William Cleland

A Treatise on the Geometry of the Circle

and some Extensions to Conic Sections by the Method of Reciprocation

William Cleland

A Treatise on the Geometry of the Circle
and some Extensions to Conic Sections by the Method of Reciprocation

ISBN/EAN: 9783743335134

Manufactured in Europe, USA, Canada, Australia, Japa

Cover: Foto ©berggeist007 / pixelio.de

Manufactured and distributed by brebook publishing software (www.brebook.com)

William Cleland

A Treatise on the Geometry of the Circle

A TREATISE

ON THE

GEOMETRY OF THE CIRCLE.

A TREATISE

ON THE

GEOMETRY OF THE CIRCLE

AND SOME EXTENSIONS TO

CONIC SECTIONS BY THE METHOD OF RECIPROCATION

WITH NUMEROUS EXAMPLES

BY

WILLIAM J. M'CLELLAND, M.A.

PRINCIPAL OF THE INCORPORATED SOCIETY'S SCHOOL, SANTRY, DUBLIN

𝔏𝔬𝔫𝔡𝔬𝔫

MACMILLAN AND CO.

AND NEW YORK

1891

PREFACE.

My object in the publication of a treatise on Modern Geometry is to present to the more advanced students in public schools and to candidates for mathematical honours in the Universities a concise statement of those propositions which I consider to be of fundamental importance, and to supply numerous examples illustrative of them.

Results immediately suggested by the propositions, whether as particular cases or generalized statements, are appended to them as Corollaries.

The Examples are printed in smaller type, and are classified under the Articles containing the principal theorems required in their solution.

The more difficult ones are fully worked out, and in most cases hints are given to the others.

The reader who is familiar with the first six books of Euclid with easy deductions and the elementary formulæ in Plane Trigonometry will thus experience little difficulty in mastering the following pages.

I have dwelt at length in Chap. II. on the Theory of Maximum and Minimum.

Chap. III. is devoted to the more recent developments of the geometry of the triangle, initiated in 1873 by Lemoine's paper entitled "Sur quelques propriétés d'un point remarquable du triangle."

The study of the Brocardian Geometry is appropriate at this stage, as I have shown that the deductions of M. Brocard and of other geometers, both in England and on the Continent, are simple and direct inferences of the well-known property of Art. 19, which has been called the Point O Theorem.

Chap. IX. gives an account of the researches of Neuberg and Tarry on Three Similar Figures.

A feature of the volume is the application of Reciprocation to many of the best known theorems by which the corresponding properties of the Conic are ascertained. This method and that of Inversion are pursued as far as is admissible within the scope and limits of an elementary treatise on Geometry.

In the preparation of the book, I consulted chiefly the writings of Mulcahy, Cremona, Catalan, Salmon, and Townsend, and hereby acknowledge my indebtedness for the valuable stores of information thus placed at my disposal.

Many of the Examples are from the Dublin University Examination Papers, and more especially from those set by Mr. M'Cay.

I have as far as possible indicated my additional sources of information, and given the reader references to the original memoirs from which extracts have been taken.

WILLIAM J. M'CLELLAND.

SANTRY SCHOOL,
1st November, 1891.

CONTENTS.

CHAPTER I.

INTRODUCTION.

CHAPTER II.

MAXIMUM AND MINIMUM.

SECTION I.

INTRODUCTION.

SECTION II.

METHOD OF INFINITESIMALS.

SECTION III.

SECTION IV.

CHAPTER III.

RECENT GEOMETRY.

SECTION I.

THE BROCARD POINTS AND CIRCLE OF A TRIANGLE.

CHAPTER V.

COLLINEAR POINTS AND CONCURRENT LINES.

CHAPTER VI.

INVERSE POINTS WITH RESPECT TO A CIRCLE.

CHAPTER VII.

POLES AND POLARS WITH RESPECT TO A CIRCLE.

SECTION I.

CONJUGATE POINTS. POLAR CIRCLE.

SECTION II.

SALMON'S THEOREM.

SECTION III.

RECIPROCATION.

CHAPTER VIII.

COAXAL CIRCLES.

Section I.

COAXAL CIRCLES.

Section II.

ADDITIONAL CRITERIA OF COAXAL CIRCLES.

Section III.

CIRCLE OF SIMILITUDE.

CHAPTER IX.

THEORY OF SIMILAR FIGURES.

SECTION I.

TWO SIMILAR FIGURES.

SECTION II.

THREE SIMILAR FIGURES.

CHAPTER X.

CIRCLES OF SIMILITUDE AND OF ANTISIMILITUDE.

SECTION I.

SECTION II.

CIRCLES OF ANTISIMILITUDE.

CHAPTER XI.

INVERSION.

SECTION I.

INTRODUCTORY.

SECTION II.

ANGLES OF INTERSECTION OF FIGURES AND OF THEIR INVERSES.

CONTENTS.

SECTION III.

ANHARMONIC RATIOS UNALTERED by INVERSION.

CHAPTER XII.

GENERAL THEORY OF ANHARMONIC SECTION.

SECTION I.

ANHARMONIC SECTION.

SECTION II.

ANHARMONIC SECTION OF AN ANGLE.

CHAPTER XIII.

INVOLUTION.

CHAPTER XIV.

DOUBLE POINTS.

CHAPTER I.

INTRODUCTION.

Definitions.—Right lines passing through a point are called a *Concurrent System.*

The point is the *Vertex* of the system, and the lines are a *Pencil of Rays.*

Collinear points are those which lie on a right line.

Symmetry. Convention of Positive and Negative.—

1. The letters A, B, C, ..., are generally used to denote points and positions of lines, and a, b, c, *lengths*, e.g., the vertices of a triangle are A, B, C, and the opposite sides a, b, c.

By AB is meant the distance from A to B measured from A towards B, and by BA the same distance measured in the *opposite* direction.

Thus $AB = -BA$ or $AB + BA = 0$.

Similarly for three collinear points A, B, C:

$$AB + BC = AC = -CA, \text{ therefore } BC + CA + AB = 0.$$

2. If four points A, B, C, D, be taken in alphabetical order on a circle, we have by Ptolemy's Theorem

$$BC \cdot AD + AB \cdot CD = BD \cdot AC = -CA \cdot BD,$$

ᴄ   ᴀ

the six linear segments being measured from left to right, or we shall say *positively*, in figure;

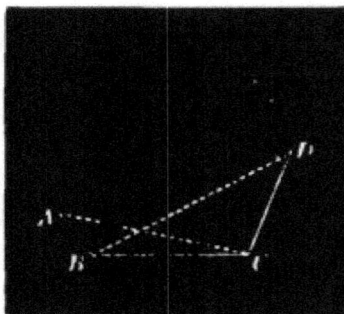

hence, by transposing,

$$BC . AD + CA . BD + AB . CD = 0.$$

Again, since each chord is proportional to the sine of the angle it subtends at any fifth point O on the circle, this equation reduces to

$$\sin BOC \sin AOD + \sin COA \sin BOD + \sin AOB \sin COD = 0,$$

a result which is therefore true for any pencil of four lines, and is deduced directly from Ptolemy's Theorem by describing a circle of any radius through its vertex.

In this equation it is implied that AOC denotes the magnitude of the angle measured from A towards C, and that therefore $\sin AOC = -\sin COA$.

3. Let $O . ABCD$ denote a system of lines concurrent at O; A, B, C, D, the points in which a line L meets it; and p the distance of the vertex O from L.

Then $\qquad 2BOC = BC . p = OB . OC \sin BOC,$

and $\qquad 2AOD = AD . p = OA . OD \sin AOD;$

by multiplication

$$BC . AD . p^2 = OA . OB . OC . OD . \sin BOC . AOD; \quad \dots (1)$$

similarly

$$CA \cdot BD \cdot p^2 = OA \cdot OB \cdot OC \cdot OD \cdot \sin COA \cdot \sin BOD ; \quad (2)$$

dividing (1) by (2) we have

$$BC.AD : CA.BD = \sin BOC . \sin AOD : \sin COA . \sin BOD. (3)$$

The student will observe that three pairs of angles are formed by taking any pair of rays with the remaining or *Conjugate* pair.

Thus BOC and AOD may be conveniently denoted by α and α', COA and BOD by β and β', and AOB and COD by γ and γ'.

With this notation (3) is written

$$BC.AD : CA.BD = \sin\alpha\sin\alpha' : \sin\beta\sin\beta',$$

and generally we infer from symmetry that

$$BC.AD : CA.BD : AB.CD = \sin\alpha\sin\alpha' : \sin\beta\sin\beta' : \sin\gamma\sin\gamma'. (4)$$

COR. 1. If we draw four parallels to the rays of the pencil, we in general obtain a triangle and a transversal to its sides. Moreover, if we denote the angles of the triangle by α, β, γ, those made by the transversal with its sides are the opposites α', β', γ'; hence for any triangle and transversal we have always

$$\sin\alpha\sin\alpha' + \sin\beta\sin\beta' + \sin\gamma\sin\gamma' = 0.$$

COR. 2. Let the line $ABCD$ be divided *harmonically* or such that $AB/BC = AD/CD$, then $BC.AD = AB.CD$; hence by (3) the pencil is divided harmonically, *i.e., the angle COA is divided internally in B and externally in D in the same ratio of sines.*

Defs. The three ratios and their reciprocals on the left side of (3) are termed the *Anharmonic Ratios* of the four points on the line; and those on the right the Anharmonic Ratios of the pencil $O.ABCD$.

Their equivalence is expressed thus:—A variable line drawn across a pencil is cut in a constant anharmonic ratio ; or any pencil and transversal to it are *Equianharmonic.*

The foot of the perpendicular from a point on a line is the *Projection of the point* on the line, and the perpendicular is called its *Projector.*

If A' and B' be the projections of A and B on a line L, $A'B'$ is called the *Projection* of AB, and is equal to $AB \cos \theta$, where θ is the angle between AB and L.

EXAMPLES.

1. The sum of the projections of the sides of a polygon on any right line $=0$; and generally if lines be drawn equally inclined and proportional to the sides of a polygon, the sum of their projections is zero.

2. $\quad \cos a + \cos\left(a + \dfrac{2\pi}{n}\right) + \cos\left(a + \dfrac{4}{n}\right) + \dots \cos\left(a + \dfrac{2(n-1)\pi}{n}\right) = 0,$

and the sum of the sines of the series of angles is also equal to 0.

[For they are proportional to the projections of the sides of a regular polygon on two lines at right angles.]

3. In any quadrilateral whose sides are a, b, c, d, to prove that

$$d^2 = a^2 + b^2 + c^2 - 2bc \cos \widehat{bc} - 2ca \cos \widehat{ca} - 2ab \cos \widehat{ab},$$

where $\widehat{bc}$ denotes the angle between the sides b and c.

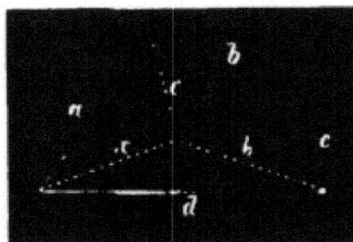

[For completing the parallelogram whose sides are b and c and drawing x we have $\quad d^2 = b^2 + x^2 + 2bx',$

where x' is the projection of x on the parallel b; but by **Ex. 1.**

$$x' = a \cos \widehat{ab} + c \cos \widehat{bc},$$

substituting for x' its value and for x^2, $a^2+c^2-2ac\cos\widehat{ac}$, the above result is obtained.]

4. **Euler's Theorem.**[*]—For three collinear points A, B, C and any fourth P to prove the relation

$$BC . AP^2 + CA . BP^2 + AB . CP^2 = - BC . CA . AB.$$

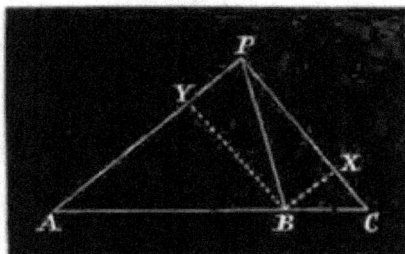

[By Euc. II. 12, 13, $AP^2 = AB^2 + BP^2 - 2AB . BP \cos B$............(1)

and $CP^2 = BC^2 + BP^2 + 2BC . BP \cos B$..........(2)

multiplying (1) by BC and (2) by AB and adding to eliminate $\cos B$, the above follows on reduction.]

4A. Having given the base c of a triangle and $la^2 + mb^2 = \text{const.}$, find the locus of the vertex, l and m being given quantities.

5. If APC is a right angle the relation in Ex. 4 is equivalent to

$$BC^2 . AP^2 + AB^2 . CP^2 = AC^2 . BP^2.$$

[This follows from Ex. 4 or is obtained directly thus; let fall perpendiculars BX and BY on CP and AP, then

$$XY^2 = BP^2 = BX^2 + BY^2 = BC^2\sin^2C + AB^2\sin^2A;$$

multiplying the equation $BP^2 = BC^2\sin^2C + AB^2\sin^2A$ by AC^2; therefore, etc.].

6. If the transversal to a harmonic pencil is parallel to one ray D, the intercept AC is bisected by B the conjugate of D.

---

[*] "Catalan's Théorèmes et Problèmes de Géométrie Élémentaire," 1879, p. 141.

7. If a line L turn around a fixed point P and meet two fixed lines OA and OB in A' and B'; the locus of the harmonic conjugate Q of P with respect to $A'B'$ is a line passing through O; and

$$\frac{1}{PA'}+\frac{1}{PB'}=\frac{2}{PQ}\quad\ldots\ldots\ldots\ldots\text{(By Ex. 6.)}$$

Note. By Euc. VI. 2 if the variable PQ is bisected at Q' the locus of Q' is a parallel to OQ and

$$\frac{1}{PA'}+\frac{1}{PB'}=\frac{1}{PQ'}.$$

Hence for any three lines A, B, C we find in the same manner that

$$\frac{1}{PA'}+\frac{1}{PB'}+\frac{1}{PC'}=\frac{1}{PQ'},$$

where Q' describes a right line.

8. For any system of lines A, B, C, D ... the locus of Q' such that

$$\frac{1}{PA'}+\frac{1}{PB'}+\frac{1}{PC'}+\ldots=\frac{1}{PQ'}\left(\text{or }\ \Sigma\frac{1}{PA'}=\frac{1}{PQ'}\right)$$

is a right line. [See Exs. 6 and 7.]

9. For a regular cyclic polygon, if P coincides with the centre

$$\Sigma\frac{1}{PA'}=0.$$

[Through P draw the line parallel to one of the sides, etc.]

10. If parallels be drawn through any point O to the four lines in Ex. 4, the relation may be written

$$\frac{\sin\beta'\sin\gamma'}{\sin\beta\sin\gamma}+\frac{\sin\gamma'\sin\alpha'}{\sin\gamma\sin\alpha}+\frac{\sin\alpha'\sin\beta'}{\sin\alpha\sin\beta}=1.$$

11. From the formula $BC.AD+CA.BD+AB.CD=0$, prove that if A, B, C be three collinear points and P any fourth point $BC\cot A+CA\cot B+AB\cot C=0$, the angles being all measured in the same aspect; and hence find the locus of the vertex, having given the base c and $l\cot A+m\cot B=$const.

4. **Limiting Cases.** 0 *and* ∞.

Def. The *Angle of intersection of two circles* is that between the tangents drawn to them at either point of intersection; it is therefore equal to the angle between the radii drawn to either common point.* (Euc. III. 19.)

If the circles touch *Internally* this angle is 0°, if *Externally* 180°. They are said to intersect *Orthogonally* when the angle is 90°.

The *Angle made by a line and circle* is that between the line and the tangent to the circle at its intersection.

EXAMPLES.

1. To find the angles between the circum- and ex-circles of a triangle ABC.

[Since $\delta_1{}^2 = R^2 + 2Rr_1$, etc., we easily obtain $2\cos\frac{1}{2}\theta_1 = \sqrt{\dfrac{r_1}{R}}$; with similar expressions for θ_2 and θ_3.]

2. To find the angle of intersection of the in- and circum-circles.

[$\delta^2 = R^2 - 2Rr$, therefore $2\sin\frac{1}{2}\theta = \sqrt{\dfrac{ir}{R}}$ where $\sqrt{-1} = i$.]

3. If two concentric circles cut orthogonally one is real and the other imaginary, and their radii are of the forms ρ, $i\rho$.

---

* If O_1, r_1; O_2, r_2, be the circles, δ the distance O_1O_2, θ the angle of intersection, and t the direct common tangent, we have

$$\delta^2 = r_1{}^2 + r_2{}^2 - 2r_1r_2\cos\theta$$
$$= (r_1 - r_2)^2 + 4r_1r_2\sin^2\tfrac{1}{2}\theta ;$$

hence $\qquad \delta^2 - (r_1 - r_2)^2 = 4r_1r_2\sin^2\tfrac{1}{2}\theta, \dots\dots\dots\dots\dots\dots(1)$

or $\qquad\qquad \dfrac{t^2}{4r_1r_2} = \sin^2\tfrac{1}{2}\theta.$

Similarly $\qquad \delta^2 - (r_1 + r_2)^2 = -4r_1r_2\cos^2\tfrac{1}{2}\theta,$

hence if t' be the transverse common tangent,

$$t'^2 = -4r_1r_2\cos^2\tfrac{1}{2}\theta \dots\dots\dots\dots\dots\dots(2)$$

Multiplying (1) and (2) and reducing we have

$$tt' = 2i \cdot r_1r_2\sin\theta .. \dots\dots\dots\dots\dots\dots(3)$$

where $\sqrt{-1} = i$; also if γ denote the length of the common chord, of the circles (real or imaginary) since $2r_1r_2\sin\theta = \gamma\delta$, $t \cdot t' = i \cdot \gamma \cdot \delta$.

It is obvious that either the transverse common tangent to the circles or their angle of intersection is imaginary.

Let AX be a variable chord passing through a fixed point A at which a tangent is drawn. According as the

chord AX and angle TAX diminish in magnitude X approaches the tangent. When X is indefinitely near to A, AX is said to have reached its *limiting position* and may then be considered to coincide with the tangent.

Hence a *tangent to a circle is in the direction of the infinitesimal chord at its point of contact, or is the chord joining two indefinitely near points.*

Again, let the tangent T and its point of contact be fixed and the chord AX given in length. As the radius of the circle increases the curvature diminishes, and the point X obviously approaches the tangent. Hence X may be made to move as near as we please to the tangent by continually increasing the value of the radius of the circle.

In the limit, when the latter is indefinitely great, the distance of X from T is so very small that we may consider the point to lie on the line. Hence a finite *portion of a circle of indefinitely great radius opens out into a right line*, the remainder being, of course, at a distance infinitely great, *i.e.*, at infinity.

5. **Envelopes.**—Let a variable line turn around a fixed point O and meet any fixed line.

According as its angle of inclination to the perpendicular OM increases, the segments OA, OB, OC continue to increase and the angles A, B, C to diminish. In the

limit it reaches a position at right angles to OM. Here the angle between it and the fixed line vanishes, and their point of section is at infinity. In this case the lines are parallel (Euc. I. 28); hence

Parallel lines may be regarded as having angles of inclination $= 0°$ or lines intersecting at infinity. Thus a system of parallels is a pencil of rays whose vertex is at infinity.

6. Let A and X be any two points on a curve of which A is fixed and X variable, and TA and TX tangents. It appears as before that as X approaches A the chord AX and the base angles A and X of the triangle TAX gradually diminish and ultimately vanish.

But as the base angles diminish the vertex T approaches the base and *a fortiori* the *element of curve AX*. Hence in the limiting position, *i.e.*, when the tangents are consecutive, their point of intersection is on the curve.

A curve touched by a variable line is called the *Envelope* of the line. *Thus the envelope of a line which varies according to any law is the locus of the intersection of its consecutive positions.*

EXAMPLES.

1. The envelope of equal chords in a circle is a concentric circle (Euc. III. 14).

2. **Bobillier's Theorem.**—If two sides of a given triangle touch fixed circles the third side also touches, or envelopes, a circle.

[Let ABC be the given triangle. Through O_1 and O_2, the centres of the given circles, draw parallels to the sides meeting the base in A' and B' and each other in C'. Describe a circle O_1O_2C', and draw $C'O_3$ parallel to AB.

Since $O_2C'O_3$ is a given angle $(=A)$, O_3 is a fixed point. But $A'B'C'$ is given in all respects save position; hence the distance p of O_3 from $A'B'$ is a known quantity. The envelope of the base AB is therefore a circle whose centre is O_3 and radius $= p$.]

3. To find the radius (ρ) of a circle which touches the sides AC and BC of a triangle and the circum-circle ABC.

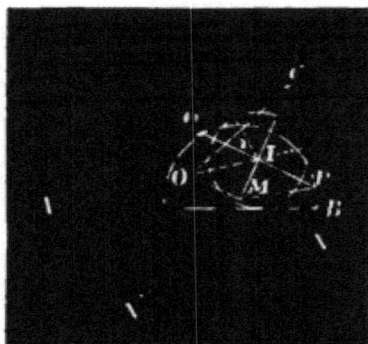

[Let I denote the in- and O the circum-centre of the triangle; M the centre of the circle whose radius is required is on the line CI. Then $OM=R-\rho$, $OI^2=R^2-2Rr$, $IC=r/\sin\frac{1}{2}C$, $MC=\rho/\sin\frac{1}{2}C$, and $MI=(\rho-r)/\sin\frac{1}{2}C$.

Also, since C, I, M are three points in a line and O any fourth point, by *Euler's Theorem* we obtain on reducing

$$r=\rho\cos^2\tfrac{1}{2}C\dotfill(1)$$

Again, if the circle M, ρ has external contact with the circum-circle, it can be similarly proved that

$$r_3 = \rho \cos^2\tfrac{1}{2}C \dots\dots\dots\dots\dots\dots\dots\dots(2)$$

Note.—The relation (1) is otherwise expressed :—

Since $\qquad r/\rho = \cos^2\tfrac{1}{2}C, \ (\rho - r)/\rho = \sin^2\tfrac{1}{2}C.$

But $\qquad (\rho - r)/\rho = MI/MC$ and $\rho^2/MC^2 = \sin^2\tfrac{1}{2}C,$

hence $\qquad\qquad MI \cdot MC = \rho^2 \dots\dots\dots\dots\dots\dots\dots(3)$

or the chord of contact PQ of the circle M, ρ with the sides of the triangle passes through the centre of the inscribed circle.]

4. **Mannheim's Theorem.**—Having given the vertical angle and radius of the in- or corresponding ex-circle, the envelope of the circum-circle is a circle.

[By Ex. 3.]

7. We shall conclude the present chapter with the following useful property, of the common tangents to four circles which touch a fifth, due to the late Dr. Casey.

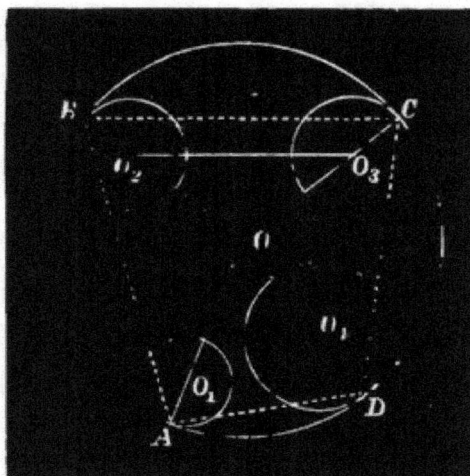

Denote the circle whose centre is O and radius r by O, r; and let the four circles O_1r_1, O_2r_2, O_3r_3, O_4r_4, touch a fifth O, R at the points A, B, C, D. Let the distance O_2O_3 be δ_{23}, and the direct common tangent to the corresponding circles be $\overline{23}$.

Then $\qquad\qquad \overline{23}^2 = \delta_{23}{}^2 - (r_2 - r_3)^2.$

In the triangle OO_2O_3 we have

$$O_2O_3{}^2 = OO_2{}^2 + OO_3{}^2 - 2OO_2 . OO_3 \cos BOC$$
$$= (OO_2 - OO_3)^2 + 4OO_2 . OO_3 \sin^2\tfrac{1}{2}BOC ;$$

or $\delta_{23}{}^2 - (r_2 - r_3)^2 = 4OO_2 . OO_3 \sin^2\tfrac{1}{2}BOC ;$

or $\overline{23}^2 = 4OO_2 . OO_3 \sin^2\tfrac{1}{2}BOC = OO_2 . OO_3 . BC^2/R^2.$

Similarly

$$\overline{14}^2 = OO_1 . OO_4 . AD^2/R^2 ;$$

hence by multiplication and reduction

$$\overline{23} . \overline{14} = (OO_1 . OO_2 . OO_3 . OO_4)^{\frac{1}{2}} BC . AD/R^2,$$

and by Ptolemy's Theorem

$$\overline{23} . \overline{14} + \overline{31} . \overline{24} + \overline{12} . \overline{34} = 0 \dots\dots\dots\dots\dots(1).$$

The contacts in the figure are similar, or all of the same kind, but it will be observed that if the fifth circle touches any two with contacts of opposite species, their transverse common tangents must be substituted in (1).

We let $\overline{12'}$ denote the transverse tangent to O_1, r_1 and O_2, r_2; then

$$\overline{12'}^2 = \delta_{12}{}^2 - (r_1 + r_2)^2.$$

For example, if the circle O_1, r_1 is external and the remaining circles internal to O, R the relation is written

$$\overline{23}\ \overline{14'} + \overline{31'}\ \overline{24} + \overline{12'}\ \overline{34} = 0,$$

with analogous expressions for all other cases.

NOTE.—The student must carefully observe that of the three terms of the equation two are positive and one negative ; the latter corresponding to the pairs of circles whose contacts are alternate. Thus in the figure, O_1, r_1 and O_3, r_3 have alternate contacts with the given circle, therefore the term $\overline{31} . \overline{24}$ is negative, and taking the absolute values only the equation is

$$\overline{23} . \overline{14} + \overline{12} . \overline{34} = \overline{31} . \overline{24}.$$

This is of great importance, and should be borne in mind in the following Examples.

EXAMPLES.

1. What does the general property reduce to when the circles become points ?   Ptolemy's Theorem.

2. Express a condition that the circum-circle of a given triangle may touch another circle.

[If a, b, c be the sides and t_1, t_2, t_3 the tangents from the vertices to the other circle we have $at_1 + bt_2 + ct_3 = 0$.]

3. **Feuerbach's Theorem.**—The nine points circle of a triangle touches the in- and ex-circles.

[The middle points of the sides and the in-circle are four circles satisfying the equation of Ex. 2.   For $\overline{23} = \tfrac{1}{2}a$ and $\overline{14} = \tfrac{1}{2}(b-c)$; therefore $\Sigma\overline{23}.\overline{14} = \tfrac{1}{4}\Sigma a(b-c) = 0*$.]

4. If a, b, c be the sides of a triangle inscribed in a circle, and λ, μ, ν the distances of its vertices from any tangent, show that the equation in Ex. 2 reduces to

$$a\sqrt{\lambda} + b\sqrt{\mu} + c\sqrt{\nu} = 0.\dagger$$

5. More generally if λ, μ, ν denote the distances from any line, give the geometrical interpretation of the equation

$$a\sqrt{\lambda - x} + b\sqrt{\mu - x} + c\sqrt{\nu - x} = 0,$$

and hence find a relation connecting the sides of a triangle with the distances of its vertices from a given line.

[The roots of the quadratic in x are the distances from the line of the tangents to the circle parallel to it, etc.]

6. **Hart's Extension of Feuerbach's Theorem.**—If the sides of a triangle be replaced by three circles, and four circles corresponding to the in- and ex-circles of the triangle described to touch them ; the group of four is touched by a circle.

[Let the triangle formed by the circles be ABC, and let $a<b<c$. Then $s-a>s-b>s-c$.   If the in- and ex-circles are numbered

---

* This proof is an application of the *converse* of Dr. Casey's relation.

† This result may be otherwise shown as follows :—Let P be the point of contact of the tangent.   Then $BC.AP + CA.BP + AB.CP = 0$. But $AP^2 = 2r\lambda$, $BP^2 = 2r\mu$, and $CP^2 = 2r\nu$, substituting these values ; therefore, etc.

1, 2, 3, 4 respectively, the side a is touched by the four circles and the transverse tangents are drawn to 2 ; also the order of the contacts is 3, 1, 2, 4 ; hence the equation is

$$- \overline{23}' . \overline{14} + \overline{31} . \overline{24}' + \overline{12}' . \overline{34} = 0 \dots\dots\dots\dots\dots\dots (1)$$

For the side b the transverse tangents are drawn to 3, and the order of the contacts is 2, 1, 3, 4 ; hence

$$- \overline{23}' . \overline{14} + \overline{31}' . \overline{24} + \overline{12} . \overline{34}' = 0 \dots\dots\dots\dots\dots\dots(2)$$

For the side c the transverse tangents are drawn to 4, and the order of the contacts is 3, 4, 1, 2 ; hence

$$\overline{23} . \overline{14}' - \overline{31} . \overline{24}' + \overline{12} . \overline{34}' = 0 \dots\dots\dots\dots\dots\dots(3)$$

Adding (1) and (3) and subtracting (2) we get

$$\overline{23} . \overline{14}' - \overline{31}' . \overline{24} + \overline{12}' . \overline{34} = 0,$$

showing that 2, 3, 4 have similar and 1 opposite contacts with a circle which touches all four.]

CHAPTER II.

MAXIMUM AND MINIMUM—INTRODUCTION.

8. When the base and vertical angle of a triangle are given the locus of the vertex is a segment of a circle described on the base, containing an angle equal to the vertical angle. (Euc. III. 21.) Let a number of triangles be constructed satisfying the given conditions, and it will be observed that as the vertex recedes from either extremity of the base the altitude and area both increase up to a certain point, after which they begin to diminish.

This point is obviously the middle point of the segment—the vertex of the isosceles triangle with the given parts—or the point at which the tangent to the arc is parallel to the base.

Here the area and altitude are said to have attained their *maximum* values.

Again since the rectangle under the sides AC and BC is equal to the rectangle under the diameter of the circum-circle and altitude ($ab = dp$); ab and p are maxima simultaneously.

Also since $$a^2 + b^2 = 2(\tfrac{1}{2}c)^2 + 2\beta^2,$$

where β is the median to the side c; when β is a maximum or minimum, $a^2 + b^2$ is maximum or minimum.

15

And if N be the middle point of the arc of the circle below the base, then, since $AN = BN$ ($=x$ say) by Ptolemy's Theorem, we have

$$ax + bx = c \cdot CN,$$

or $$x(a + b) = c \cdot CN,$$

from which it appears that $a + b$ and CN are maxima together; that is when the vertex C is at the middle point M of the arc AB.

On the other hand it is manifest that the difference of base angles $(A - B)$ and difference of sides $(a - b)$ both diminish as the vertex C approaches M and vanish at that point; and after C passes through this point each difference begins to increase. At C they are said to have their *minimum* values, though this need not necessarily be nothing.

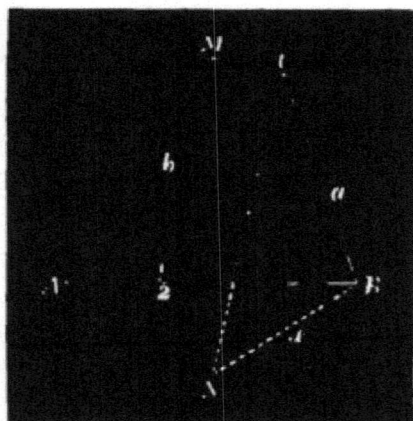

Thus generally:—a variable quantity which, under certain conditions, increases up to a definite limit and then begins to diminish, is said to have attained its maximum value at the limit; and if, after diminishing, it again begins to increase, it attains a minimum value at the stage where it has ceased to diminish.

The foregoing remarks may be thus summed up:—Of all triangles having a given base and vertical angle the isosceles has the following maxima—area, altitude, rectangle under sides, sum of sides, bisector of base, and sum of squares of sides.*

EXAMPLES.

1. The triangle of greatest area and perimeter inscribed in a circle is equilateral.

[For each vertex must lie mid-way between the other two, or the area and perimeter would both be increased by removing any vertex to the middle point.]

2. A regular polygon of n sides inscribed in a circle has a greater area and perimeter than any other inscribed polygon of the same order. [By Ex. 1.]

9. **Theorem.**—*If two sides AC and AB of a triangle are given in length the area of the triangle ABC is a maximum when they contain a right angle.*

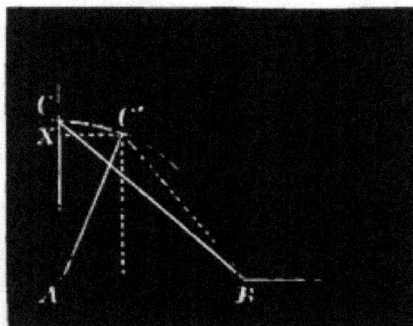

Let ABC denote the right-angled triangle, and ABC' any other triangle formed with the given sides. Draw $C'X$ perpendicular to AC.

Since $AC = AC'$ and $AC' > AX$; therefore $AC > AX$, hence (Euc. I. 41) the triangle $ABC > ABC'$, and similarly for any other position; therefore, etc.

---

* The vertical angle is supposed to be acute.

EXAMPLES.

1. If the ends of a string of given length are joined, the area of the figure enclosed is a maximum when it takes the form of a semi-circle.

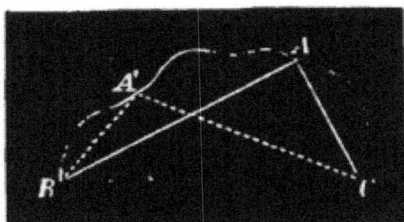

[Take any point A on the string ABC and join AB and AC. Consider the segments into which the string is divided at A to be rigidly attached to the lines AB and AC. If the angle at A is not right, by rotating AC around A until it is perpendicular to AB, the area of the triangle ABC, and therefore also of the whole figure, is increased.

Similarly for any other point A'; hence the area enclosed is a maximum when the joining line BC subtends a right angle at every point on the string.]

2. A closed curve of given perimeter is of greatest area when its form is a circle.

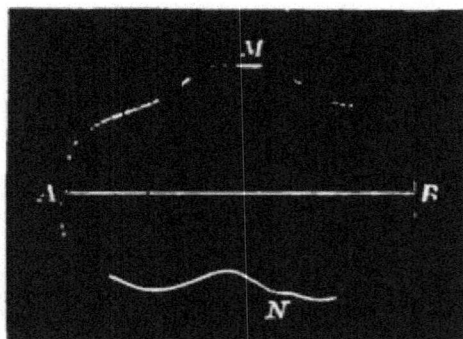

[Let A be any point on the curve, and take B such that AMB and ANB are equal in length. Then the areas AMB and ANB are each a maximum when AB is the diameter of semicircles on opposite sides ; therefore, etc.]

3. Having given the four sides a, b, c, d of a quadrilateral, its area is a maximum when it is cyclic.

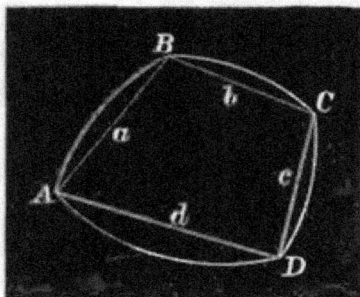

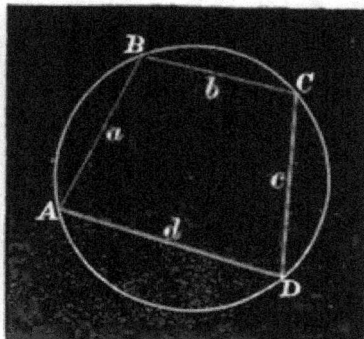

[Let $ABCD$ be the cyclic quadrilateral with the given sides, and consider the segments on the sides to be rigidly attached to them.* If then the figure be distorted in any way into a new position

* The construction of the cyclic quadrilateral whose four sides are given is as follows :—

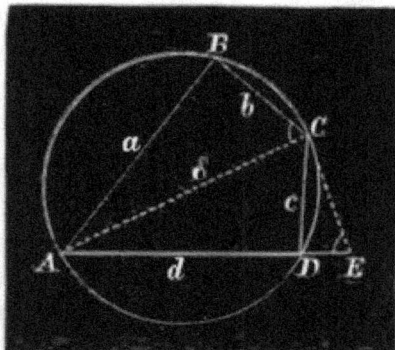

Draw CE making $\angle DCE = \angle BAC$. Since by Euc. iii., 22, $\angle CDE = \angle ABC$, the triangles ABC and CDE are similar ; therefore $DE : c = b : a$ (Euc. vi. 4) ; hence DE is known and E is a fixed point.

Again, $AC : CE = a : c$; therefore in the triangle ACE we have the base AE and ratio of sides ; the locus of C is therefore a circle (Euc. vi. 3) ; this locus intersects the circle described with D as centre and c as radius at the point C ; therefore, etc.

$A'B'C'D'$, the area of the circle $ABCD > A'B'C'D'$ (Ex. 2), but the segments $AB = A'B'$, $BC = B'C'$, etc.: take away these equal parts and there remains the quadrilateral $ABCD$ greater than $A'B'C'D'$.]*

4. If three sides a, b, c, of a quadrilateral are given in magnitude, the area is a maximum when the fourth side d is the diameter of the circle through the vertices; and generally,

When all the sides but one of a polygon of any order are given in magnitude, the area is a maximum when the circle on the closing side as diameter passes through the remaining vertices.†

[Proof as above.]

5. Having given of a quadrilateral the diagonals δ and δ' and a pair of opposite sides BC and AD, its area is a maximum when BC is parallel to AD.

[Take any position of the quadrilateral and through C draw CE parallel and equal to δ. Join BE and AE.

The triangles BDE and BCD are equal (Euc. I. 37); to each add ABD, therefore $ABCD = ABED$.

---

* The student should learn the proof of the Trigonometrical expression for the area of any quadrilateral in terms of the four sides and the sum of either pair of opposite angles.

$$(\text{Area})^2 = (s-a)(s-b)(s-c)(s-d) - abcd\cos^2\tfrac{1}{2}(A+C).$$

(Casey's *Plane Trig.*, art. 152, cors. 3, 4.)

† To construct the quadrilateral. Let θ be the angle between a and b, and $AC = x$.

Then $\qquad\qquad d^2 = c^2 + x^2 = a^2 + b^2 + c^2 - 2ab\cos\theta;$ ....(1)

but $\qquad\qquad\qquad \cos\theta = -c/d;$

substituting in (1) and simplifying we have the following expression for d:— $\qquad d^3 - d(a^2 + b^2 + c^2) - 2abc = 0,$

an equation which has only one positive root. (Burnside and Panton's *Theory of Equations*, Art. 13.)

In the particular case when $a = b = c$, the equation for d reduces to

$$(d - 2a)(d + a)^2 = 0;$$

hence $\qquad\qquad\qquad\qquad d = 2a,$

thus showing that the quadrilateral is half the regular inscribed hexagon.

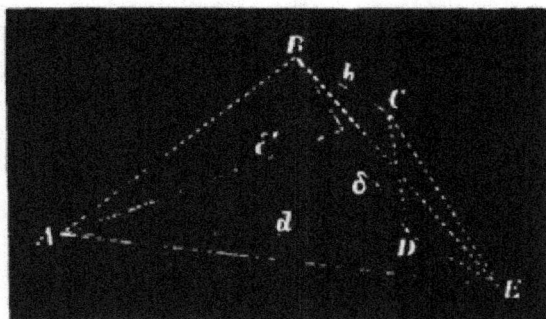

Now, *ABDE* is a maximum when *AD* and *DE* are in the same straight line; hence *ABCD* is a maximum when *BC* is parallel to *AD*.]

6. The diagonals of a quadrilateral are 9 and 10 feet and two opposite sides 5 and 3 feet; find when its area is a maximum.

10. **Theorem.**—*Having given the base AB of a triangle and the locus of the vertex a line L meeting the base produced, the sum of the sides AC+BC is a minimum when L is the external bisector of the vertical angle.*

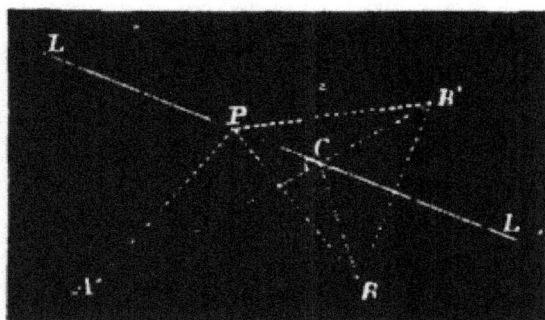

Let fall a perpendicular *BL* and make *B'L=BL*. Join *AB'* and let *C* be its intersection with *L*. Take any other point *P* on the line and join *AP* and *B'P*.

The triangles BCL and $B'CL$ are equal in every respect (Euc. I. 4); hence $BC=B'C$. Similarly $BP=B'P$. Hence since (Euc. I. 20) $AP+B'P > AB'$ it follows that $AP+BP > AC+BC$.

Cor. 1. If the line L cuts the base internally the difference of the sides $(AC-BC)$ is a maximum when it bisects internally the angle C.

EXAMPLES.

1. The triangle of minimum perimeter inscribed in a given one is formed by joining the feet X, Y, Z of the perpendiculars* let fall from the vertices on the opposite sides.

[For the joining lines are equally inclined to sides on which they intersect (Euc. III. 21).]

2. The polygon of least perimeter that can be inscribed in a given one is that whose angles are bisected externally by its sides. (By Ex. 1.)

3. The base and area of a triangle being given, the perimeter is least when the triangle is isosceles.

[For the line L is parallel to the base.]

4. If from O, the point of intersection of the diagonals of a cyclic quadrilateral, perpendiculars are drawn to the sides and their feet P, Q, R, S joined, the quadrilateral $PQRS$ is of minimum perimeter.

4a. If points P', Q', R', S' be taken on the sides of the given quadrilateral, such that $P'Q'$, $Q'R'$, $R'S'$ are parallel to PQ, QR, RS, then $P'S'$ is parallel to PS and the perimeters of the quadrilaterals are equal. [Euc. VI. 2 and I. 5.]

5. The value of the minimum perimeter of the indeterminate inscribed quadrilateral in Ex. 4 is $2\delta\delta'/D$, where D is the diameter of the circum-circle.

6. Given a triangle ABC, find a point O such that
$$OA + OB + OC \text{ is a minimum.}$$
[Where $BOC = COA = AOB = 120°$.]

---

* These are generally known as the *Perpendiculars of the Triangle*, and XYZ as the *Pedal Triangle* of ABC.

11. Problem.—*Given an angle C of a triangle and a point P on the base, construct the triangle of minimum area.*

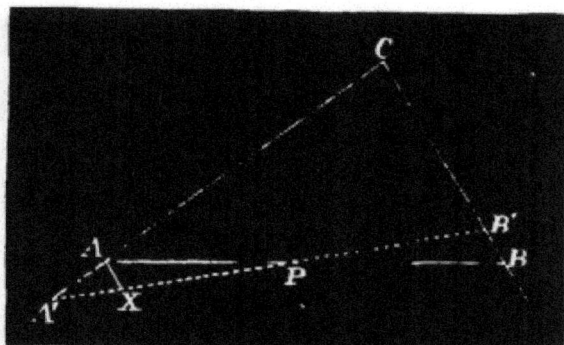

Through P draw APB such that $AP = BP$. The triangle ABC is less than any other $A'B'C$.

For draw AX parallel to BB'. Then the triangles APX and BPB' are equal in all respects (Euc. I. 4); hence $AA'P > BB'P$. To each add $APB'C$, therefore $A'B'C > ABC$; hence *the triangle of least area is that whose base is bisected at this point.*

12. Theorem.—*Given an angle and any curve concave to its vertex C. The tangent AB which forms with the sides of the angle a triangle ABC of minimum area is bisected at its point of contact (P).*

For this tangent cuts off a less area than any other line $A'B'$ through P, because it is bisected at P. Now draw any other tangent XY, and let $PA'B'$ be parallel to it. Since the curve is concave to C, $A'B'C < XYC$; *a fortiori* $ABC < XYC$.

Cor. 1. In the particular case when the curve is a circle whose centre is at C the triangle is isosceles. This

property may be stated otherwise.   *When the vertical
angle and altitude of a triangle are given, the base and
area are both minima when the triangle is isosceles.*

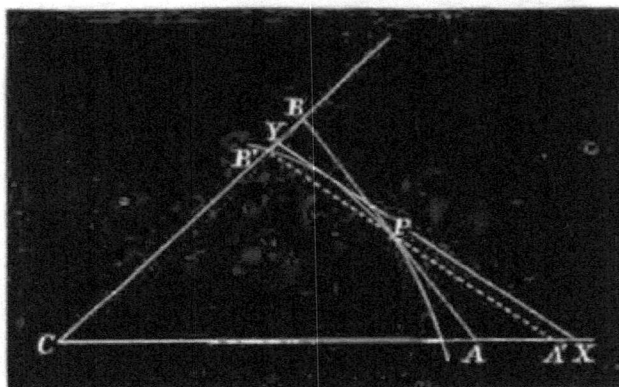

On account of its importance an independent proof of this property of the isosceles triangle is given.

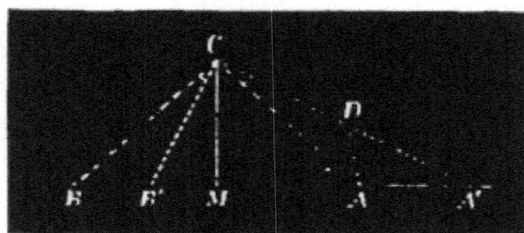

Let ABC be an isosceles triangle and $A'B'C$ any other, having
the same vertical angle and altitude CM.

Now $BC > B'C$ (Euc. III. 8), but $BC = AC < A'C$, hence $A'C > B'C$.
Let $CD = B'C$, join AD. The triangles ACD and $BB'C$ are equal
in every respect (Euc. I. 4), hence $AA'C > BB'C$; therefore
$A'B'C > ABC$; therefore, etc.

Cor. 2. When the curve is a circle touching the sides
of the angle the tangent AB and area ABC are each
minima when the triangle is isosceles.

COR. 3. If we consider the portion of the circle in Cor. 2, which is convex to C, the intercept of a variable tangent made by the sides of the angle subtends a constant angle a at the centre of the circle $(2a = \pi - C)$.

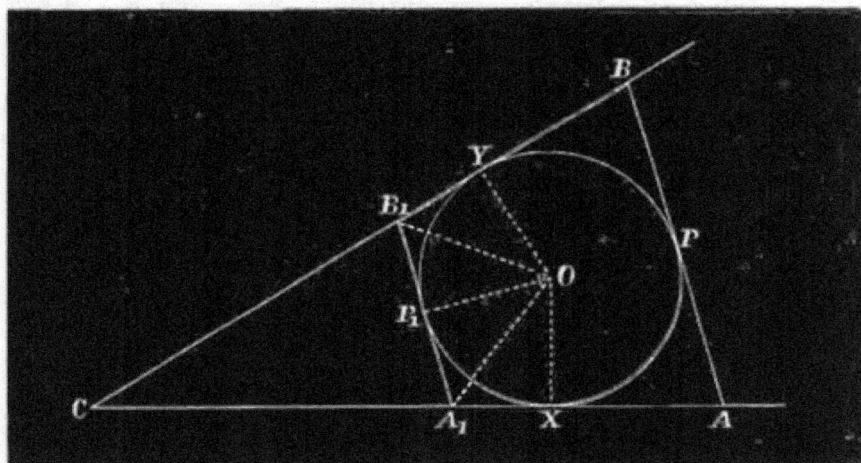

Hence the variable triangle A_1B_1O has a constant vertical angle (a) and altitude (γ), and therefore its base and area are minima when $A_1O = B_1O$. In this case the point of contact P_1 is the middle point of A_1B_1. Therefore, having given a circle and two fixed tangents, the portion of a variable tangent intercepted by the fixed tangents becomes a minimum in two positions, viz., when its point of contact bisects the arc XY internally or externally.

In the latter case the area cut off (ABC) is a minimum but in the former a maximum;

For $\qquad A_1B_1C = CXOY - 2A_1B_1O$;

therefore, since $CXOY$ is constant, when A_1B_1O is a minimum, A_1B_1C is a maximum.

EXAMPLES.

1. The triangle of least area and perimeter escribed to a circle is equilateral.

[For the point of contact of each side bisects the arc between the other; cf. Art. 8, Ex. 1.]

2. The polygon of least area and perimeter escribed to a circle is regular. (By Ex. 1.)

3. Having given the vertical angle C of a triangle in position and magnitude, and the in- or corresponding ex-circle, to prove that the line LM joining the middle points of the sides forms with the centre of the circle a triangle of constant area.

[For the ex-circle: if p be the perpendicular of ABC drawn from C to the base, and r_3 the radius, we have $2OLM = \frac{1}{2}c(\frac{1}{2}p + r_3)$ $= \frac{1}{2}ABC + AOB = \frac{1}{2}OCXY = $ const., etc.]

13. Problem.—*Given an angle O of a triangle and a point P on the base, construct it such that $AP \cdot BP$ is a minimum.*

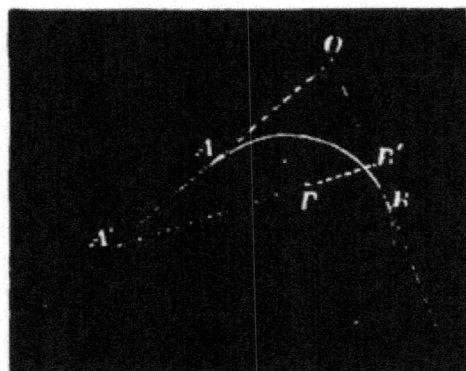

Through P draw AB so that the triangle ABO is isosceles. Describe a circle touching the sides of the angle at A and B, and draw any other line $A'PB'$.

It is evident that $AP \cdot PB < A'P \cdot B'P$, and is therefore a minimum.

1. Through the point of intersection P of two circles draw a line APB such that $PA \cdot PB$ is a minimum.

[This reduces to describe a circle touching the two given ones at A and B such that A, B and P are in a line.

It will be afterwards seen that this line passes through a point Q, on the line of centres $O_1 O_2$ of the circles where $QO_1/QO_2 =$ the ratio of the radii.]

14. Theorem.—*If a right line be divided into any two parts a and b, their rectangle is a maximum when the line is bisected.*

For Euc. (II. 5) $\quad ab + \left(\dfrac{a-b}{2}\right)^2 = \left(\dfrac{a+b}{2}\right)^2 = \text{const.},$

hence ab is a maximum when $a - b = 0$ or when $a = b$.

COR. The continued product of the segments of a line is a maximum when the parts are equal.

1. Through any point P on the base of a triangle parallels PX and PY are drawn to the opposite sides ; the area of the parallelogram $PXCY$ is a maximum when the base AB is bisected at P.

[For the triangles APX and BPY are constant in species, hence $PX \cdot PY \varpropto AP \cdot BP$. But the area of the parallelogram $= PX \cdot PY \sin C \varpropto PX \cdot PY$; therefore, etc.]*

2. The maximum rectangle inscribed in a given segment of a circle is such that if tangents BC and AC be drawn at its vertices X and Y, then $BX = CX$ and $CY = AY$.

[For NX is the maximum rectangle that can be inscribed in the triangle BCN, and therefore greater than any other $X'N$. Hence from the symmetry of the figure the rectangle on the side XY is greater than that on $X'Y'$, and a fortiori greater than that on $X''Y''$.

---

* Hence, the maximum parallelogram inscribed in a triangle is half the area of the triangle.

The construction of the maximum rectangle is as follows :—Let
BL be drawn perpendicular to OL, the diameter of the circle
parallel to AB. Join OX and let it meet BL in P. Since the
triangles OCX and BPX are equal in every respect (Euc. I. 26)

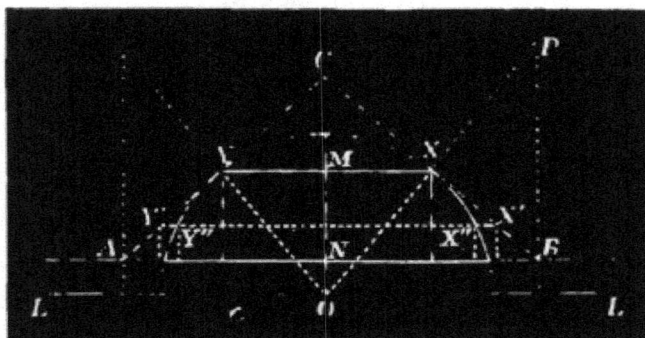

$PX = OX = r$.　　Also $OXBL$ is a cyclic quadrilateral, therefore
Euc. (III. 36),

$$PB \cdot PL = PO \cdot PX = 2r^2,$$

but $PL - PB$ is given ; hence the segment PB is known, and since it
is equal to OC, C is determined.

In the general case the line AB does not meet the circle, the segment is therefore imaginary, and the proposition may be thus
stated :—given a line AB and a circle ; construct the maximum
rectangle, having two of its vertices X and Y on the circle and the
remaining two on the line.]

3. Draw a chord XY of a circle in a given direction such that the
area of the quadrilateral $ABXY$, where AB is a given diameter, is a
maximum.

[Draw a diameter YX', and $AYBX'$ is a rectangle, hence AX' is
equal and parallel to BY. Join BX and XX', and draw BC parallel
to XX'.

Then since

triangle AXX +triangle BXY=triangle ABX',

reject the common part AMX' and let BMX be added to each,

$$BXX' = ABXY.$$

The quadrilateral is a maximum therefore when BXX' is a

maximum. It is easy to see that the latter is half the rectangle inscribed in a *given* segment *BC*.

For since *BC* is parallel to *XX'*, *AC* is perpendicular to *XX'* and therefore parallel to *PX*, hence *BAC*=a.

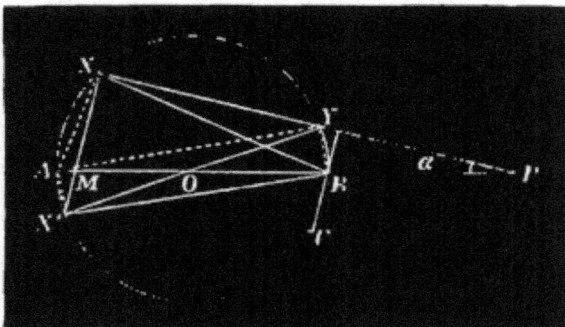

The problem is thus reducible to Ex. 2.]

4. If a given finite line be divided into any number of parts a, b, c ... ; to find when $a^\alpha b^\beta c^\gamma$ ... is a maximum, where α, β, γ are given quantities.

[This expression is a maximum when

$$\left(\frac{a}{a}\right)^\alpha \left(\frac{b}{\beta}\right)^\beta \left(\frac{c}{\gamma}\right)^\gamma \dots \text{ is a maximum.} \dots\dots\dots\dots \quad (1)$$

but a/α is one of the α equal parts into which the segment a may be divided; hence $(a/\alpha)^\alpha$ is the product of the equal subdivisions. Similarly $(b/\beta)^\beta$ is the product of the β equal subdivisions of b, and so on. Therefore (1) attains its greatest value when the subdivisions of a, b, c ... are all equal; *i.e.*, when

$$\frac{a}{\alpha} = \frac{b}{\beta} = \frac{c}{\gamma} = \dots]$$

5. Find a point O with respect to a triangle such that the product of the areas $(BOC \times COA \times AOB)$ is a maximum.

[Since $BOC + COA + AOB$ is constant, when $BOC = COA = AOB$, by Ex. 4, or when O is the centroid of the triangle.]

6. The maximum triangle of given perimeter is equilateral.

[From the formula $\Delta^2 = s(s-a)(s-b)(s-c)$; since the sum of the factors on the right hand side is constant, Δ is a maximum when $s-a=s-b=s-c$; therefore, etc.]

7. The maximum parallelogram of given perimeter and angles is equilateral.

8. If p_1, p_2, p_3 denote the perpendiculars from any point O on the sides of a triangle, the maximum value of $p_1 p_2 p_3$ is $8\Delta^3/27abc$, and O is then the centroid of the triangle. (By **Ex. 5.**)

[Otherwise thus :—Since $4abp_1p_2 \equiv (ap_1 + bp_2)^2 - (ap_1 - bp_2)^2$ for any point O on the base c, p_1p_2 is maximum when $ap_1 - bp_2$ vanishes, since $ap_1 + bp_2$ equals 2Δ. Then O is the middle point of the base. Now if p_3 be supposed constant, O is on the median through C. Similarly by regarding p_1 as constant, O would be found on the median through A ; and so on. Therefore if the three perpendiculars vary, their product is a maximum for the point of intersection of the medians.]

15. Theorem.—*If a right line be divided into any two parts a and b the sum of their squares is a minimum when the line is bisected.*

For (Euc. II. 9, 10)

$$a^2 + b^2 = 2\left(\frac{a+b}{2}\right)^2 + 2\left(\frac{a-b}{2}\right)^2.$$

Hence $a^2 + b^2$ is minimum when $a - b$ is minimum, because $a+b$ is constant ; that is when $a=b$.

Cor. The sum of the squares of the segments of a line is a minimum when the segments are equal.

16. Problem.—*If a right line be divided into any number of parts a, b, c ..., to find when*

$$\frac{a^2}{\alpha} + \frac{b^2}{\beta} + \frac{c^2}{\gamma} + \dots \text{ is minimum}$$

where α, β, γ are known quantities.

Let the segment a be divided into α equal parts ; each part is therefore a/α and the sum of squares of the parts

$$= \alpha\left(\frac{a}{\alpha}\right)^2 = \frac{a^2}{\alpha}.$$

Similarly if the segment b be divided into β equal parts the sum of squares of the subdivisions $= b^2/\beta$; and so on.

Hence the above expression denotes the sum of the squares of the subdivisions of the parts a, b, c ..., and is therefore a minimum when these are equal; *i.e.*, when

$$\frac{a}{\alpha} = \frac{b}{\beta} = \frac{c}{\gamma} = \ldots$$

EXAMPLES.

1. Divide a line into two parts a and b such that
$$3a^2 + 4b^2 \text{ is a minimum.}$$

[When $\dfrac{a^2}{4} + \dfrac{b^2}{3}$ is minimum, *i.e.*, when $\dfrac{a}{4} = \dfrac{b}{3}$, hence $3a = 4b$.]

2. To find a point P such that the sum of the squares of its distances, x, y, z, from the sides of a triangle is a minimum.

[Let Δ_1, Δ_2, Δ_3 denote twice the areas of the triangles subtended by the sides of the given one at the point. Now since $\Delta_1 = ax$, $\Delta_2 = by$, and $\Delta_3 = cz$,

$$x^2 + y^2 + z^2 = \frac{\Delta_1^2}{a^2} + \frac{\Delta_2^2}{b^2} + \frac{\Delta_3^2}{c^2} \dotfill (1)$$

and is consequently a minimum when

$$\frac{\Delta_1}{a^2} = \frac{\Delta_2}{b^2} = \frac{\Delta_3}{c^2} \dotfill (2)$$

since $\Delta_1 + \Delta_2 + \Delta_3 = \text{const.}$

From (2) it is obvious that $\quad \dfrac{x}{a} = \dfrac{y}{b} = \dfrac{z}{c} \dotfill (3)$

This result may also be seen from the identity

$$(a^2 + b^2 + c^2)(x^2 + y^2 + z^2) - (ax + by + cz)^2$$
$$\equiv (bz - cy)^2 + (cx - az)^2 + (ay - bx)^2,$$

with which the student should be familiar.]

NOTE.—This point is termed the *Symmedian Point* of the triangle, as it is obvious from (3) that the lines joining it to the vertices of the given triangle make the same angles with the sides as the corresponding medians; also since

$$\frac{x}{a} = \frac{y}{b} = \frac{z}{c} = \frac{ax + by + cz}{a^2 + b^2 + c^2} = \frac{2\Delta}{a^2 + b^2 + c^2},$$

$$x = \frac{2a\Delta}{a^2 + b^2 + c^2}, \quad y = \frac{2b\Delta}{a^2 + b^2 + c^2}, \quad z = \frac{2c\Delta}{a^2 + b^2 + c^2}.$$

3. Find a point P such that the sum of squares of its distances from the vertices of a triangle may be a minimum.

[If CP be supposed constant while AP and BP vary, the point P describes a circle around C as centre, and if M be the middle point of the base $AP^2 + BP^2 = 2AM^2 + 2MP^2$. Hence $AP^2 + BP^2 + CP^2$ is minimum when $2PM^2 + CP^2$ is minimum, since AM is constant. Therefore P is a point on the median CM such that $CP/PM = 2$, i.e., the centroid.

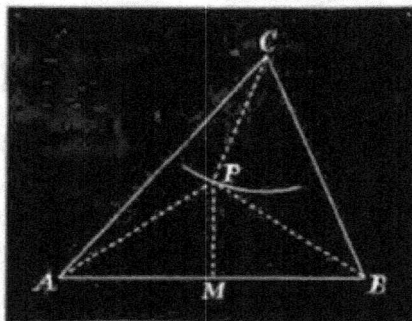

Similarly by supposing AP or BP to remain constant we find the same point. Hence the centroid is the required point when AP, BP and CP all vary.]

SECTION II.

METHOD OF INFINITESIMALS.

17. It has probably been observed in the preceding section that the positions of maximum and minimum of a quantity, varying according to a given law, are symmetrical with respect to the fixed parts of the figure. Thus when the base and vertical angle of a triangle are given, the altitude, rectangle under sides, area, etc., etc., are maxima when the triangle is *isosceles*.

In Art. 9 the triangle of maximum area is found by placing the two given sides at *right* angles:

Again, a figure of given perimeter and of maximum area is *circular*. As the variable line AB in Art. 11 rotates in a positive direction around P, according as PB recedes from the perpendicular from P on BC, the segments AP and BP approach an equality, and the triangle ABC is a minimum when $AP = BP$.

18. The several parts, of a geometrical figure which varies according to a definite law, can always be expressed in terms of the fixed parts of the figure and those quantities which are sufficient to define its position.

Take for example the figure of Art. 8. In any position of the vertex C, by assuming the triangle to be of given altitude; the variable parts, a, b, area, and other functions of the sides or angles can be found in terms of the base c, vertical angle C, and altitude.

Thus the variables may be regarded as functions of the given parts and the *co-ordinates* of their position.

It follows, then, that if the latter vary continuously those functions must do likewise.* Hence a very small change in position will cause a very slight change or *increment* in the magnitude of the function. Suppose in Art. 8 the circle to be divided into an indefinitely great number of equal parts, and let the vertex C occupy each point of section from A towards B. As the *altitude* thus receives indefinitely small increments so does the area.

Let AB be the base of a triangle and any curve CC_1C_2 the locus of its vertex.

---

* See Burnside and Panton's *Theory of Equations*, Art. 7.

c

In the figure as the vertex approaches C on the curve from left to right the intercept AX made by the perpendicular may be taken as the co-ordinate of its position, since if AX is known the position of C is also known.

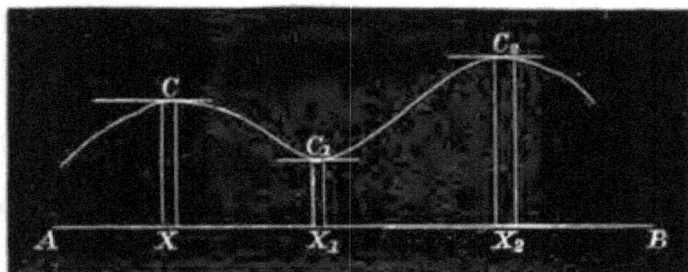

Thus while AX continues to receive positive increments, the area, altitude, and other functions of it are sometimes decreasing, as from C to C_1, and sometimes increasing, as from C_1 to C_2.

At the points C, C_1, C_2 the increments in the altitude alter in sign *and therefore consecutive values are equal.* Here also the tangents to the curve are parallel to the base AB, and at any other point C_n the increment of the variable divided by the corresponding increment in the function $= \cot a$, where a is the angle made by the tangent at C_n with AB. We have seen that if AX denote the value of a variable in any position, and CX any function of AX, when the function passes through a maximum or minimum its two consecutive values are in each case equal to one another.

Suppose, for example, that a variable chord XY of a circle moves parallel to a certain direction; it gradually increases in length as it approaches the centre and if XY

.be a diameter and $X'Y'$ a consecutive chord; since XX' and YY' are tangents to the circle, and therefore parallel, $XYX'Y'$ is a parallelogram and $XY=X'Y'$ (Euc. I. 34). Hence the diameter is the maximum chord in a circle (cf. Euc. III. 15).

EXAMPLES.

1. Having given the base and locus of vertex of a triangle; find when the area is a maximum or minimum.

[Let the locus be a curve of any order and it is readily seen (Euc. I. 39) that the tangents at the required points are parallel to the base.]

2. In Ex. 1 when is the sum of the sides a minimum or maximum?

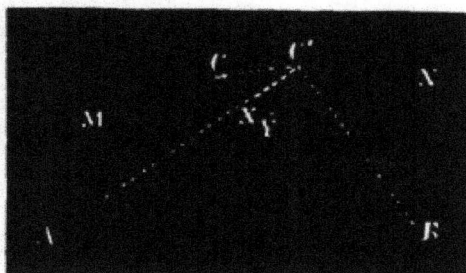

[Let C and C' be two points indefinitely near to each other on the locus MN. Draw CX and $C'Y$ perpendiculars to AC' and BC respectively.

Then since in the triangle ACX, X is a right angle and A indefinitely small, ACX is approximately a right angle and AC is nearly equal to AX. Hence in the limit

$$C'X = AC' - AX = AC' - AC.$$

Similarly CY is the increment (negative) of BC.

Therefore $C'X = CY$ and the right-angled triangles $CC'X$ and $CC'Y$ are equal in every respect, and $\angle AC'C = \angle BCC'$. But $AC'C = ACM$ when A is indefinitely small; hence the required points C on the locus are such that AC and BC are equally inclined to the curve, *i.e.*, to the tangent at their point of intersection.

It similarly follows that if A and B were upon opposite sides of the curve this relation holds when $AC - BC$ is maximum or minimum.*]

3. Given the vertex A of a triangle fixed, the angle A in magnitude and the base angles moving on fixed lines intersecting in O; to construct the triangle ABC of minimum area.

[By taking two consecutive positions as in figure, we have

$$AB \cdot AC = AB' \cdot AC' \text{ and } \angle BAB' = \angle CAC'.$$

Hence $$AB : AB' = AC' : AC,$$

and the triangles BAB' and CAC' are similar (Euc. VI. 6).]

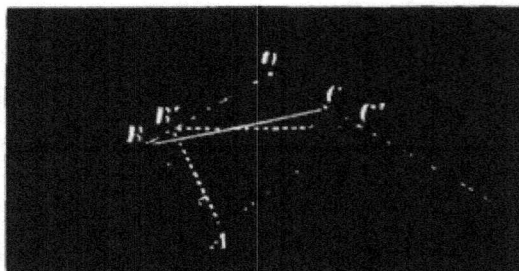

Therefore $\angle ABO = \angle AC'O = ACO$ in the limit

In the required position the sides AB and AC are *equally* inclined to the given lines. Here again we have an illustration of the symmetry of the figure when the triangle is minimum. If the angle A is 180° the property (Art. 13) follows at once.]

4. Given two sides of a triangle fixed in position and a point P on the base; when is AB a minimum?

[Taking two consecutive positions of AB and drawing perpendiculars AX and BY; as before $A'X$ is the increment of AP and $B'Y$ of BP; hence $A'X = B'Y$.

Again $\qquad A'X = AX \cot A' = AP \sin P \cdot \cot A'$.

Similarly $\qquad B'Y = BY \cot B = BP \sin P \cot B$.

Therefore in the limit

$$AP \cot A = BP \cot B.$$

---

* It follows if the curve is of such a nature that $AC + BC$ is constant then for every point on it AC and BC are equally inclined.

But if Q denote the foot of the perpendicular on the base we have

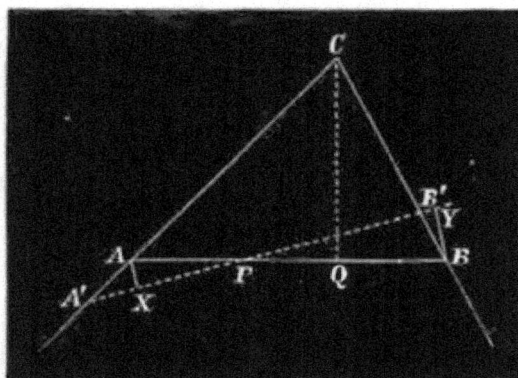

$$BQ \cot A = AQ \cot B,$$

hence
$$AP = BQ,$$

or the minimum chord is such that the *given point P and the foot of the perpendicular are equidistant from the extremities of the base.*

This is known as Philo's Line.

5. Through a given point O in the diameter produced of a semi-circle to draw a secant OBC such that the quadrilateral $ABCD$ may be a maximum.

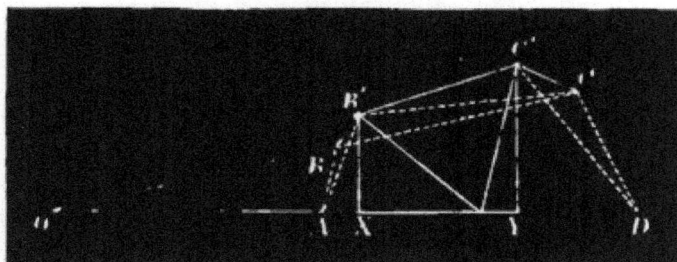

[Take two consecutive positions of the secant OBC and $OB'C'$ such that $ABCD = AB'C'D$, and join AB, AB', DC, DC', and $B'C$.

Now since $ABCD = AB'C'D$ it follows that

$$BB'CC' = ABB' + DCC'$$

or
$$BB'C + CB'C' = BB'A + CC'D.$$

Transposing we have
$$BB'C - BB'A = CC'D - CB'C',$$
or since twice the area of a triangle is the product of two sides × the sine of the included angle; in the limit this relation becomes
$$\frac{BB'(BC^2 - AB^2)}{\text{diameter}} = \frac{CC'(CD^2 - BC^2)}{\text{diameter}};$$
but from similar triangles $BB'/CC' = OB/OC$. Hence if $AB = a$, $BC = b$, $CD = c$, $AD = d$, and the angles subtended at the centre of the circle by the sides a, b, c be denoted by $2a$, 2β, 2γ, this relation may be written
$$\frac{b^2 - a^2}{c^2 - b^2} = \frac{OC}{OB},$$
which is easily reducible to
$$\cos 2a + \cos 2\gamma = 1,$$
or the *projection XY of the intercept is equal to the radius of the circle.* The construction of the chord BC will be afterwards given.]

6. Having given two opposite sides AB and CD of a quadrilateral and the diagonals CA and BD, to construct it so that the area may be a maximum.

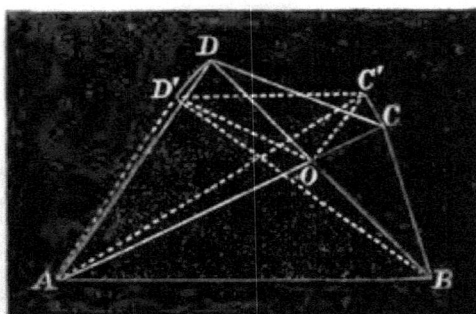

[Let AB be fixed and draw C' and D' consecutive positions of C and D. Let O be the intersection of AC and BD. Then since CC' is small compared with OC and OCC' a right angle; OCC' may be considered an isosceles triangle, and $OC = OC'$. Similarly $OD = OD'$; and since $CD = C'D'$ the triangles COD and $C'OD'$ are equal in every respect. From the equal areas $ABCD$ and $ABC''D'$ take the equals COD and $C'OD'$ and the common part AOB, and there remains $\qquad BOC + AOD = BOC' + AOD',$

or $$BOC' - BOC = AOD - AOD',$$

hence $$BO . CO = AO . DO,$$

from which it is manifest that CD and AB are parallel. Cf. Art. 9, Ex. 5.

A similar proof may be applied to show that when the four sides of a quadrilateral are given the area is a maximum when

$$CO . AO = BO . DO,$$

i.e., when the figure is cyclic. See Milne's *Companion to the Weekly Problem Papers*, 1888, p. 27.]

7. To draw a parallel to a given line meeting a semicircle in C and D such that $ABCD$ is a quadrilateral of maximum area.

[As before, when $ABCD$ is a maximum it is equal to the consecutive area $ABC'D'$.

Hence $$CC'DD' = ACC' + BDD',$$

therefore $$CC'D - CC'A = DD'B - DD'C',$$

which in the limit reduces to

$$b^2 - a^2 = c^2 - b^2 \text{ or } 2b^2 = a^2 + c^2 \quad \ldots\ldots\ldots\ldots\ldots\ldots\ldots(1)$$

Again if X and Y are the projections of C and D on the diameter d of AB we have

$$AX = a^2/d, \ BY = c^2/d \text{ and } XY = b \cos \alpha.$$

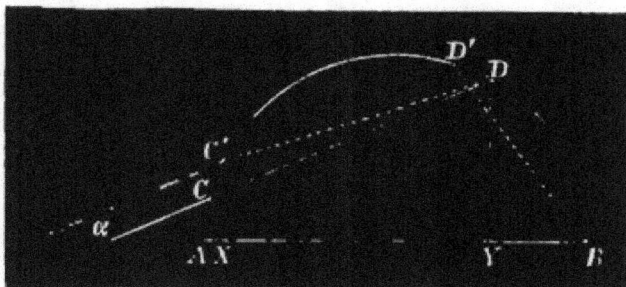

Making these substitutions in (1) we have on reducing

$$2b^2 + d \cos \alpha . b - d^2 = 0 \quad \ldots\ldots\ldots\ldots\ldots\ldots\ldots\ldots(2)$$

NOTE.—If $a = 0$ the quadrilateral is found to be one half of the inscribed hexagon.

If $a = 90$ the maximum quadrilateral is the inscribed square.

SECTION III.

THE POINT O THEOREM.

19. Theorem.—*If points P, Q, and R be taken on the sides of a triangle the circles AQR, BRP, and CPQ pass through a common point O.*

For let the circles AQR and BRP meet in O. Then since (Euc. III. 22) $QOR = \pi - A$ and $ROP = \pi - B$, we have $QOP = 2\pi - (\pi - A) - (\pi - B) = A + B = \pi - C$; therefore the quadrilateral $POQC$ is cyclic.

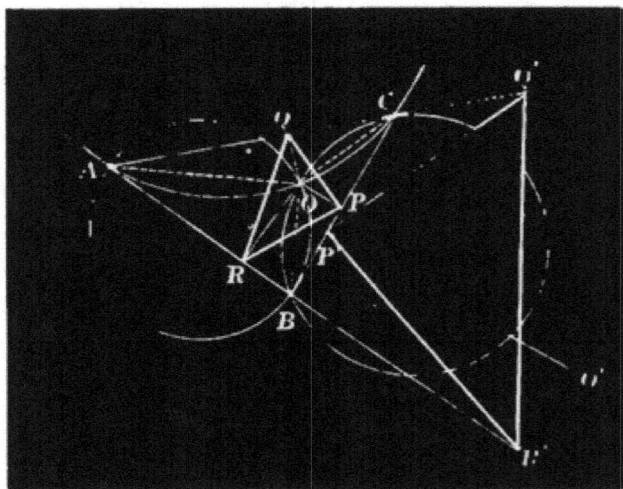

The angles BOC, COA, AOB, subtended by the sides of the given triangle at O, are respectively $A + P$, $B + Q$, $C + R$, when O is within the triangle ABC.

For, applying Euc. I. 32 to the triangles BOC and COA, it follows that $\angle AOB = C + CAO + CBO$.

But $\qquad CAO = QRO$ since $AQRO$ is cyclic,

also $\qquad CBO = PRO$ since $BPRO$ is cyclic;

therefore $\qquad AOB = C + R$ .... (α)

where R denotes an angle of the triangle PQR. Similarly for the angles BOC and COA.

If O falls outside the triangle ABC these angular relations become somewhat modified. Take for example O within the angle C.

Then from the cyclic quadrilaterals $QRAO$ and $RPBO$ we have (Euc. III. 20)

$$\angle ORP = OBP \text{ and } \angle ORQ = OAQ;$$

adding these equations

$$R = OAQ + OBP = C + AOB,$$

or $\qquad AOB = R - C.$

Again, since Euc. I. 32,

$$A + ACO = BOC + ABO,$$

by transposing

$$A - BOC = ABO - ACO \text{ .... (1)}$$

But $\qquad ABO = RPO$ since $PRBO$ is cyclic,

and $\qquad ACO = QPO$ since $PQCO$ is cyclic.

Substituting these values in (1) we have

$$A - BOC = RPO - QPO = P;$$

therefore $\qquad BOC = A - P.$

Similarly $\qquad COA = B - Q.$ .... (β)

It may be shown in the same manner that if the points P, Q, R are such that two of the angles P, Q of the triangle formed by them are greater than A and B respectively $\qquad BOC = P - A,$

$$COA = Q - B, \text{ .... (γ)}$$

and $\qquad AOB = C - R.$

Hence if a triangle PQR of given species be inscribed in a given one ABC, the circles AQR, BRP, and CPQ

pass through either of two fixed points, one of which
subtends at the sides of ABC, angles $A+P$, $B+Q$, $C+R$,
and the other $A-P$, $B-Q$, $R-C$, or $P-A$, $Q-B$, $C-R$,
according as two of the angles of the given triangle are
greater or less than the corresponding angles of the
inscribed triangle.

20. Let PQR be a triangle of given species inscribed in
ABC. We have seen that the point O is fixed, and
therefore the lines AO, BO divide the angles of ABC into
known segments. But the segments of A are equal to
the base angles of the triangle QOR; similarly of B to
the base angles of ROP, and of C to the base angles of
POQ.

Hence each of the triangles POQ, QOR, ROP are given
in species. Therefore as the inscribed triangle PQR varies
in position OQR, ORP, OPQ remain constant in species,
and $OP : OQ : OR$ are constant ratios.

Again, since the triangle OPQ is fixed in species and
one vertex O a fixed point; if P describes a line BC it
follows that the locus of Q is also a line (CA). And
generally, *when one vertex of a figure of given species is
fixed and any other vertex P or point invariably connected with it describe a locus, the remaining points Q ...
describe loci, which may be derived from P by revolving
it through a known angle POQ and increasing or
diminishing OP in the ratio of $OQ : OP$.*

The loci thus described are similar, the ratio $OP : OQ$ is
termed their *Ratio of Similitude* and the point O the
Centre of Similitude.

Thus since O is a point invariably connected with a
variable inscribed triangle PQR of given species, the

ortho-centre, circum-centre, ex-centres, median point, etc., and all other points invariably connected with the triangle, describe right lines which can at once be constructed by the above method.

Moreover, we know that if O is fixed and P describes a circle, and the variable line or *Radius Vector OP* be divided in Q, in a given ratio, the locus of Q is a circle. Now if Q be turned around O through any given angle the locus is the same circle displaced through the same angle. Therefore if *one vertex of a triangle of given species is fixed, and another vertex describe a circle, the remaining vertex and all other points invariably connected with it likewise describe circles.*

EXAMPLES.

1. Having given the diagonals and angles of a quadrilateral *ABCD*, construct it.

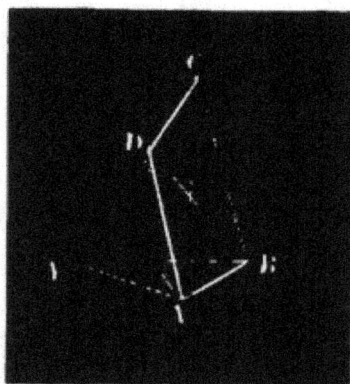

[On one diagonal *A C* describe segments of circles containing angles respectively equal to *B* and *D*. Let *ABCD* be the required quadrilateral. Produce *CD* to *Y* and *BC* to *X*. Join *BY* and *AY*.

Then since the chord *BY* of a given circle subtends a given angle *C* it is of known length. The triangle *ADY* is also given in species; hence the following construction :—On *BY* describe a segment of a circle containing an angle *C*. The triangle *ADY*, of given species, has one vertex Y fixed, another A describing the circle AYC, therefore the remaining vertex D describes a circle. Take B as centre and BD as radius, and cut this locus in the point D; therefore, etc.*]

2. Required to place a parallelogram of given sides with its vertices on four concurrent lines (M'Vicker).

[Let $ABCD$ be the parallelogram situated on the pencil $O . ABCD$. Through C and D draw parallels CP and DP to BO and AO respectively. Join OP. By Ex. 1 the diagonals and angles of the quadrilateral $CDPO$ are given; therefore, etc.]

21. When the triangle PQR is given in every respect, the triangles OPQ, OQR, ORP are completely determined; for in addition to their species we are given the sides PQ, QR, and RP, hence the sides OP, OQ, OR are easily determined. We have therefore four solutions, real or imaginary, to the problem :—

Having given two triangles ABC and PQR to place either with its vertices on the corresponding sides of the other ; for having determined the point O, the position of which depends altogether on the species of the triangles, we get the position of the vertex P by taking O as centre and OP as radius and describing a circle cutting BC.

22. When the line OP is perpendicular to BC, OQ and OR are therefore perpendiculars to CA and AB respectively, and the circle with O as centre and OP as radius touches BC. In this case the two solutions coincide, *and PQR is the minimum triangle of given species that can be inscribed in ABC.*

23. It is manifest that *a given triangle ABC may be escribed to another PQR.* For having determined the point O, the triangles BOC, COA, and AOB are given in species, and are therefore completely determined, since

---

* For other solutions see " *Mathematics from the Educational Times*," Vol. XLIV., p. 29, by D. Biddle and Rev. T. C. Simmons.

BC, CA, and *AB* are given lines. Hence any vertex (*C*) is found by describing a segment of a circle upon *PQ* containing an angle equal to *C*, and with *O* as centre and *OC* as radius describing circle. Where these circles meet is the required position of *C*.

Again in the triangle *BOC* when *BC* is a maximum *OC* is a maximum, and is therefore a diameter of the circle *OPQC*. Then *OPC* is a right angle. Hence *the maximum triangle of given species escribed to a given one is that whose sides are perpendicular to OP, OQ, OR.*

Cor. If the sides of the given escribed triangle be λ, μ, and ν, and α, β, γ the distances of *O* from *P, Q, R*,

$$\lambda\alpha+\mu\beta+\nu\gamma=\text{a minimum.}$$

Hence *required to find a point, given multiples of whose distances from three fixed points is a minimum when any two of the multiples are together greater than the third.*

Examples.

1. If *d* denote the distance of the point *O* from the circumcentre *II* of the triangle *ABC*; prove that twice the area of the minimum triangle *PQR* is $(R^2 \sim d^2) \sin A \sin B \sin C$.

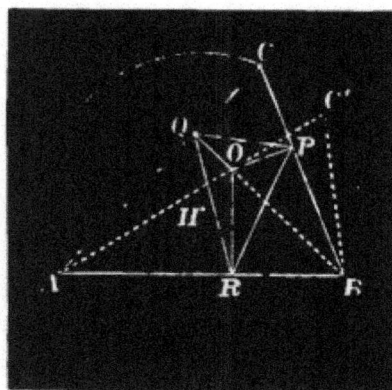

[For join *AO* and produce it to meet the circum-circle again in *C'*; join *BC'*.

Now since $\angle R = AOB - C = AOB - C' = OBC'$ (Euc. I. 32),

we have $2PQR = RP \cdot RQ \sin R = RP \cdot RQ \sin OBC'$ ....(1)

but $RP = OB \sin B$ and $RQ = OA \sin A$.

Substituting these values in (1) and putting

$$OB \sin OBC' = OC' \sin C',$$
$$2PQR = AO \cdot BO \sin A \sin B \sin OBC'$$
$$= AO \cdot OC' \sin A \sin B \sin C$$
$$= (R^2 \sim d^2) \sin A \sin B \sin C.]$$

NOTE.—If the point O is on the circum-circle $R = d$ and the area of the triangle vanishes, hence if *from any point on the circum-circle of a triangle perpendiculars be let fall upon the sides their feet lie in a line.* This is termed a *Simson Line* of the triangle, and the collinearity of the points admits of an easy direct proof.

2. If the pedal triangle PQR of a point O is constant in area the locus of the point is a circle.

[Concentric with the circum-circle by the equation of Ex. 1.]

2a. The theorem holds generally for a polygon.

3. Having given of a triangle the base c, and $ab \sin (C - a)$ where a is a given angle, find the locus of the vertex.

[In Ex. 1 we have

$$2PQR = AO \cdot BO \sin A \sin B \sin (AOB - C)$$
$$\propto AO \cdot BO \sin (AOB - C),$$

and the locus of O is in that case a circle. Hence in the triangle AOB we have the data in question; therefore the locus of the vertex is a circle concentric with H.]

4. To inscribe a quadrilateral of given species $PQRS$ in a given quadrilateral $ABCD$.

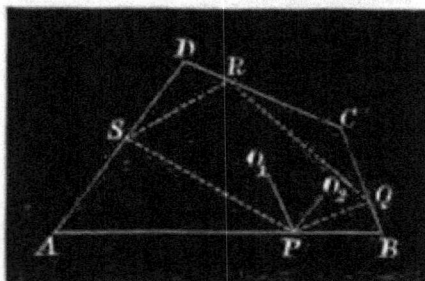

Find the point O_1 of the triangle PQR of given species inscribed in a given one, viz., that formed by three of the sides, AB, BC, CD of the quadrilateral. Similarly find O_2 of the triangle PQS inscribed in a given one. Now by Art. 19, since the species of each of the triangles O_1PQ and O_2PQ is given, we have $\angle O_1PO_2 = O_2PQ \sim O_1PQ =$ a known quantity; therefore the point P is determined.

5. To escribe a quadrilateral $ABCD$ of given species to a given one $PQRS$.

[Take any quadrilateral $abcd$ of the same species as $ABCD$. Inscribe in it by Ex. 4 a quadrilateral $pqrs$ of the species $PQRS$. It is obvious that $\angle SPA = spa$, since the figures are similar, hence the problem reduces to drawing lines in known directions through $P, Q, R, S.$

Otherwise thus :—

Upon a pair of opposite sides PQ and RS describe segments of circles containing angles equal to B and D respectively. Find a point M such that the arcs PM and QM subtend angles equal to ABD and CBD respectively. Similarly find N such that CDN and ADN may be equal to the known segments of the angle C. Join MN; where it meets the circles in B and D are two of the required vertices of the quadrilateral $ABCD$.]

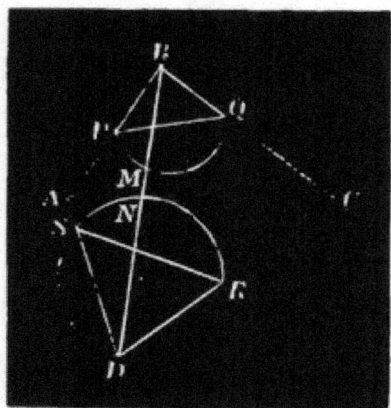

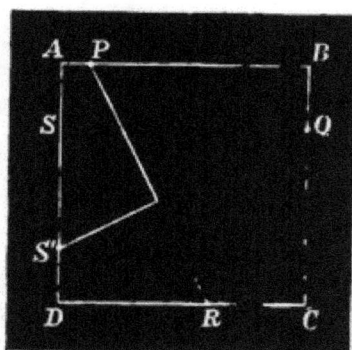

6. To escribe a square $ABCD$ to a quadrilateral $PQRS$.

[By Ex. 5 or simply thus :—Join PR and let fall a perpendicular from Q upon it. Make $QS' = PR$. SS' is a side of the required square. This construction depends upon the property that *any two rectangular lines terminated by the opposite sides of a square are equal to one another* (Mathesis).

7. From any point P on the base of a triangle perpendiculars PX and PY are drawn to the sides, find the locus of the middle point N of XY.

[Bisect CP in M, join MX, MY and MN. It is easy to see that MXY is an isosceles triangle of given species, each of its base angles being the complement of C; and since its vertices X, M, Y move on fixed lines, any point N invariably connected with it describes a line. By taking P to coincide alternately with A and B the locus is seen to be the line joining the middle points of the perpendiculars from the extremities of the base of the triangle ABC.]

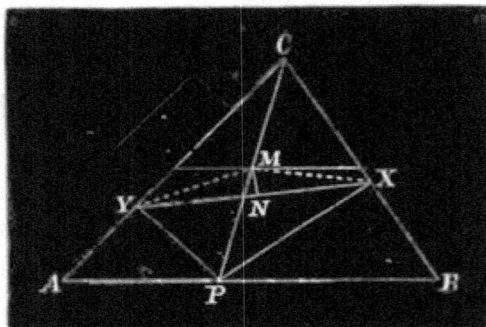

8. The sides of the pedal triangle PQR are in the ratios
$$a \cdot AO : b \cdot BO : c \cdot CO.$$
[For $QR = AO \sin A \varpropto a \cdot AO$, etc.]

9. **Extension of Ptolemy's Theorem.**—If the three pairs of opposite connectors of four points be denoted by a, c; b, d; δ, δ' to prove the relation
$$\delta^2 \delta'^2 = a^2 c^2 + b^2 d^2 - 2abcd \cos (\theta + \theta'),$$
where $\theta + \theta'$ is the sum of a pair of opposite angles of the quadrilateral.

[Let A, B, C, O be the four points. From any one of them O let fall perpendiculars OP, OQ, OR on the sides of the triangle ABC formed by the remaining three; then since
$$PQ^2 = QR^2 + RP^2 - 2QR \cdot RP \cos R,$$
substituting for PQ, QR, RP the values in Ex. 8, and reducing, the above equation follows at once (M'Cay).]

9a. What does this theorem reduce to for the quadrilateral $ABCP$ in the figure of Ex. 7? Deduce the relation of Art. 3, Ex. 5, as a further particular case.

10. A variable circle passes through the vertex of an angle and a second fixed point ; find the locus of the intersection of tangents at the extremities of its chord of intersection.

11. If a, β, γ denote the distances of any point O from the sides of a triangle ; to prove that

$$a\beta\gamma = \frac{SS'}{2R}$$

where S and S' are the rectangles under the segments of a variable chord through O^* of the circum-circles of ABC and of the pedal triangle of the point O (M'Vicker).

[In Ex. 1 let K be the point where RO meets the circum-circle of PQR ; then $\gamma = S'/OK = S' \sin P/\beta \sin OQK$.

But $\sin OQK = \sin(A + P) = \sin BOC$; $\therefore \beta\gamma = S' \sin P/\sin BOC$. Also $a = OB . OC \sin BOC/a$, therefore $a\beta\gamma = S' . OB . OC \sin P/a$.

Again $OB = RP/\sin B$, etc. ... therefore by substitution

$$a\beta\gamma = \frac{S' . RP . PQ \sin P}{a \sin B \sin C} = \frac{S' . PQR}{\Delta/2R} = \frac{SS'}{2R}.]$$

12. In the particular cases when O coincides with the in- or ex-centres of the triangle ABC, the formula in Ex. 11 reduces to

$$\delta^2 = R^2 - 2Rr \text{ or } \delta_1{}^2 = R^2 + 2Rr, \text{ etc.}$$

24. Theorem.—*When three points P, Q, R are taken collinearly on the sides of a triangle, the circles circumscribing the four triangles QRA, RPB, PQC, ABC meet in a point.*

This theorem may be easily proved directly, but it is obviously a particular case of Art. 19, for the circles QRA, RPB, PQC meet in a point O (Art. 19) such that $COA = Q - B$, which in this case is $180 - B$; therefore, etc. Euc. III. 22.

The transversal PQR to the sides of ABC is the limiting case of a triangle inscribed in ABC, the angles at P and

---

* The constant rectangle under the segments of a variable chord of a circle passing through a fixed point has been termed by Steiner the *Power of the Point* with respect to the circle.

R being each $0°$ and $Q = 180°$. The species of the limiting triangle is determined by the ratios $QR : RP : PQ$, or their equivalents $a . AO : b . BO : c . CO$. (Art. 23, Ex. 8.)

Hence if a transversal is drawn to a triangle such that the ratios of its segments made by the sides is constant; the ratios $AO : BO : CO$ are known and with them the point O. As in the general case, the triangles QOR, ROP, POQ are constant in species.

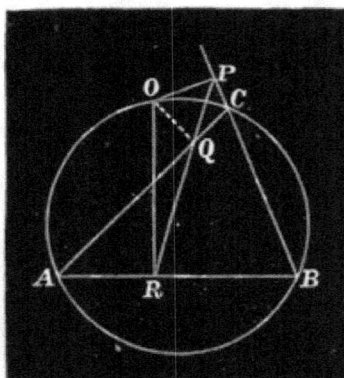

It follows then that if P, Q, R be the feet of the perpendiculars from O on the sides of ABC, *and the lines OP, OQ, OR rotated through any angle in the same direction, P, Q, R will always remain collinear and the ratios PQ : QR : RP are constant.**

Cor. **Ptolemy's Theorem.**—Since $QR:RP:PQ = a . AO : b . BO : c . CO$, and $PQ + QR = PR$;

therefore $a . AO + c . CO = b . BO$.

Examples.

1. Place a given line PQ divided in any point R such that the points P, Q, R may lie in an assigned order on the sides of a given triangle.

---

* Chasles' Géométrie supérieure, p. 281.

2. Draw a line across a quadrilateral, meeting the sides in $PQRS$ such that the ratios $PQ : QR : RS$ may be given.

3. The line joining O to the orthocentre of ABC is bisected by the Simson line PQR, and intersects it on the nine points circle.

4. The angle subtended by any two points O_1 and O_2 on the circle is equal to the angle between their Simson lines.

5. The Simson lines of two points diametrically opposite intersect at right angles on the nine points circle. (By Ex. 4.)

25. **Theorem**. — For *three positions*, PQR, $P_1Q_1R_1$, $P_2Q_2R_2$, *of the triangle of given species inscribed in a given one ABC; to prove that*

$$PP_1 : PP_2 = QQ_1 : QQ_2 = RR_1 : RR_2.$$

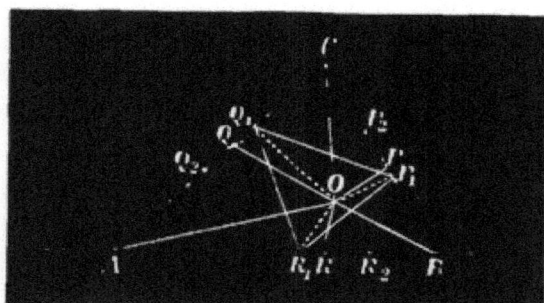

Since the triangles OPQ, OP_1Q_1, OP_2Q_2 are similar, we have $OP : OP_1 = OQ : OQ_1$, also $\angle POP_1 = QOQ_1$ since $\angle POQ = P_1OQ_1$; therefore the triangles POP_1 and QOQ_1 are similar. Hence

$$PP_1 : QQ_1 = OP : OQ.$$

Similarly $QQ_1 : RR_1 = OQ : OR;$

therefore $PP_1 : QQ_1 : RR_1 = OP : OQ : OR.$

Similarly $PP_2 : QQ_2 : RR_2 = OP : OQ : OR;$

therefore, etc.

Now if $P_1Q_1R_1$ and $P_2Q_2R_2$ denote two fixed positions of the variable inscribed triangle PQR of constant species, and PQR any arbitrary position, it follows that *a variable line PQ, dividing similarly two linear segments P_1Q_1 and P_2Q_2, subtends a constant angle POQ at a fixed point O.*

The point O is determined by the intersection of the loci of the vertices of the triangles P_1Q_1O and P_2Q_2O, whose bases P_1Q_1 and P_2Q_2 are given and ratio of sides $(=P_1P_2 : QQ_1)$, *or the intersection of the circles CP_1Q_1 and CP_2Q_2.*

Since P_1P_2 and Q_1Q_2 form similar triangles with O, this point is termed the *Centre of Similitude* of the segments. Thus the centre of similitude of two segments AB and CD is the intersection of the circles passing through the two pairs of non-corresponding extremities and the intersection O of the given lines. Or it may be regarded as the common vertex of two similar triangles described on the sides.

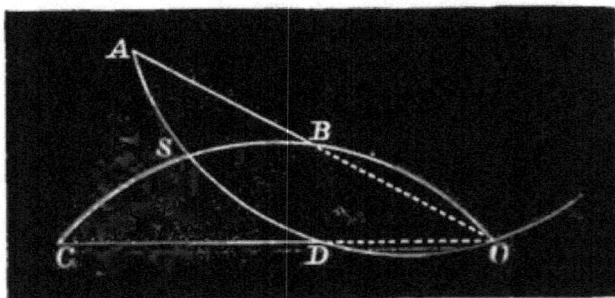

If the points B and D coincide, O coincides with them, and the circle ADO meeting CD in coincident points D and O therefore touches CD. In the same case the circle BCO touches AB.

Cor. The centres of similitude of the sides of a triangle taken in pairs are therefore found by describing circles on BC and AC touching the sides AC and BC respectively. The second point of intersection of these circles is a centre of similitude of AC and BC; similarly for each of the remaining pairs of sides.

EXAMPLES.

1. Draw a line L dividing three linear segments A_1A_2, B_1B_2 and C_1C_2 in the same ratio. (*Dublin Univ. Exam. Papers.*)

[Let the required line intersect the segments in P, Q and R respectively, O_1 and O_2 the centres of similitude of the pairs of lines A_1A_2, B_1B_2 and B_1B_2, C_1C_2. Then in the triangle O_1QO_2 we know the base O_1O_2 and vertical angle, since it is equal to $180 - O_1QP - O_2QR$; therefore, etc.]

2. The centres of similitude of the sides of a triangle taken in pairs are the middle points of the symmedian chords of the circumcircle.

[Let X, Y, Z denote the middle points of the sides of the triangle ABC; CD and CE the median and symmedian chords of the circle respectively; M the middle point of CE. Join ZE, AM and BM.

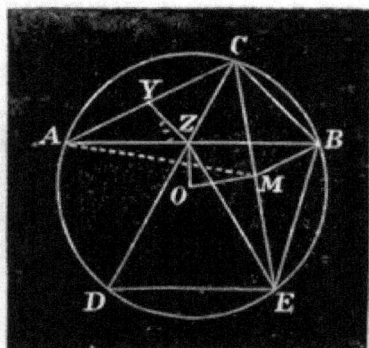

Then since $\angle ACD = BCE$ and $\angle CAZ = CEB$, the triangles ACZ and ECB are similar, and Y and M being the middle points of a pair of corresponding sides, CYZ and CMB are therefore similar. Hence $\angle CBM = CZY = BCZ = ACM$. Similarly $\angle CAM = BCM$; therefore the triangles BCM and CAM are similar.]

3. Prove the following results from **Ex. 2** :—

    1°. $CEZ =$ difference of base angles $(B-A)$.

    2°. The triangles ADZ and BEZ equal in every respect.

    3°. $CZ \cdot CE = ab$.

    4°. $CM = ab / \sqrt{a^2 + b^2 + 2ab \cos C}$.

    5°. $BMC = CMA = \pi - C$.

    6°. The circum-circle of ABM passes through the centre of the circle ABC.

4. Having given the base (c) bisector of base (CZ) and difference of base angles $(B-A)$; construct the triangle.

[The triangle CEZ is readily constructed ; therefore, etc.]

5. Having given the bisector of base (CZ) rectangle under sides (ab) and difference of base angles $(B-A)$; construct the triangle.

[As in Ex. 4.]

6. Having given the base, median, and symmedian of a triangle; construct it.

SECTION IV.

MISCELLANEOUS PROPOSITIONS.

26. Prop. I.—*Through a point P to draw a line across*

an angle such that the intercepted segment MN may subtend at a fixed point Q a triangle of maximum area.

The transversal PMN such that the parallels OM and ON to the sides of the angle intersect on PQ is the required line.

For draw any other line $PM'N'$. Join $M'N$. Then the triangles MON and $M'ON$ are equal (Euc. I. 37), but $M'ON > M'ON'$; therefore $MON > M'ON'$.

But $\dfrac{MON}{M'ON'} = \dfrac{MQN}{M'QN'}$ because $\dfrac{MON}{MQN} =$ ratio of the altitudes $= PO/PQ$. Similarly $\dfrac{M'ON'}{M'QN'} = PO/PQ$; therefore $MQN > M'QN'$.

To find the point O. Evidently by similar triangles

$$PA/PO = PM/PN = PO/PB\,;$$
therefore
$$PA \,.\, PB = PO^2.$$

Prop. II.—*On the sides BC and CA of a triangle, to find points M and N such that if the lines AM and BN meet in O the triangle MON may be a maximum.*

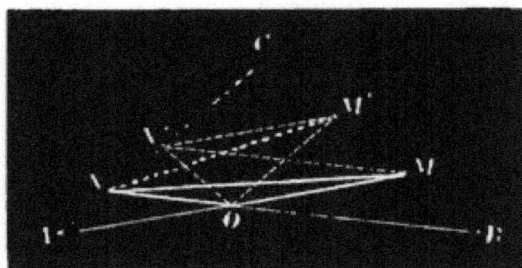

Regarding A as a point on the base produced of BCN and AOM a transversal to the sides, MON is maximum when ON' and MN' parallels to these sides respectively meet on AC. Similarly since B is on the base produced of ACM and BON a transversal to the sides, OM' and NM' parallels to the sides meet on the base.

Then we have $ANM'O$ and $CN'OM'$ equal parallelograms (Euc. I. 36), therefore $AN = CN'$, also $BM = CM'$. But by Prop. I. $AN \cdot AC = AN'^2$, therefore $AN \cdot AC = CN^2$; similarly $BM \cdot BC = CM^2$, or *the sides of the triangle ABC are divided in extreme and mean ratio, the greater segments being measured from the vertex.*

Prop. III. *Through one extremity A of the diameter APB of a semicircle draw a chord AMN to meet a perpendicular through P to the diameter AB in M and the circle in N, such that the triangle MBN may be a maximum.*

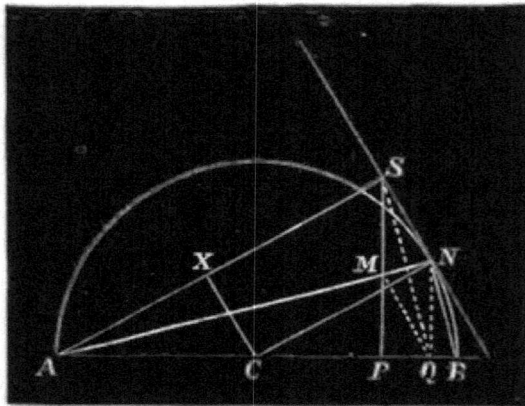

Suppose a tangent is drawn at the required point N. Let it meet PM in S. Join AS. From the centre C let fall CX perpendicular on AS. Join CN.

By Prop. I. the parallels MQ to the tangent and NQ to PS meet on AB, for then with respect to the angle PSN the triangle MBN is maximum; therefore *a fortiori* it is the maximum triangle whose vertex N lies on the circle ANB.

$\angle ANS = ABN = ANQ$. Hence since MN, the diagonal of a parallelogram $MSNQ$, bisects the angle N, the figure is a rhombus, and $NQ = NS$. Then the triangles ANS and ANQ are equal in every respect (Euc. I. 4), therefore ASN is a right angle; hence $CNSX$ is a rectangle, and SX is equal to the radius of the circle.

Also $CPSX$ is a cyclic quadrilateral, therefore

$$AS \cdot AX = AC \cdot AP$$

which is known. Therefore we have the rectangle and difference of AS and AX, from which data these lines are at once determined. Then we can construct the right-angled triangle ACX, which fixes the point X; therefore, etc.

Cor. In the particular case when PMS is a vertical radius, if SN meet the tangent AT in T and AB in T', we have $AS \cdot AX = r^2$, therefore by parallels $AT' \cdot AC = CT'^2$.

Similarly $TT' . TS = T'S^2$, but $TT' . TS = AT^2 = TN^2$; therefore $TN = T'S$, and $TS = T'N$.

But when a line TT' is divided in extreme and mean ratio in S and from the greater segment a part $T'N$ is taken equal to the less TS, $T'S$ is divided into extreme and mean ratio.

Ex. Draw the transversal AMN such that the quadrilateral $MNBP$ may be a maximum.

Prop. IV.* *Through a given point O in the tangent at C to a circle draw a secant AB such that the triangle ABC may be of maximum area.*

Draw tangents at A and B to meet in T. The required triangle is such that the parallels through A and B to the tangents at these points meet on OC in P.

For since O and T, O and C, are pairs of conjugate points with respect to the circle, CT is the polar of O.

---

* This Proposition may be omitted on the first reading.

Let OC', the second tangent from O, meet PT in C', Since $PBTA$ is a rhombus, AB is at right angles to PT; also since $TC'MP$ is a harmonic row, we have

$$TC'/C'M = TP/PM = 2;$$

therefore $\qquad TM$ or $PM = 3MC'$.

Then a given angle COC' is divided by the required line AB, such that the ratio of the tangents of its segments is known; therefore, etc.

Ex. If a, b, c denote the sides of the maximum triangle ABC, prove that

$$(1) \quad \frac{OA}{OB} = \frac{c^2 - a^2}{b^2 - c^2}.$$

$$(2) \quad c^2 = \frac{a^4 + b^4}{a^2 + b^2}.$$

CHAPTER III.

RECENT DEVELOPMENTS OF POINT O THEOREM.

SECTION I.

THE BROCARD POINTS AND CIRCLE OF A TRIANGLE.

27. **Brocard Points Ω, Ω'.**—In Art. 20 if the inscribed triangle PQR is similar to ABC and $P = A$, $Q = B$, $R = C$, then $BOC = A + P = 2A$, similarly $COA = 2B$ and $AOB = 2C$; therefore O is the centre of the circum-circle.

Secondly, let $P = B$, $Q = C$ and $R = A$. Then
$$BOC = A + P = A + B = \pi - C\,;$$
similarly $\qquad COA = B + Q = B + C = \pi - A,$
and $\qquad AOB = \pi - C.$

Thirdly, let $P = C$, $Q = A$ and $R = B$. It follows as in the last case that $BOC = \pi - B$, $COA = \pi - C$ and $AOB = \pi - A$.

Thus we see that a triangle PQR similar to a given one may be inscribed in the latter in three different ways; and that the point O in each case may be found as in the general method by describing segments of circles on two of the sides containing given angles.

In the second and third positions the points of inter-section of the circles are usually denoted by the letters Ω and Ω'. They are termed the *Brocard Points* of the

60

triangle ABC, and are distinguished as Positive (Ω) and Negative (Ω').

28. **Brocard Angle** (ω).—Since $B\Omega C$ is the supplement of C, $\Omega BC + \Omega CB = C$ or $\Omega BC = \Omega CA$. For a similar reason $\Omega CA = \Omega AB$,

hence $\qquad \Omega BC = \Omega CA = \Omega AB = \omega$ (say).

The angle ω is called the *Brocard Angle* of the triangle ABC.

We may remark that the angle subtended at Ω by the base c is the supplement of B, the angle at the right extremity of AB, and at Ω' equal to the supplement of A, the angle at the other extremity of AB.

The same relations hold for the sides a and b; hence the names *Positive* and *Negative* Brocard points.

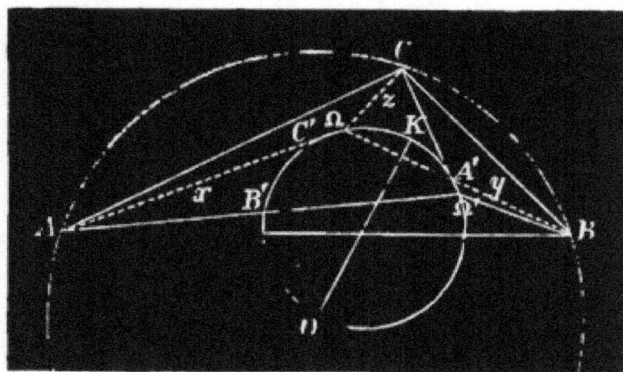

The value of ω as a function of the sides or angles is thus found.

Let x, y, z denote the lengths of $A\Omega$, $B\Omega$ and $C\Omega$ respectively. Then in the triangle $B\Omega C$

$$\cot \omega = \frac{\cos \omega}{\sin \omega} = \frac{a^2 + y^2 - z^2}{2ay \sin \omega} = \frac{a^2 + y^2 - z^2}{4B\Omega C}.$$

Similarly in the triangles $C\Omega A$ and $A\Omega B$

$$\cot \omega = \frac{a^2+y^2-z^2}{4B\Omega C} = \frac{b^2+z^2-x^2}{4C\Omega A} = \frac{c^2+x^2-y^2}{4A\Omega B}$$

$$= \frac{a^2+b^2+c^2}{4ABC} \dots\dots\dots\dots\dots\dots\dots\dots\dots\dots(1)$$

It is proved in like manner for Ω' that

$$\Omega'CB = \Omega'AC = \Omega'BA,$$

and that the value of these angles is also given by (1).

Again $\cot A = \dfrac{b^2+c^2-a^2}{2bc\sin A} = \dfrac{b^2+c^2-a^2}{4\Delta}$ with similar values

for $\cot B$ and $\cot C$.   Hence

$$\cot A + \cot B + \cot C = \Sigma\frac{b^2+c^2-a^2}{4\Delta} = \frac{a^2+b^2+c^2}{4\Delta},$$

$$\text{or } \cot \omega = \cot A + \cot B + \cot C \dots\dots\dots(2)$$

EXAMPLES.

1. Prove that

        (1) $\operatorname{cosec}^2\omega = \operatorname{cosec}^2 A + \operatorname{cosec}^2 B + \operatorname{cosec}^2 C.$

        (2) $\sin^2\omega = \dfrac{4\Delta^2}{b^2c^2 + c^2a^2 + a^2b^2}.$

        (3) $\cos^2\omega = \dfrac{(a^2+b^2+c^2)^2}{4(b^2c^2 + c^2a^2 + a^2b^2)}.$

2. The distances of Ω from the sides of ABC are $2R\sin^2\omega\dfrac{c}{b}$, $2R\sin^2\omega\dfrac{a}{c}$, $2R\sin^2\omega\dfrac{b}{a}$; and of Ω', $2R\sin^2\omega\dfrac{b}{c}$, $2R\sin^2\omega\dfrac{c}{a}$, $2R\sin^2\omega\dfrac{a}{b}$.

[For let the distances of Ω be denoted α, β, γ.   Then

$$\alpha = y\sin \omega = \frac{c\sin^2\omega}{\sin B}; \text{ therefore, etc.}$$

The *ratios of the distances** are evidently as follows :—

        $\alpha : \beta : \gamma = c^2a : a^2b : b^2c,$

and $\alpha' : \beta' : \gamma' = ab^2 : bc^2 : ca^2,$

and also $\alpha\alpha' = \beta\beta' = \gamma\gamma' = 4R^2\sin^4\omega.]$

---

  * Or Trilinear Co-ordinates of the points with respect to the triangle, which is also called the *Triangle of Reference.*

3. AD is the bisector of the angle A of a triangle ABC, and ω_1, ω_2 the Brocard angles of the triangles ABD and ACD respectively ; prove that $\cot \omega_1 + \cot \omega_2 = 2 \operatorname{cosec} A + \cot A + \cot \omega$, with similar expressions for the triangles formed by the bisectors of the angles B and C.

4. If ω_1 and ω_2 denote the Brocard angles of the triangles CAD and BAD, where AD is the median to the side BC,

$$\cot \omega_1 - \cot \omega_2 = \frac{b^2 \sim c^2}{2\Delta},$$

with similar expressions for the medians BE and CF.

5. Hence prove that $\cot \omega_1 + \cot \omega_3 + \cot \omega_5 = \cot \omega_2 + \cot \omega_4 + \cot \omega_6$,

$$\text{and } \Sigma \cot \omega_1 = \frac{2(a^2 + b^2 + c^2)}{\Delta}.$$

6. If ABC is divided as in the previous exercises by the symmedians, prove that $\Sigma(b^2 + c^2)(\cot \omega_1 - \cot \omega_2) = 0$.

7. Ω and Ω' are Brocard points of their pedal triangles PQR and $P'Q'R'$. (Euc. III. 21.)

8. The triangles PQR and $P'Q'R'$ are equal in area.
[For $\Omega P'Q$ and ΩBC are similar ; hence (Euc. VI. 19)

$$\Omega PQ : \Omega BC = \Omega P'^2 : \Omega B^2 = \sin^2\omega \; ;$$

similarly $\Omega QR : \Omega CA = \Omega RP : \Omega AB = \sin^2\omega \; ;$

therefore $PQR = P'Q'R' = ABC \cdot \sin^2\omega.$]

9. The Brocard points are equidistant from the circum-centre.
[By Ex. 8 and Art 23, Ex. 1.]

10. If A', B', C' be the points of intersection of the pairs of lines $y, z' : z, x' : x, y'$, prove that the six points A', B', C', O, Ω, Ω' lie on a circle.
[For the triangles BCA', CAB' and ABC' are isosceles and similar, their base angles each being equal to ω, hence OA', OB', OC' are the bisectors of their vertical angles. In the quadrilateral $O\Omega\Omega'A'$ we have $O\Omega = O\Omega'$ and OA' the bisector of the angle $\Omega A'\Omega'$; therefore O is a point on the circum-circle of $\Omega A'\Omega'$, and the quadrilateral is therefore cyclic. Similarly B' and C' are on the circum-circle of the triangle $O\Omega\Omega'$.]

DEF. This is called the *Brocard Circle*, and $A'B'C'$ the *First Brocard Triangle* of ABC.

11. To find the distance of the Brocard points from the circum-centre ($O\Omega = O\Omega' = \delta$).

[By Art. 23, Ex. 1, $2PQR = (R^2 - \delta^2)\sin A \sin B \sin C$,

but (Ex. 8) $\qquad PQR = ABC \sin^2\omega = 2R^2 \sin A \sin B \sin C \sin^2\omega,$

hence $\qquad R^2 - \delta^2 = 4R^2 \sin^2\omega$ or

$$\delta = R\sqrt{1 - 4\sin^2\omega}.]$$

12. The angle subtended at the circum-centre by $\Omega'\Omega = 2\omega$.
(By Ex. 10 and Euc. III. 22.)

13. To find the distance $\Omega\Omega'$ between the Brocard points.
[Since $O\Omega\Omega'$ is an isosceles triangle,

$$\Omega\Omega' = 2O\Omega \sin \omega = 2R \sin \omega \sqrt{1 - 4\sin^2\omega}, \text{ by Ex. 11.}]$$

14. The diameter of the Brocard circle is equal to

$$R \sec \omega \sqrt{1 - 4\sin^2\omega}.$$

[For it equals $\delta/\sin 2\omega$; therefore, etc.]

15. The altitudes of the similar isosceles triangles BCA', CAB', ABC'' are equal to the distances of the symmedian point (K) from the sides.

[For $\qquad C'Z = \tfrac{1}{2}c \tan \omega = \dfrac{2c\Delta}{a^2 + b^2 + c^2};$

therefore, etc., by Art. 28, (1).]

16. The circle on OK as diameter is the Brocard circle.

[For KA' is parallel and OA' perpendicular to BC, hence OK subtends a right angle at A'; similarly for the points B' and C''; therefore, etc.]

17. Brocard's first triangle is *Inversely Similar* to ABC; *i.e.*, by rotation in the plane of the paper their sides cannot be brought into a position of parallelism with each other.

[For $B'C''$ subtends equal angles at A' and K, but KB' and KC'' are respectively parallel to CA and AB, and therefore contain an angle A; similarly the angles B' and C'' are equal to B and C.]

18. Having given the base c and Brocard angle ω of a triangle ABC, find the locus of the vertex (Neuberg).

[Let ρ be the median CZ and θ the angle between it and PZ. Since $\cot \omega = (a^2 + b^2 + c^2)/2c$. CR and $a^2 + b^2 = \tfrac{1}{2}c^2 + 2\rho^2$, we have

$$2\rho^2 + \tfrac{3}{2}c^2 = 2c \cot \omega \,.\, CR = 2c \cot \omega \,.\, \rho \cos \theta,$$

or
$$\rho^2 - c \cot \omega \,.\, \rho \cos \theta + \tfrac{3}{4}c^2 * = 0.$$

NOTE.—Comparing this result with the standard form of the equation in the footnote we have by equating coefficients

$$c \cot \omega = 2d \text{ and } d^2 - r^2 = \tfrac{3}{4}c^2,$$

or
$$d = \tfrac{1}{2}c \cot \omega \text{ and } r^2 = \tfrac{1}{4}c^2 \cot^2 \omega - \tfrac{1}{4}c^2.$$

It is evident that the locus is a curve symmetrical with respect to the perpendicular bisector of the base, as to each position of the vertex C there is a corresponding one, C''' of the inversely similar triangle ABC''' described on the base.

The distance of C', a vertex of Brocard's first triangle, from $c = \tfrac{1}{2}c \tan \omega$; therefore $ZC' \,.\, ZO = (\tfrac{1}{2}c)^2$ where O is the centre of the required locus.

This example is a particular case of :—*Having given the base c and* $(la^2 + mb^2 + nc^2)/\Delta$ *to find the locus of the vertex;* a solution of which is similarly obtained.

18a. Six similar triangles are constructed on a given base and on the same side of it. Prove that their vertices C_1, C_2, ... C_6 are concyclic. (Mathesis, t. 2, p. 94.)

---

* This is known by Analytical Geometry to be the *Polar Equation of a Circle.* If we take any point Z and draw a variable line (*Radius Vector*) to a given circle (O, r) and let $d = ZO$, the equation connecting ρ and θ

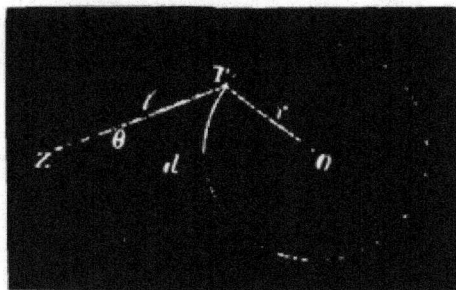

is for all points on the circle $\rho^2 - 2\rho d \cos \theta + d^2 - r^2 = 0$; and ρ and θ are called the *Polar Co-ordinates* of the point P.

19. Having given the base c, and Brocard Angle ω, find the locus of the centroid of ABC.

[A circle whose equation is formed from that in Ex. 18 by changing ρ into 3ρ; hence

$$12\rho^2 - 4c \cot \omega \,.\, \rho \cos \theta + c^2 = 0.$$

It has many important properties, which will be found in the *Transactions of the Royal Irish Academy*, vol. XXVIII. xx, where M'Cay names it the " C " circle of the triangle ABC.]

20. The lengths of the tangents drawn from A, B, C to the Brocard Circle are inversely proportional to a, b, c, and the sum of their squares $= 2\Delta \operatorname{cosec} 2\omega$.

SECTION II.

THE SYMMEDIANS OF A TRIANGLE.

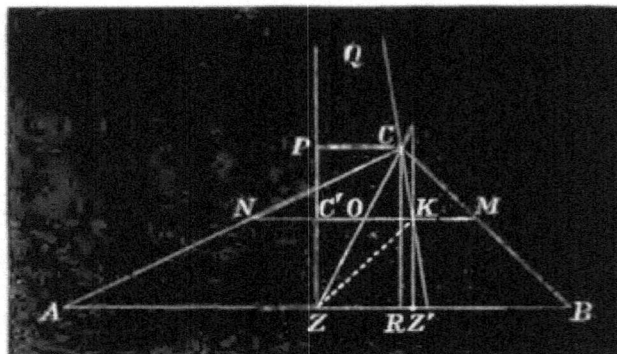

29. Let K be the symmedian point of ABC, a' and β' the distances of Z' from BC and CA respectively. Then $a'/\beta' = a/b = BZ' \sin B / AZ' \sin A$, hence

$$\frac{AZ'}{BZ'} = \frac{b^2}{a^2} \dots\dots\dots\dots\dots\dots\dots(1)$$

or *the symmedians divide each side in the duplicate ratio of the remaining two.*

Again from (1) $AZ'/c = b^2/(a^2+b^2)$ or $AZ' = b^2c/(a^2+b^2)$;
similarly $\qquad\qquad BZ' = a^2c/(a^2+b^2)$ ....(2)

Also $\quad CZ'/CK = a'/a = (a^2+b^2+c^2)/(a^2+b^2)$, hence

$$\frac{CK}{KZ'} = \frac{a^2+b^2}{c^2} \quad ....(3)$$

Cor. If $C = 90°$ then $CK = KZ'$ (Euc. I. 47) and K is the middle point of the perpendicular on the hypotenuse.

30. **The length of the symmedian CZ' is found as follows :—**

In the formula $b^2BZ' + a^2AZ' = cAZ' \cdot BZ' + cCZ'^2$ substitute the values in (2) and reduce. We easily obtain

$$CZ' = \frac{\sqrt{a^2+b^2+2ab\cos C}}{a/b+b/a}$$

with similar expressions for the lines through A and B.

EXAMPLES.

1. The symmedian is divided harmonically at K, and Q its point of intersection with the perpendicular to the base of the triangle at its middle point Z.

$$\left[\, ZZ' = \frac{b^2c}{a^2+b^2} - \frac{c}{2} = \frac{c^2}{a^2+b^2}ZR; \text{ hence} \right.$$

$$\frac{CP}{ZZ'} = \frac{ZR}{ZZ'} = \frac{a^2+b^2}{c^2} = \frac{CK}{KZ'} \quad (\text{Art. 29 (3)});$$

$$\left. \text{therefore } CQ/QZ' = CK/KZ' = (a^2+b^2)/c^2. \right]$$

2. Since $Z \cdot CKZ'Q$ is an harmonic pencil any line through K is cut harmonically by its rays, hence if KC' is parallel to one ray, it is bisected at O by the conjugate ray CZ. Also the parallel through K to PL is bisected at K.

3. The vertices of Brocard's first triangle and the symmedian point are equidistant from the extremities of the parallels through K to the sides of ABC.

[Let O be the middle point of MN. Since $OM = ON$ and (Ex. 2) $OK = OC'$, subtracting these results ; therefore, etc.]

4. The lines joining the middle points of the sides of ABC to the middle points of the perpendiculars on them meet in a point.

[By **Ex.** 2 the point of concurrence is the symmedian point. The ratios of the segments into which the joining lines are divided at K are easily seen to be $bc \cos A/a^2$, etc., etc.]

5. Prove that $\cot KBC + \cot KCA + \cot KAB = 3 \cot \omega$.

6. The sides of the pedal triangle of K are at right angles to the medians of ABC.

ANTIPARALLELS.

Def. A straight line meeting the sides a and b of a triangle at angles A and B is parallel to the base. If a line meet these sides at angles A and B respectively it is said to be *Antiparallel* to c.

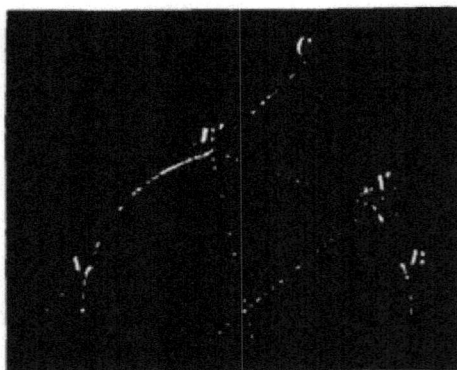

31. The following are the fundamental and obvious properties of antiparallels to the sides of any triangle :—

(1) Antiparallels to the sides a and b meet c at equal angles (C).

(2) They are parallels to the sides of the pedal triangle.

(3) Or to the tangents at A, B, C to the circum-circle.

(4) The locus of the middle point of a variable anti-parallel to a side, *c*, is the corresponding symmedian chord *CK*.

(5) Antiparallels through *K* to each side are bisected at the point, and are equal to one another. The latter part follows from (1).

(6) The median and symmedian to *c* of the triangle *ABC* are respectively the symmedian and median of the triangle *A'B'C* cut off by any antiparallel *A'B'*.

(7) The extremities of a parallel and antiparallel to any side of a triangle are concyclic.

THE PEDAL TRIANGLES OF THE BROCARD POINTS.

32. From Ω let fall perpendiculars on the sides and denote their feet as in figure by $A'B'C'$.

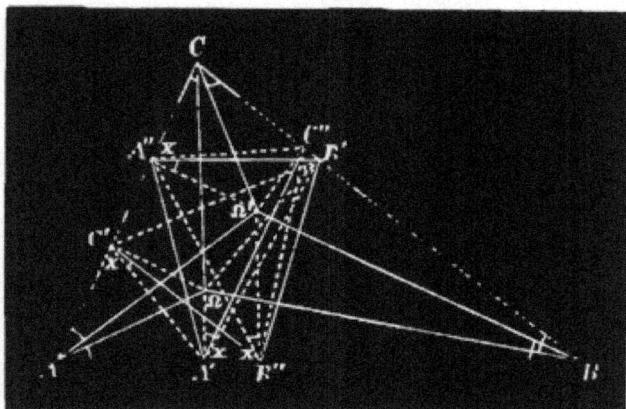

It follows conversely since $A\Omega B$ is the supplement of *B* (Art. 28), and is equal to $C + A'$ (Art. 19) that $A' = A$; similarly $B' = B$ and $C' = C$. Also A'', B'', C'' are respectively equal to *A*, *B* and *C*.

33. Theorems I. Ω *is the common positive Brocard point of ABC and A'B'C'.*

Since $AC'A'\Omega$ is a cyclic quadrilateral $\Omega AB = \Omega C'A' = \omega$ (Euc. III. 21); similarly $\Omega B'C'$ and $\Omega A'B'$ are each equal to ω.

It follows also that Ω' is the common negative Brocard point of ABC and $A''B''C''$.

II. *The sides of A'B'C' and A''B''C'' are equally inclined to the corresponding sides of ABC.*

For by (1) $CB'C' = AC'A' = BA'B' = 90 - \omega$,
and $\qquad\qquad BC''B'' = AB''A'' = CA''C'' = 90 - \omega$.

III. *The six points A', B', C', A'', B'', C'' are concyclic.*

For the angles $AC'A' = AB''A''$, therefore $A'A''B''C'$ is cyclic (Euc. III. 22).

Similarly $B'B''C''A'$ and $C'C''A''B'$ are cyclic. But generally if three pairs of points on the sides of a triangle are such that every two pairs are cyclic, the six points lie on a circle.* For if they do not the tangents to the three circles from A, B and C are easily seen to be equal, which is impossible.

IV. *The lines B''C', C''A', A''B' are parallel to the sides a, b, c respectively.*

We know that each pair of sides of ABC with Ω and Ω' form similar triangles, *i.e.*, $B\Omega C$ and $A\Omega'C$, $C\Omega A$ and $B\Omega'A$, $A\Omega B$ and $C\Omega'B$ are similar; hence the perpendiculars (or other corresponding lines) through Ω and Ω' divide the opposite sides similarly. In the triangles $C\Omega A$

---

* For example, if $A'B'C'$ be the middle points of the sides and $A''B''C''$ the feet of the perpendiculars, it follows immediately that $A'B'C'A''B''C''$ is a cyclic hexagon since each pair of points AA' and BB' form a cyclic quadrilateral. ("Nine Points" Circle.)

and $B\Omega'A$ we have therefore $AC'/AC = AB''/AB$, or $B''C'$ is parallel to a.

V. *Hence also* $A'A''$, $B'B''$, $C'C''$ *are antiparallels to the sides* a, b, c. (Euc. III. 22.)

<h2 style="text-align:center">SECTION III.</h2>

<h3 style="text-align:center">TUCKER'S CIRCLES.</h3>

34. By Art. 24 if the inscribed triangle $A'B'C'$ is given in species only it may be conceived to vary its position by rotating around the point Ω which is fixed. Let it revolve in a positive direction through any angle θ and also let $A''B''C''$ revolve in the opposite direction through an equal angle.

Then each of the equal angles of inclination of the sides of $A'B'C'$ and $A''B''C''$ are diminished by θ, therefore for all values of θ the sides are equally inclined and the vertices of the two triangles are always concyclic.

The circles thus described are called the *Tucker Circles* of the triangle.

Thus the lines $B''C'$ and $A'A''$, etc., are always parallel and antiparallel respectively to the opposite side a, and therefore remain constant in direction.

Now since the point Ω is fixed and the triangle $A'B'C'$ of constant species; since the vertices move on given lines all points fixed relatively to the figure describe lines. The locus of the centre of the system of Tucker's circles is therefore a line. (Art. 20.)

By taking particular positions of the triangle we find points on the line of centres. In the case where $\theta = 0$ the vertices of ABC and $A'B'C'$ coincide, and the circum-circle is thus seen to be one of Tucker's circles. The line of centres thus passes through the circum-centre of ABC.

Similarly the loci of the other Brocard points of the triangle $A'B'C'$ and $A''B''C''$ are lines.

35. Let the vertices of the triangle formed by the parallels $B''C'$, $C''A'$, $A''B'$ to the sides of ABC be denoted by X, Y, Z.

Then $AA'A''X$ is a parallelogram, as are also $BB'B''Y$, $CC'C''Z$; and since the diagonals bisect each other AX bisects the antiparallel $A'A''$. AX, BY, CZ are the symmedians of ABC.

Hence the following construction for Tucker's circles:

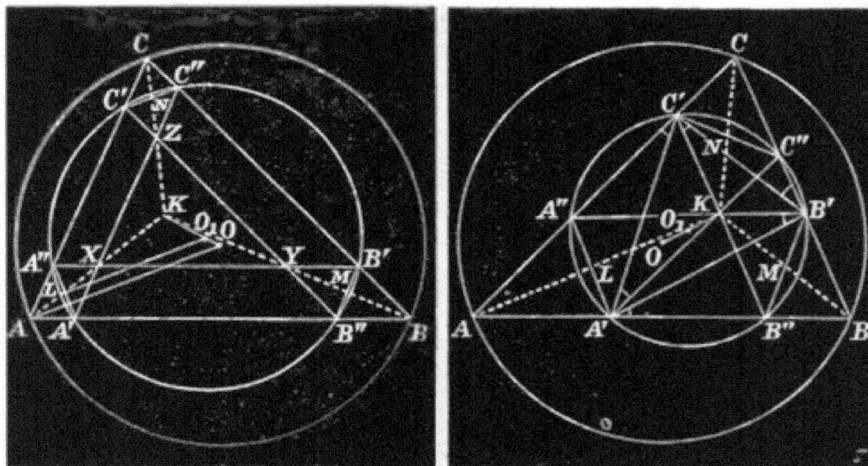

Let K be the symmedian point of ABC. Join AK, BK, CK. Take any point X on AK and draw parallels through it to the sides b and c. Let them meet BK and CK in Y and Z respectively. YZ is parallel to a, and the hexad of points in which the sides of ABC are cut by these parallels lie on one of the required circles.

36. *The antiparallels $A'A''$, $B'B''$, $C'C''$ are equal.* For since $A''B'$ is parallel to c, and $A'A''$ and $B'B''$ are equally inclined to c (at an angle C), $A'A''=B'B''$; therefore, etc.; or they are the chords of a Tucker circle intercepted by parallel lines.

37. Theorem. *The line OK is the locus of the centre of Tucker's system of circles.*

For let L be the middle point of the chord $A'A''$ of one of the system. Draw LO_1 at right angles to it meeting OK in O_1. Join AO.

Since the tangent at A to the circum-circle is anti-parallel to a, AO and LO_1 are parallel lines.

But $AK/AX = BK/BY = CK/CZ$ (Euc. VI. 2); therefore $AK/AL = BK/BM = CK/CN = OK/OO_1$, or O_1 is the centre of the Tucker circle.

38. Since Ω is the positive Brocard point of the triangles ABC and $A'B'C'$, and ΩAB and $\Omega A'B'$ a pair of similar triangles; if θ be the inclination of the sides of $A'B'C'$ to those of ABC, we have

$$\frac{\Omega A'}{\Omega A} = \frac{\sin\omega}{\sin(\theta+\omega)}\quad\ldots\ldots\ldots(1)$$

This ratio is the *Ratio of Similitude* of the triangles, and is the constant relation between all *corresponding* lines of $A'B'C'$ and ABC.

For example, if ρ be the radius of Tucker's circle for any value of θ,
$$\frac{\rho}{R} = \frac{\sin\omega}{\sin(\theta+\omega)}\quad\ldots\ldots\ldots(2)$$

In (2) we have the following particular cases:—

when $\theta=0°$ $\rho=R$............(circum-circle);

„ $\theta=\omega$ $\rho=\frac12 R\sec\omega$.........(T. R. circle);

„ $\theta=90°$ $\rho=R\tan\omega$...........(cosine circle).

Also area $A'B'C' : ABC = \sin^2\omega : \sin^2(\theta+\omega)$ (Euc. VI. 19).

Section IV.

Tucker's Circles, Particular Cases.

39. I. Cosine Circle. As a particular case of the
general theorem (Art. 33 v.) we shall consider the antiparallels $A'A''$, $B'B''$, $C'C''$ to pass through K. The points
L, M, N will therefore coincide with K, which is also the
centre of the corresponding Tucker's circle.

It is otherwise evident that the six segments KA', KA'',
etc., of antiparallels through K to the sides are equal.
(Art. 31 (5)).

Also $B'C'B''C''$, $C'A'C''A''$, $A'B'A''B''$ are rectangles since
their diagonals are equal.

Again because $A'B'B''$ is a right-angled triangle
$$A'B'' = B'B'' \cos A'B''B' = B'B'' \cos C,$$

or $A'B'' = 2\rho \cos C$, with similar expressions for $B'C''$ and $C'A''$. Hence

The segments intercepted by the circle on the sides of ABC are proportional to the cosines of the opposite angles. [*]

It is from this property the circle derives its name.

40. The middle point M of $A''B'$ is on the median through C to the opposite side c; hence the perpendicular through K to this side passes through M, or as has been shown otherwise (Art. 30, Ex. 4). If a perpendicular be drawn through K to the base meeting it in N and the median in M, $MK = NK$, from which it follows *that the lines joining the middle points of the sides to the middle points of the corresponding perpendiculars meet at the symmedian point* (Hain).

41. The sides of the triangles $A'B'C'$ and $A''B''C''$ are perpendicular to the corresponding sides of ABC. The cosine circle may therefore be obtained by rotating the two inscribed triangles in opposite directions until $\theta = 90^{\circ}$. (Art. 39.)

The ratio of similitude of $A'B'C'$ and $ABC = \tan\omega$.

42. II. **Triplicate Ratio Circle.**—Let the parallel in figure of Art. 35 pass through K.

Then L, M, N are the middle points of $AK, BK,$ and CK, since $AA'A''K$, etc., etc., are parallelograms; and the centre O of the corresponding Tucker circle bisects OK.

The sides of $A'B'C'$ are inclined to those of ABC at an angle $= \omega$. For consider the angles in the equal segments $A'A''$, $B'B''$, $C'C''$, and it is obvious (Euc. III. 21) that $A'B'A'' = A'C'A'' = B'C'B'' = B'A'B'' = C'A'C'' = C'B'C''$.

---

[*] See Mathesis, t. i., p. 185 :—

" Sur le centre des Médianes Antiparallèles," Neuberg (1881).

Hence K *is the negative Brocard point of* $A'B'C'$.
Similarly it is the positive Brocard point of $A''B''C''$.

It follows generally that the locus of the negative Brocard point of $A'B'C'$ is a line passing through K.

43. The ratio of similitude of $A'B'C'$ and ABC is $\sin \omega/\sin 2\omega$ since $\theta = \omega$; hence

$$\rho = \tfrac{1}{2}R \sec \omega \dots\dots\dots\dots\dots\dots\dots(1)$$

44. The intercepts $B'C''$, $C'A''$, $A'B''$ made by the circle on the sides are thus determined:—The triangles $A'KB''$ and ABC are similar, therefore $A'B''/c$ = ratio of altitudes

$$= \frac{2c\Delta}{a^2+b^2+c^2} \Big/ \frac{2\Delta}{c} = \frac{c^2}{a^2+b^2+c^2};$$

hence

$$A'B'' = \frac{c^3}{a^2+b^2+c^2} \dots\dots\dots\dots\dots\dots\dots (1)$$

with similar expressions for $B'C''$ and $C'A''$. The general property of the circle may be thus stated:—*Parallels through the symmedian point meet the non-corresponding sides in six points which lie on a circle;* and the intercepts made on each side are in the ratios $a^3 : b^3 : c^3$. From the latter property the circle takes its name. For the sake of brevity it is often written "T.R." Circle.*

45. III. Taylor's Circle.—Let the antiparallels $A'A''$, $B'B''$, $C'C''$, which, it will be remembered, are always parallel to the sides of the pedal triangle (PQR) of ABC, pass through the middle points a, β, γ of the sides of PQR.

Consider the segments into which $A'A''$ is divided by β and γ. We have $\beta\gamma = \tfrac{1}{2}QR$, $\gamma A'' = \tfrac{1}{2}PQ$ (Euc. I. 5), and

---

* An account of the circle will be found in Mathesis in the article by Neuberg already referred to (Art. 39). See also *Nouvelles Annales*, 1873, p. 264.

for the same reason $\beta A' = \frac{1}{2}RP$; therefore $A'A''$ is equal to the semiperimeter of PQR

$$= \tfrac{1}{2}(a \cos A + b \cos B + c \cos C) = 2R \sin A \sin B \sin C.$$

Hence generally

$$A'A'' = B'B'' = C'C'' = 2R \sin A \sin B \sin C \ldots\ldots (1)$$

Again, since $B''aC'$ is an isosceles triangle, the perpendicular to the chord $B''C'$ of Tucker's circle at the middle point bisects the vertical angle a and passes through the in-centre of $a\beta\gamma$. Similarly for the chords $C''A'$ and $A''B'$. Hence

The centre of the circle coincides with the in-centre of the median triangle $(a\beta\gamma)$ of PQR.

Many properties of this circle are proved in Neuberg's article in Mathesis, t. 1, p. 185, but it was described independently in England by Mr. H. M. Taylor, and now bears his name. (*Proc. Lond. Math. Society*, vol. xv. p. 122.)

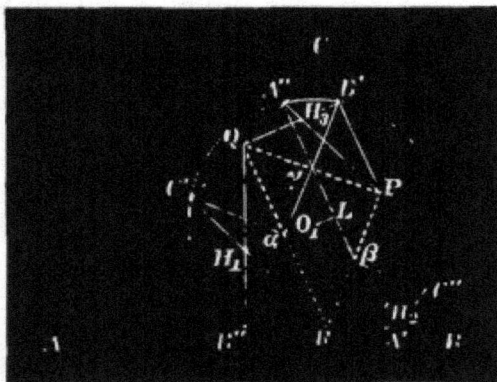

46. Since $aQ = aR = aB'' = aC'$, the circle on QR as diameter passes through B'' and C' and $RB''Q = RC'Q = 90°$; or B'' and C' are the projections of Q and R on the sides AB and AC; hence

The six projections of the vertices of the pedal triangle on the sides of ABC lie on Taylor's circle.

47. The triangle $B'aC''$ is isosceles, therefore O_1a the bisector of its vertical angle a is at right angles to BC; hence generally

The lines O_1a, $O_1\beta$, $O_1\gamma$ are perpendiculars to the sides of ABC.

Let H_3 denote the orthocentre of CPQ; then QH_3 and O_1a are parallel; similarly PH_3 and $O_1\beta$ are parallel; hence the triangles PQH_3 and $a\beta O_1$ are similar, their ratio of similitude being $= \frac{1}{2}$, or H_3R is bisected at O_1.

Similarly PH_1 and QH_2 are each bisected at O_1; and therefore the triangles $H_1H_2H_3$ and PQR are equal in all respects.

48. Theorem.—*Taylor's circle of the triangle ABC is the common orthogonal circle of the ex-circles of PQR.*

In the triangle $AA'A''$ we have by rule of sines

$$AA'' = A'A'' \sin C/\sin A = 2R \sin B \sin^2 C \text{ (Art. 45 (1))},$$

also $\qquad AC' = AR\cos A = b\cos^2 A$;

multiplying these results and reducing

$$AA'' . AC' = 4R^2 \sin^2 B \sin^2 C \cos^2 A,$$

but $AQ = c\cos A$; substituting we obtain

$$AA'' . AC' = AQ^2 \sin^2 B,^*$$

or the square of the perpendicular from A on QR. Hence the tangent from A an ex-centre of PQR to Taylor's circle

---

* Otherwise from the right-angled triangle $AA''P$ and ACP we have
$AA'' = b\sin^2 C$; and from the triangles ACR and $AC'R$,
$AC' = b\cos^2 A$; therefore $AA'' . AC' = b^2 \sin^2 C \cos^2 A$.

is equal to the radius of the ex-circle; similarly for the ex-centres B and C; therefore, etc.*

1. To find the value of the radius ρ of a circle cutting the ex-circles of a triangle PQR orthogonally.

[In figure of Art. 45 $\rho^2 = O_1A'^2$. But if a perpendicular be drawn from O_1 to $\beta\gamma$ it is equal to the radius of the in-circle of the triangle $a\beta\gamma$ or half the radius ($\frac{1}{2}r$) of PQR; and the distance of its foot from A' is equal to the semiperimeter of $a\beta\gamma$—*i.e.*, $\frac{1}{2}s$ of PQR.

Hence (Euc. I. 47) $\rho^2 = \frac{1}{4}(r^2 + s^2)$.

Similarly for the radii ρ_1, ρ_2, ρ_3 of the circles cutting two escribed and the inscribed of PQR orthogonally we obtain

$$\rho_1{}^2 = \tfrac{1}{4}(r_1{}^2 + \overline{s - a^2}),$$

$$\rho_2{}^2 = \tfrac{1}{4}(r_2{}^2 + \overline{s - b^2}),$$

$$\rho_3{}^2 = \tfrac{1}{4}(r_3{}^2 + \overline{s - c^2}),$$

and by adding these results we have, on reducing,

$$\rho^2 + \rho_1{}^2 + \rho_2{}^2 + \rho_3{}^2 = 4R^2$$

or,

the sum of the squares of the radii of the four circles cutting orthogonally the inscribed and escribed circles of any triangle taken in threes is equal to the square of the diameter of the circum-circle.

---

* In the triangle PQR since perpendiculars PA' and QB'' are let fall from the extremities of the base PQ on the external bisector AB of the vertical angle R, by a well-known property $\gamma A' = \gamma B'' = \frac{1}{2}$ sum of sides. But the distance of the middle point of any side from the points of contact of the ex-circles which touch it externally $= \frac{1}{2}$ sum of sides. Hence if a circle be described with γ as centre and $\gamma A' = \gamma B''$ as radius, it cuts the ex-circles of PQR whose centres are at A and B orthogonally. It follows that the locus of the centre of a circle cutting these two orthogonally is the line γO_1, since it is perpendicular to the line of centres; similarly aO_1 and βO_1 are the loci for the centres of circles orthogonal to the remaining pairs of ex-circles, whose centres are at B and C, C and A respectively.

Therefore O_1 is the centre and $O_1A' = O_1B'' =$ etc., the radius of the common orthogonal circle, *i.e.*, Taylor's circle.

2. To find the radius ρ of Taylor's circle of a triangle ABC.

[Taylor's circle for the triangle ABC is the circle in **Ex. 1** for PQR; hence we have to express r and s of the latter triangle * in terms of the parts of ABC. We easily obtain

$$\rho^2 = 4R^2(\sin^2\!A\,\sin^2\!B\,\sin^2\!C + \cos^2\!A\,\cos^2\!B\,\cos^2\!C)$$

also $\qquad \rho_1{}^2 = 4R^2(\sin^2\!A\,\cos^2\!B\,\cos^2\!C + \cos^2\!A\,\sin^2\!B\,\sin^2\!C)$

with similar values for $\rho_2{}^2$ and $\rho_3{}^2$.

From these expressions we have the result given in **Ex. 1** : $\Sigma\rho^2 = 4R^2$.]

3. The lines $B''C'$, $C''A'$, $A''B'$, parallels to the sides of ABC, are the chords of contact of the ex-circles of PQR with its sides.†

[Let $A''B'$ meet PR in the point Q'. Then $B'B''RQ'$ is a parallelogram, therefore $RQ' =$ semiperimeter of PQR, etc.]

4. Employing the notation of Art. 35, prove that the lines joining the corresponding vertices of the two triangles PQR and XYZ are concurrent at the circum-centre of the latter.

[Let p and q be the perpendiculars from R on the sides YZ and ZX of the triangle XYZ. Then $p/q = RB''\sin B/RA'\sin A$. But $RB''/RA' = QR/RP = a\cos A/b\cos B$. Substituting and reducing we have $p/q = \cos A/\cos B$.

But if Z be joined to the circum-centre of XYZ, the joining line is the locus of a point such that perpendiculars from it on the sides are in this ratio; hence ZR passes through the circum-centre of XYZ.‡ And similarly for the lines PX and QY.]

---

5. The Simson lines of the median triangle LMN of a given one ABC with respect to the vertices P, Q, R of the pedal triangle pass through the centre of Taylor's circle.*

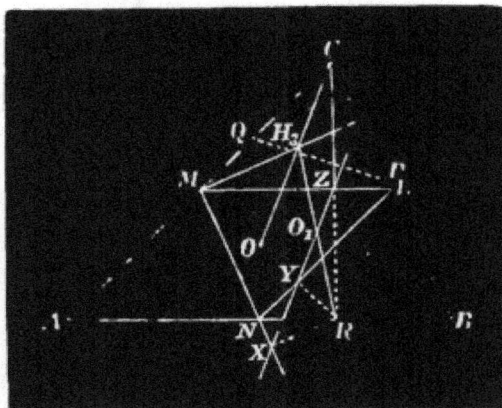

[The circum-centre O of ABC is the orthocentre of LMN. Hence RO is bisected by the Simson line XYZ of R. Also $CZ = RZ$; therefore the line XYZ is parallel to OC. But the centre of Taylor's circle O_1 is (Art. 47) the middle point of RH_3; therefore, etc.]

6. The Simson lines of PQR, whose poles are L, M, N, pass through O_1.

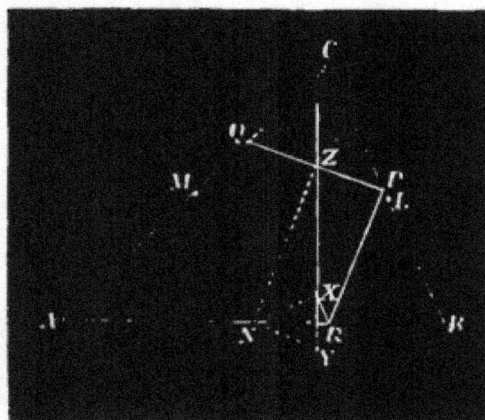

---

* The point on the circum-circle from which perpendiculars or other isoclinals are let fall on the sides of an inscribed triangle is called the *Pole* of the Simson line.—*V*. Mathesis, t. 2, p. 106, "Sur la Droite de Simson," par M. Barbarin.

[For the perpendicular NZ from N on PQ bisects it (Euc. III. 3); and the perpendiculars NX and NY are equally inclined to AB (Euc. I. 26), hence the line XYZ is a perpendicular to AB through the middle point of PQ; therefore, etc. (Art. 47.)]

7. Prove that the common inclination (θ) of the sides of the triangles $A'B'C'$ and $A''B''C''$ to those of ABC is given by the equation

$$\tan \theta = - \tan A \tan B \tan C. \qquad \text{(Taylor)}$$

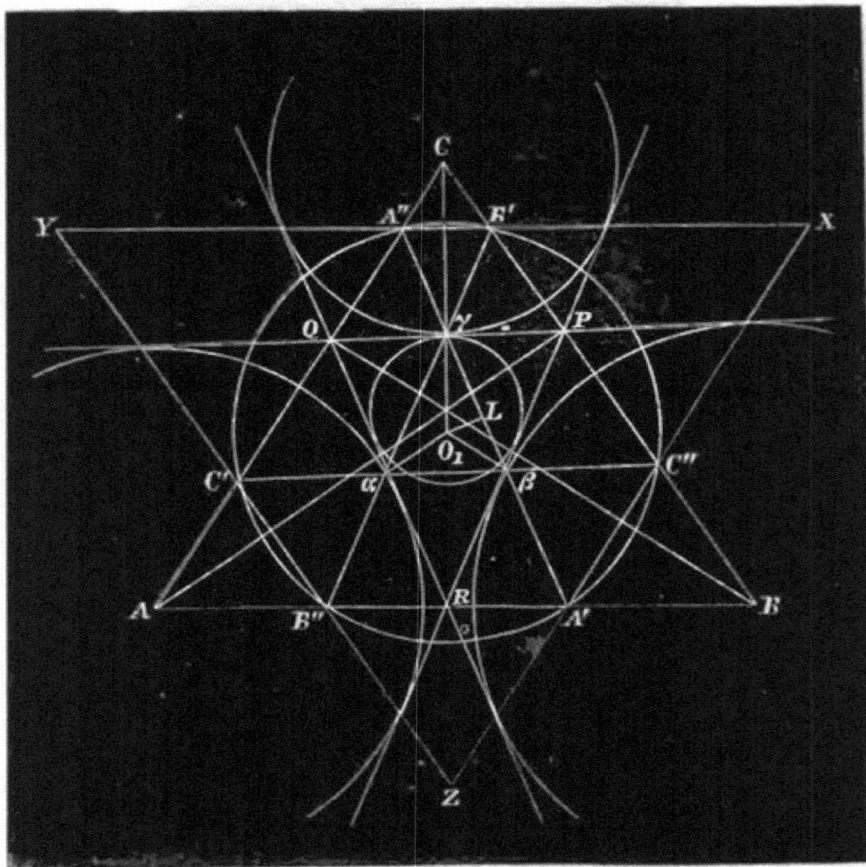

8. The intercepts made by Taylor's circle on the sides are $a \cos A \cos (B - C)$, $b \cos B \cos (C - A)$, $c \cos C \cos (A - B)$.

$$[A'B'' = A'R + RB'' = (a \cos A + b \cos B) \cos C = \text{etc.}]$$

9. The circum-centre of a triangle, its symmedian point, and the orthocentre of its pedal triangle are collinear. (Tucker.)

[The orthocentre of the pedal triangle has been shown to be (Ex. 4) the circum-centre of XYZ, and K is the centre of similitude of ABC and XYZ; therefore, etc.]

10. The circum-centre and the orthocentre of its pedal triangle are equidistant from, and collinear with, the centre of Taylor's circle. (Neuberg.)

[For CH_3 and ZR are parallel, since both are at right angles to PQ; also RH_3 is bisected at O_1 (Art. 47), therefore, etc., by Art. 37.]

CHAPTER IV.

GENERAL THEORY OF THE MEAN CENTRE OF A SYSTEM OF POINTS.

49. We now proceed to the discussion of the general linear relation connecting the distances of a system of points from a given line.

Let A, B, C, D ... be the system of points, AL, BL CL ... their distances from any line L, and $\Sigma(a \cdot AL)$ the algebraic sum

$$a \cdot AL + b \cdot BL + c \cdot CL + \ \dots$$

where a, b, c ... are given quantities.

By $\Sigma(a \cdot AL)$ is therefore meant the sum of given multiples of the distances of the system of points from the line ; perpendiculars from points on opposite sides of L being taken with opposite signs.

50. **Theorem**.—*For any two lines M and N and systems of points A, B, C ... and multiples a, b, c ... having given*

$$\Sigma(a \cdot AM) = 0 \text{ and } \Sigma(a \cdot AN) = 0$$

to prove that

$$\Sigma a \cdot AL = 0,$$

where L is any line passing through O the intersection of M and N.

Join AO and let this line be denoted by R. Then since $LMNR$ is a concurrent system of lines we have
$$\sin MN \cdot \sin LR + \sin NL \cdot \sin MR + \sin LM \cdot \sin NR = 0,$$
but, by Art. 2,
$$\sin LR : \sin MR : \sin NR = AL : AM : AN;$$
therefore
$$\sin MN \cdot AL + \sin NL \cdot AM + \sin LM \cdot AN = 0.$$
Similarly for the points $B, C \ldots$ we have
$$\sin MN \cdot BL + \sin NL \cdot BM + \sin LM \cdot BN = 0$$
$$\sin MN \cdot CL + \sin NL \cdot CM + \sin LM \cdot CN = 0.$$
Multiplying these equations respectively by $a, b, c \ldots$ and adding
$$\sin MN \Sigma(a \cdot AL) + \sin NL \Sigma(a \cdot AM) + \sin LM \Sigma(a \cdot AN) = 0,$$
hence if $\Sigma(a \cdot AM) = 0$ and $\Sigma(a \cdot AN) = 0$, it follows that
$$\Sigma a \cdot AL = 0.$$

Def. The point O which satisfies the relation $\Sigma(a \cdot AL) = 0$ for every line L passing through it is termed the *Mean Centre* of the system of points $A, B, C \ldots$ for the system of multiples $a, b, c \ldots$.

51. **Theorem.**—*The position of the mean centre for a given system of multiples is either unique or indeterminate.*

For let O_1 and O_2 be two of its positions, and O any point whatever. Join O_1O and O_2O, and denote these lines by M and N.

Since $\Sigma(a \cdot AM) = 0$ and $\Sigma(a \cdot AN) = 0$, it follows by Art. 50 that any line L through O, *i.e.* any line whatever, satisfies the equation
$$\Sigma a \cdot AL = 0.$$
It is obvious, in the general case, that when all the points of the system, and all save one of the multiples are

given ; by assigning a definite value to the last multiple, the position of the mean centre is determinate ; and conversely *any point* whatever is the mean centre of a given system for multiples, all of which save two may be arbitrarily chosen.

EXAMPLES.

1. The middle point of a right line is the mean centre of its extremities (Euc. I. 26).

2. The mean centre O of two points A and B for the multiples a and b divides the line AB inversely as the multiples, *i.e.*,
$$AO : BO = b : a.$$
The mean centre of the same points for the multiples $a, -b$, divides the line *externally* such that
$$AO : BO = b : a.$$

3. The mean centre O of a linear system of points A, B, C ... for multiples each $=1$ satisfies the equation $\Sigma AO = 0$.

4. The bisectors L, M, N of the sides of a triangle ABC are concurrent.

[For $\Sigma AL = 0, \Sigma AM = 0$ and $\Sigma AN = 0,$

hence each line passes through the mean centre (centroid or centre of gravity) of the vertices.]

5. The lines joining the middle points of the three pairs of opposite connectors BC and AD, CA and BD, AB and CD of four points A, B, C, D are concurrent, and each is bisected at the point of concurrence.*

---

* In the particular case when the fourth point D coincides with the orthocentre O of the triangle ABC we infer at once the well-known property :—

The lines joining the middle points of the sides of a triangle with those of the segments towards the angles of the corresponding perpendiculars meet in a point and bisect each other. From this it follows immediately (Euc. I. 4) that the six segments are equal, and that the circle passing through the middle points of the sides passes through the feet of the perpendiculars and bisects the segments of the latter towards the angles. This is the fundamental property of the *Nine-Points-Circle.*

6. The geometrical centre O of a regular polygon is the mean centre of the vertices A, B, C ....

[Join AO and BO. If the polygon be of an even order these lines (L and M) will pass through the opposite vertices, and the perpendiculars from the remaining vertices are equal in pairs and opposite in sign ; and if the polygon be of an odd order L and M bisect the opposite sides at right angles ; therefore, etc.]

7. $ABCD$ ... is a regular cyclic polygon and L any line passing through its centre O ; prove that

$$AL + BL + CL + \ldots = 0.$$

52. Theorem.—*Any point O is the mean centre of the vertices of a triangle ABC for multiples proportional to the areas BOC, COA, AOB.*

For letting L coincide with AOX and applying the relation $\Sigma a AL = 0$ we have

$$b \cdot BL + c \cdot CL = 0,$$

or disregarding signs $BL/CL = c/b$.

Also since the triangles COA and AOB are upon the same base AO, $BL/CL = AOB/COA$; equating these values,

therefore $\qquad \dfrac{b}{c} = \dfrac{COA}{AOB}.$

Similarly $\qquad \dfrac{c}{a} = \dfrac{AOB}{BOC}.$

Hence $\qquad a : b : c = BOC : COA : AOB.$

If the point O is outside the triangle, and within the angle A, the multiples are proportional to

$$-BOC, \; COA \text{ and } AOB,$$

with similar results when O is within the angles B or C.

Examples.

1. The in-centre of a triangle is the mean centre of the vertices for multiples proportional to the sides.

2. The ex-centres are the mean centres for systems of multiples $-a$, b, c ; a, $-b$, c ; a, b, $-c$; or quantities proportional to them.

3. If O, O_1, O_2, O_3 denote the in- and ex-centres of a triangle, each is the mean centre of the remaining three for multiples,

$$s-a,\ s-b,\ s-c\ ;\ s-b,\ s-c,\ -s,\ \text{etc.}$$

[For the areas in the first case are O_2O_3O, O_3O_1O, O_1O_2O, and these are obviously proportional to $s-a$, $s-b$, $s-c$. Similarly for each of the ex-centres. Thus generally since $-s:s-a:s-b:s-c = -1/r:1/r_1:1/r_2:1/r_3$; for the points O, O_1, O_2, O_3 each is the mean centre of the remaining three for the corresponding multiples of the system $-1/r,\ 1/r_1,\ 1/r_2,\ 1/r_3$.]

4. Prove the following points are the mean centres of the vertices for the system of multiples written opposite to them.

Circum-centre	$\begin{cases} a\cos A,\ b\cos B,\ c\cos C, \\ \sin 2A,\ \sin 2B,\ \sin 2C. \end{cases}$
Orthocentre	$\tan A,\ \tan B,\ \tan C.$
Symmedian Point	$a^2,\ b^2,\ c^2.$
Brocard Points	$\dfrac{1}{b^2}\ \dfrac{1}{c^2}\ \dfrac{1}{a^2}\ ;\ \dfrac{1}{c^2}\ \dfrac{1}{a^2}\ \dfrac{1}{b^2}.$

"Nine-Points" Centre $a\cos(B-C)$, $b\cos(C-A)$, $c\cos(A-B)$.*

5. The lines drawn from the vertices of a triangle to the points of contact of the in-circle are concurrent at the mean centre of the vertices for multiples $r_1,\ r_2,\ r_3$.

6. The lines drawn to the internal points of contact of the three ex-circles meet at the mean centre of the vertices for multiples

$$1/r_1,\ 1/r_2,\ 1/r_3.$$

7. If a point O be the mean centre of the vertices for multiples l, m, n, its *Isotomic Conjugate* † is the mean centre for multiples the reciprocals of l, m, n.

7a. The *Isogonal Conjugate* † of O is the mean centre for multiples a^2/l, b^2/m, c^2/n.

---

* From this it is evident that the sides of the triangle ABC meet the Nine-Points-Circle at angles $B-C$, $C-A$, $A-B$.

† Two points X and X' equidistant from the extremities of a line BC are called *Isotomic Conjugates* with respect to the line. It is easy to see, and it will be afterwards proved, that if the sides of a triangle ABC be divided isotomically in the pairs of points X, X'; Y, Y'; Z, Z'; such that AX, BY and CZ are concurrent at a point O; then AX',

8. Any point O on the segment AB of the circum-circle of an equilateral triangle ABC is the mean centre of the vertices for multiples $1/OA$, $1/OB$, $-1/OC$.

9. The mean centre of O, O_1, O_2, O_3 is in Ex. 3 the circum-centre of the triangle.

10. The centre of Taylor's circle is the mean centre of the vertices of the pedal triangle of ABC for multiples
$$a\cos(B-C),\ b\cos(C-A),\ c\cos(A-B).$$

11. The mean centre O of the vertices of ABC for multiples l, m, n is the mean centre of the vertices of the pedal triangle PQR of O for multiples a^2/l, b^2/m, c^2/n.

[From the figure of Art. 23, Ex. 1, we have
$$QOR:ROP:POQ=OQ.OR\sin A:OR.OP\sin B:OP.OQ\sin C$$
$$=a/OP:b/OQ:c/OR \dots\dots\dots\dots(1)$$
But $\qquad OP:OQ:OR=BOC/a:COA/b:AOB/c$
$$=l/a:m/b:n/c.$$
Substituting these values in (1); therefore, etc.]

12. The symmedian point O of any triangle is the centroid of the pedal triangle of O.

[For $BOC:COA:AOB=a^2:b^2:c^2$ by Art. 16, Ex. 2 (2).]

13. The lines joining A, B, C to the corresponding vertices of Brocard's first triangle are concurrent, and the point of concurrence is the mean centre of the vertices of ABC for multiples the reciprocals of a^2, b^2, c^2.

[For it has been shown that it is the isotomic conjugate of the symmedian point, Art. 30, Ex. 3.]

14. If perpendiculars be let fall from any point P on the sides of a regular polygon; the mean centre of their feet lies on the line joining P to the circum-centre.

BY', CZ' are also concurrent at O'. The points O and O' are termed *Isotomic Conjugates with respect to the triangle ABC.*

If the pairs of lines AX, AX', etc., are equally inclined to the sides b and c, etc., they are *Isogonal Conjugates* with respect to the angles; and if AX, BY, CZ are concurrent, AX', BY', CZ' are also concurrent. The points of concurrence are *Isogonal Conjugates with respect to the triangle.*

[Through O draw OAA' parallel and PA' perpendicular to p_1. The projection of p_1 on $OP=$ projection of AA'; but A, B, C, ... and

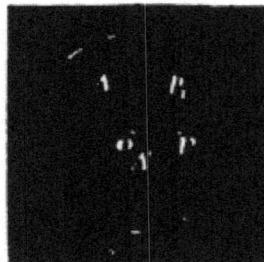

A', B', C', ... are the vertices of regular polygons, whose mean centres are both on OP. Therefore the sum of the projections of p_1 ... on $OP=0$.]

53. **Theorem.**—*For any line L to prove that*
$$\Sigma a . AL=\Sigma(a)OL.$$
Draw M through O parallel to L.

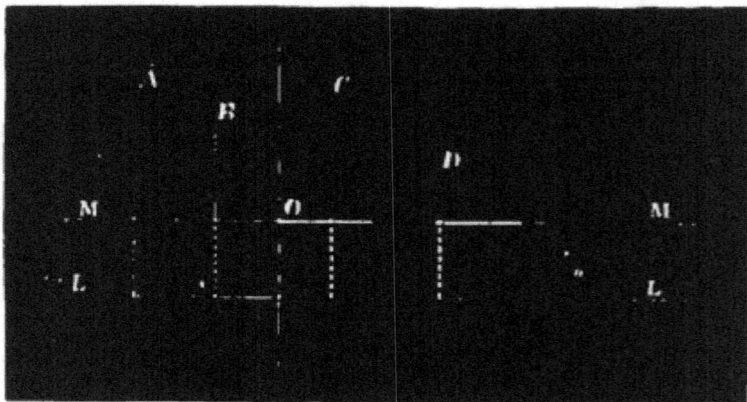

Then
$$AL=AM+OL,$$
$$BL=BM+OL,$$
$$CL=CM+OL, \text{ etc.}$$
Multiplying these equations respectively by a, b, c, ... and adding, we have
$$\Sigma(a . AL)=\Sigma(a . AM)+\Sigma(a)OL;$$

but $\Sigma(a.AM)=0$ since M passes through the mean centre; therefore, etc.

This property enables us *to find the mean centre.* For by taking a line L in an arbitrary position and calculating $\Sigma(a.AL)/\Sigma(a)$ we have for the locus of O a line parallel to L at this distance from it. Again, take a line in another position and construct the locus of O as before. The intersection of these loci is the point required.

Cor. 1. If $\Sigma a.AL$ is a constant, the line L touches, or envelopes, a circle concentric with O.

Cor. 2. If the multiples are all equal $\Sigma AL=n.OL$, where n denotes the number of points in the system.

Cor. 3. For systems of points and multiples and their mean centres

$$A_1B_1C_1 \ldots, a_1b_1c_1 \ldots, O_1,$$
$$A_2B_2C_2 \ldots, a_2b_2c_2 \ldots, O_2,$$
$$\ldots \ldots \ldots \ldots$$
$$A_nB_nC_n \ldots, a_nb_nc_n \ldots, O_n,$$

the mean centre O of all the points and their corresponding multiples is the mean centre of $O_1, O_2, \ldots O_n$ for the multiples $\Sigma(a_1), \Sigma(a_2), \ldots \Sigma(a_n)$.

[For since $\Sigma a_1A_1L=\Sigma(a_1)O_1L$, $\Sigma a_2A_2L=\Sigma(a_2)O_2L$, etc., on adding these equations

$$\Sigma a_1A_1L+\Sigma a_2A_2L+ \ldots +\Sigma a_nA_nL=\Sigma(a_1)O_1L+ \ldots$$
$$=\sum(\Sigma a_1)OL.]$$

Hence the mean centre of a system of points can be found as follows :—Find the mean centre O_1 of two of the points A and B; next find the mean centre of O_1 and C for multiples $a+b$, c. Denote this by O_2, and find the mean centre of O_2 and D for multiples $a+b+c$, d, and so on. When the entire system has thus been exhausted the last mean centre found is that of the system.

EXAMPLES.

1. The sum of the distances of the vertices of a triangle from any line is equal to three times the distance of its centroid from the line.

2. Draw a tangent to a circle such that $\Sigma a . AL$ may be a maximum, minimum, or have any given value.

[The extremities of the diameter passing through the mean centre are obviously the points of contact in the extreme cases. The general case reduces to draw a common tangent to two circles.]

3. If L touches the in-circle $\Sigma a . AL = 2\Delta$ when the multiples are equal to the sides of the triangle.

3*a.* For the ex-circle to the side c the equation becomes
$$aAL + bBL - cCL = 2\Delta.$$

4. The projection of the mean centre on any line is the mean centre of the projections of the system of points on the line.

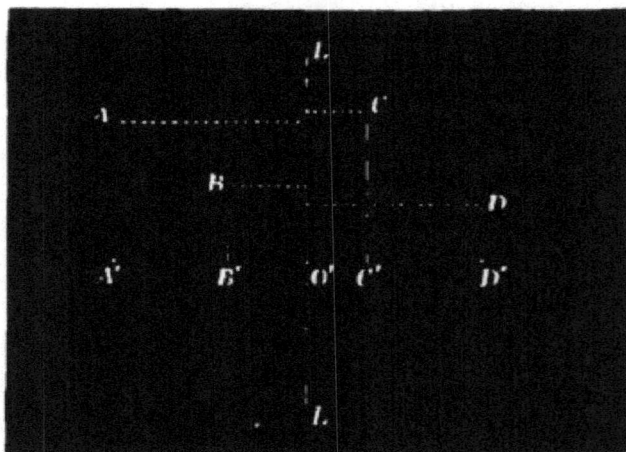

[Let the projections be denoted by O', A', B', C' ... and L the line OO'. Then $A'O' = AL$, $B'O' = BL$, etc. Hence
$$\Sigma a . A'O' = \Sigma a . AL = 0 ; \text{ therefore, etc.]}$$

5. If O, O_1, O_2, O_3 denote the in- and ex-centres of a triangle,
$$(s-a)O_1L + (s-b)O_2L + (s-c)O_3L = s . OL.*$$

---

* This relation may be otherwise written :—
$$\frac{O_1L}{r_1} + \frac{O_2L}{r_2} + \frac{O_3L}{r_3} = \frac{OL}{r}.$$

[For O is the mean centre of the remaining points for multiples $s-a$, $s-b$, $s-c$ (Art. 52, Ex. 3), and since
$$\Sigma(s-a)=s \; ; \text{ therefore, etc.}]$$

6. Let three similar triangles BCA', CAB' and ABC' be described on the sides of ABC in the same aspect ; to prove that the mean centres of the triangles ABC and $A'B'C'$ coincide (Brocard).

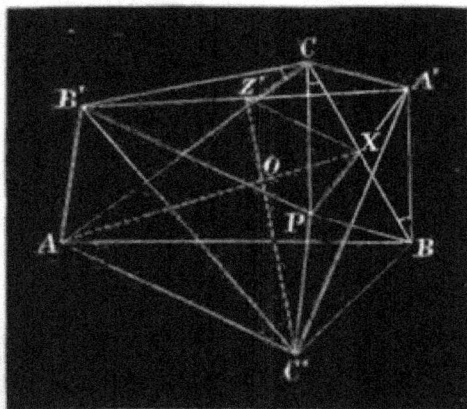

[Let X be the middle point of BC and Z' of $A'B'$. Complete the parallelogram $BA'CP$. Join AX, $C'Z'$, $Z'X$ and PB'. The triangles BPC and $B'CA$ are similar, therefore $CP/CB=B'C/AC$ (Euc. VI. 4), or by alternation $B'C/CP=AC/BC$; also the angles $B'CP$ and ACB are equal, therefore the triangles $B'PC$ and ABC are similar (Euc. VI. 6); hence $CB'/B'P=CA/AB$; alternately $CB'/CA=PB'/AB$; but $CB'/CA=C'A/AB$ (hyp.) ; therefore $PB'/AB=C'A/AB$ from which
$$PB'=AC'.$$
Again $\angle PB'C=\angle BAC$, to these add the equals ACB' and BAC'' respectively ; therefore PB' and AC' are parallel. But $Z'X$ is parallel and equal to half of PB' ; therefore it is parallel and equal to half of AC'. Hence the medians AX and $C'Z'$ trisect each other.* Otherwise thus : †—Let another triangle ABC''' be described below the base AB symmetrically equal to ABC'. It is easy to see that

---

* For another proof see Milne's *Companion to the Weekly Problem Papers*, Art. 123.

† *Educational Times.* Reprint. Vol. liv., p. 102.

the triangles ABA' and CBC'' are equal in area; similarly ABB' and CAC'' are equal. By addition we have $ABA' + ABB' = ABC + ABC''$ or $ABA' + ABB' - ABC'' = ABC$, *i.e.* the algebraic sum of the perpendiculars on AB from A', B', C' = the perpendicular from C on AB. Similar results are obtained for the sides BC and CA; therefore, etc. Syamadas Mukhopadhyay.]

7. If two points A and B be displaced to new positions A' and B', their mean centre M for any multiples is displaced to M' found by the following construction :—

Through M draw lines MP and MQ equal and parallel to AA' and BB' respectively. Join PQ and divide it in M' such that $PM'/QM' = AM/BM$.

[For since $AA'PM$ and $BB'QM$ are parallelograms, $A'P = AM$ and $B'Q = BM$; therefore by similar triangles $PA'M'$ and $QB'M'$,

$$\frac{A'P}{B'Q} = \frac{A'M'}{B'M'} = \frac{PM}{QM} : \text{therefore, etc.}]$$

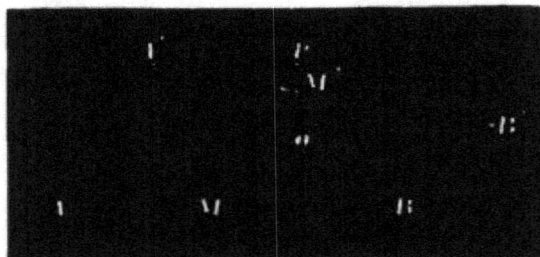

8. If three points A, B and C be displaced to new positions A', B' and C'', their mean centre M is displaced to M' found by the following construction :—

Through M draw lines MP, MQ and MR equal and parallel to the displacements AA', BB' and CC' respectively; M' is the mean centre of P, Q, R.

[For let X denote the mean centre of A and B, X' which is found by Ex. 7 of A' and B'. Draw MO equal and parallel to XX'. Join OX', RC' and $X'C'$.

It is evident by parallels that O is the mean centre of P and Q; also $MX = OX'$ and $MC = RC'$; therefore in the similar triangles $OM'X'$ and $RM'C$, $\dfrac{M'X'}{M'C} = \dfrac{OX'}{RC'} = \dfrac{MX}{MC}$ ....(1)

hence M' is the mean centre of X' and C', that is of A', B' and C', for the same multiples that M is of A, B and C.

But each of the ratios in (1) is equal to $M'O/M'R$; therefore M' is the mean centre of O and R, that is of P, Q and R for the same set of multiples.

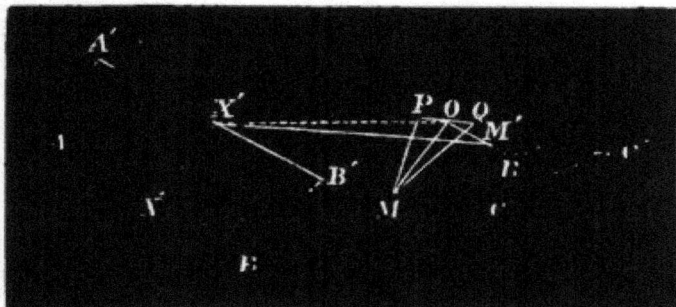

NOTE.—The construction for the displaced mean centre may in the same manner be extended to the quadrilateral and generally to a polygon of any number of sides.

Hence for two systems of points A, B, C, ... and A', B', C', ... and their mean centres M and M' for the same set of multiples a, b, c ... if we draw through M parallels MP, MQ, MR, ... equal to AA', BB', CC', ... respectively, the mean centre of the third system P, Q, R, ... for the same multiples coincides with M'.

9. If through any point M are drawn MP, MQ and MR parallel and proportional to the sides of a triangle ABC, the mean centre of P, Q and R for multiples each equal to unity coincides with M.

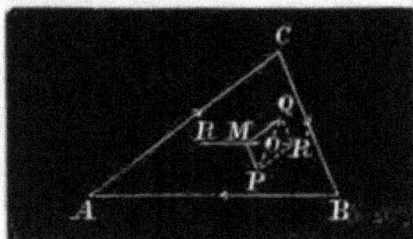

[By Ex. 8, or thus :—Complete the parallelograms $PMQR'$ and draw MR'.

Since $\dfrac{PM}{PR} = \dfrac{PM}{QM} = \dfrac{a}{b}$ and the angles at P and C equal, the triangles PMR' and ABC are similar, hence $MR = MR' = 2MO$, and O

is the mean centre of P and Q, and therefore M is the mean centre of P, Q, R.]

10. Prove the similar property for the quadrilateral; and generally :—

If through any point M lines are drawn parallel and proportional to the sides of a polygon; the mean centre of their extremities for multiples each $= 1$ coincides with M.

11. If a system of points $A, B, C, \ldots$ be displaced to $A', B', C', \ldots$ such that $AA', BB', CC', \ldots$ are parallel and proportional to the sides of a polygon, the mean centre of the system remains a fixed point.

[By aid of Exs. 8 and 10.]

12. **Weill's Theorem.**—A variable polygon is inscribed to one circle and escribed to another; to prove that the mean centre of the points of contact of its sides with the latter circle is a fixed point.

[Let $ABC\ldots$ denote the polygon, $A'B'C'\ldots$ a consecutive position, T and T' the points of contact of AB and $A'B'$ with the circle of radius r; θ the small angle between AB and $A'B'$, and X their intersection.

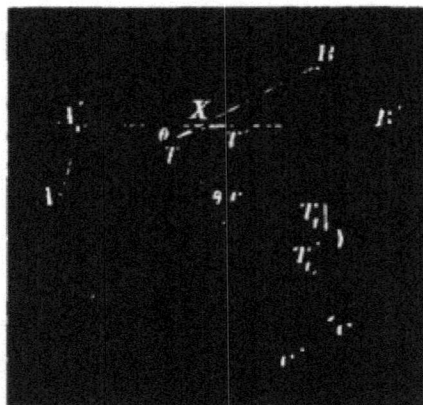

The triangles $AA'X$ and $BB'X$ are similar, hence $BB'/AA' = BX/A'X$

and
$$\frac{BB'}{AA' + BB'} = \frac{BX}{BX + A'X} = \frac{BX}{BX + AX} \text{ in the limit} = \frac{BX}{AB} \quad ......(1)$$

Similarly
$$\frac{BB'}{BB' + CC'} = \frac{BY}{BC} \quad ....(2)$$

Also, since AB and $A'B'$ are indefinitely near to one another, X is indefinitely near to the point of contact T, and BX and BY are therefore equal because they are tangents from the same point to a circle.

Dividing (2) by (1)

$$\frac{AB}{BC} = \frac{AA' + BB'}{BB' + CC'} \quad\dots\dots\dots\dots\dots\dots\dots\dots\dots(3)$$

Again

$$\frac{AA'}{AX} = \frac{BB'}{BX} = \frac{\theta}{\sin A'} \qquad \text{(Rule of Sines)},$$

hence

$$\frac{\theta}{\sin A'} = \frac{AA' + BB'}{AB} \; ;$$

but $AB = (\text{diameter of } ABC) \times \sin A'$

and $TT' = 2r\theta \; ;$

therefore $TT' \propto AA' + BB' \propto AB$ (by 3).

Thus as the polygon $ABC\dots$ varies, its points of contact are displaced for each consecutive position in the direction of its sides, and proportional to them; therefore the mean centre is a fixed point.

Note.—If the side BC is a variable tangent to a third circle of radius r, the result of dividing (2) by (1) is

$$\frac{AB}{BC} = \frac{AA' + BB'}{BB' + CC'} \cdot \frac{BX}{BY} \; ;$$

therefore if the three circles are so related that BX/BY is a constant ratio k,

$$\frac{AB}{BC} = k \cdot \frac{AA' + BB'}{BB' + CC'}$$

and $TT'/T_1 T_1' = r/kr' \cdot AB/BC.]$

13. The mean centres of the vertices of any polygon and of similar triangles similarly described on its sides coincide (M'Cay).

[Let the vertices of the triangles on the sides AB, BC, $CD \dots$ be A', B', $C' \dots$ respectively.

Since $AA' : BB' : CC' \dots = AB : BC : CD \dots$ and are inclined to the sides of the polygon at the same angle; we may regard the vertices of the given polygon displaced to $A'B'C' \dots$ distances proportional and parallel to its sides turned through that angle (cf. Ex. 6).]*

---

* The proofs of Examples 11-13 were communicated to the Author by Mr. Charles M'Vicker.

14. Through the centre O of a regular polygon any line is drawn meeting the sides in A', B', C', ... to prove that $\Sigma \dfrac{1}{OA'} = 0$.

[Let M be the middle point of one side, then $MA'O$ is a right-angled triangle, and if a perpendicular MM' be let fall on the hypotenuse we have

$$OA' \cdot OM' = r^2 \text{ or } \Sigma \frac{1}{OA'} = \frac{1}{r^2} \Sigma OM' = 0. \quad \text{Art. 50.} \quad \text{See Art. 3, Ex. 9.]}$$

54. Theorem.—*For any system of points A, B, C, ... their mean centre O, and any line L; to prove that*

$$\Sigma a \cdot AL^2 = \Sigma a \cdot AL'^2 + \Sigma(a)OL^2,$$

where L' is the line through O parallel to L.

For $AL = AL' + OL$; $\therefore$ $AL^2 = AL'^2 + OL^2 + 2AL' \cdot OL$;

$\quad BL = BL' + OL$; $\therefore$ $BL^2 = BL'^2 + OL^2 + 2BL' \cdot OL$;

....

Multiplying these equations by a, b, c, ... respectively and adding results,

$$\Sigma a \cdot AL^2 = \Sigma a \cdot AL'^2 + \Sigma(a)OL^2 + 2OL\Sigma(a \cdot AL'),$$

but $\Sigma a \cdot AL' = 0$ (Art. 50); therefore, etc.

Cor. 1. When the multiples are equal

$$\Sigma AL^2 = \Sigma AL'^2 + nOL^2,$$

also since $\Sigma AL = n \cdot OL$; OL is the arithmetical mean of the several lines AL, BL, CL ..., and AL', BL' ... the several differences between each and their mean.

Hence, *the sum of the squares of n quantities $= n$ times the square of their mean value $+$ the sum of squares of the n differences ;* or if the quantities are the segments of a line this property may be stated : *the sum of the squares of the unequal parts $=$ the sum of the squares of the equal parts $+$ the sum of the squares of the n differences.* This property is obviously an extension of Euc. II. 9, 10.

Cor. 2. For any two parallel lines L and M,

$$\Sigma a \cdot AL^2 - \Sigma a \cdot AM^2 = \Sigma(a)(OL^2 - OM^2).$$

55. Theorem.—*For any point P to prove that*
$$\Sigma a \cdot AP^2 = \Sigma a \cdot AO^2 + \Sigma(a)OP^2.$$

Project the system of points on the line OP and denote their projections by A', B', C', ...

Then (Euc. III. 12-13)

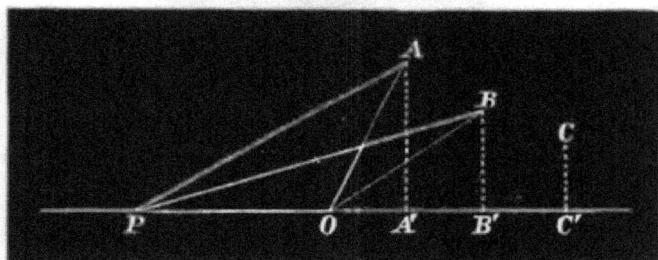

$$AP^2 = AO^2 + OP^2 + 2OP \cdot OA'.$$

Similarly $\qquad BP^2 = BO^2 + OP^2 + 2OP \cdot OB'$ etc.

Multiplying these equations by a, b, c ... and adding the results,

$$\Sigma a \cdot AP^2 = \Sigma a \cdot AO^2 + \Sigma(a)OP^2 + 2OP\Sigma a \cdot OA',$$

but O is the mean centre of the system A', B', C' ... (Art. 53, Ex. 4); therefore $\Sigma a \cdot OA' = 0$.

COR. 1. If the n multiples are equal
$$\Sigma AP^2 = \Sigma AO^2 + n \cdot OP^2.$$

COR. 2. For a regular cyclic polygon the sum of the squares of the distances of any point on the circle from the n vertices is constant and $= 2nR^2$.

COR. 3. If $\Sigma a \cdot AP^2$ is constant, the locus of P is a circle concentric with O the square of whose radius is equal to $\qquad \dfrac{\Sigma a \cdot AP^2 - \Sigma a \cdot AO^2}{\Sigma(a)}$.

COR. 4. $\Sigma a \cdot AP^2$ is a minimum when P coincides with O. See Art. 16, Ex. 3.

EXAMPLES.

1. $ABCD$ ... is a regular cyclic polygon, O the centre, R the radius, and P any point on the circle to prove that the sum of the squares of the perpendiculars from P on the radii OA, OB, OC ... $= \frac{1}{2}nR^2$.

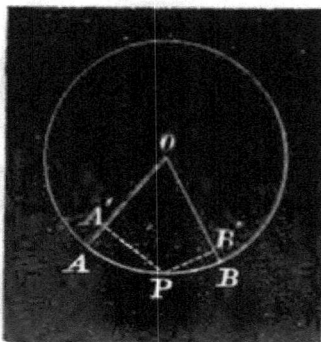

[Denote the feet of the perpendiculars by A', B', C' ... The circle on OP as diameter passes through these points (Euc. III. 31); also since $A'B'$, $B'C'$, ... subtend equal angles $(2\pi/n)$ at O, a point on the circle, $A'B'C'$... is a *regular cyclic polygon*. Hence (Cor. 2)

$$\Sigma PA'^2 = 2n(\tfrac{1}{2}OP)^2 = \tfrac{1}{2}nR^2.$$

Similarly $\qquad \Sigma OA'^2 = \tfrac{1}{2}nR^2.$]

2. For any line L passing through O, $\Sigma AL^2 = \tfrac{1}{2}nR^2$.

[Let L coincide with OP. By similar triangles

$$AL = PA', \quad BL = PB', \text{ etc.}$$

therefore $\qquad \Sigma PA'^2 = \Sigma AL^2 = \tfrac{1}{2}nR^2$ by Ex. 1.]

3. The sum of the squares of the perpendiculars $p_1, p_2, p_3 \ldots p_n$

from *any* point P upon the sides of the polygon is equal to $n(r^2+\frac{1}{2}\delta^2)$, where r is the radius of the in-circle and $\delta = OP$.

[Through O draw parallels OA', OB', OC' ... to the sides of the polygon meeting the corresponding perpendiculars from P in A', B', C', ... As before $A'B'C'$... is a regular cyclic polygon inscribed in the circle on OP as diameter.

Since the sum of the perpendiculars is constant and $=nr$

$$\Sigma p_1{}^2 = nr^2 + \Sigma PA'^2 \quad \dots\dots\dots\dots\dots\text{(Art. 54) (1)}$$

but $\qquad \Sigma PA'^2 = \frac{1}{2}n\delta^2 \qquad\qquad\qquad\qquad$ (Ex. 1),

substituting this value in (1); therefore, etc.]

4. In Ex. 3 if P is on the in-circle $\Sigma p_1{}^2 = \frac{3}{2}nr^2$.

5. If π_1, π_2, π_3, ... denote the distances of the vertices from any line L and $\delta = OL$, $\qquad \Sigma\pi_1{}^2 = n(\delta^2 + \frac{1}{2}R^2)$.

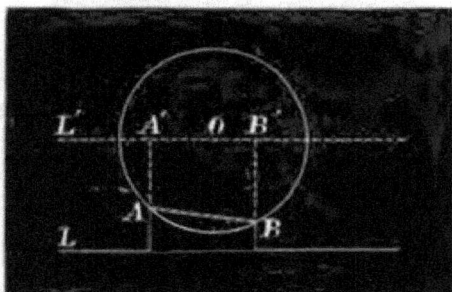

[Through O draw L' parallel to L and let A', B', C' be its intersections with AL, BL, CL ... respectively.

Since $\qquad\qquad \Sigma AL = nOL \qquad\qquad\qquad$ (Art. 53),

$\qquad\qquad\qquad \Sigma AL^2 = n \cdot OL^2 + \Sigma AA'^2 \qquad$ (Art. 54),

but $\qquad\qquad \Sigma AA'^2 = \frac{1}{2}nR^2 \qquad\qquad\qquad$ (Ex. 2);

therefore by substitution $\Sigma AL^2 = n(OL^2 + \frac{1}{2}R^2)$,

or $\qquad\qquad \Sigma\pi_1{}^2 = n(\delta^2 + \frac{1}{2}R^2)$.

5a. If L is a tangent to the circum-circle

$$\Sigma\pi_1{}^2 = \frac{3}{2}nR^2.$$

6. If P be a point on the circum-circle of a regular polygon $ABC \dots$, $\qquad\qquad \Sigma PA^4 = 6nR^4$.

[Draw OP and produce it to meet the circle again in Q, and let A', B', C' be the projections of the vertices on this line. Since PAQ is a right-angled triangle,

$$PQ \cdot PA' = PA^2.$$

Squaring, we have $\qquad 4R^2 . PA'^2 = PA^4,$

therefore $\qquad\qquad 4R^2 \Sigma PA'^2 = \Sigma PA^4.$

But $\qquad\qquad\qquad \Sigma PA'^2 = nR^2 + \Sigma OA'^2,$ (Art. 54, Cor. 1.)

and $\qquad\qquad\qquad \Sigma OA'^2 = \frac{1}{2} nR^2.$ $\qquad\qquad$ (Ex. 2.)

Substituting $\qquad\qquad \Sigma PA^4 = 4R^2(nR^2 + \frac{1}{2} nR^2) = 6nR^4.]$

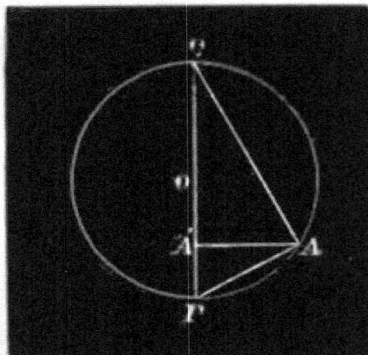

7. If a, b, c denote the sides of a triangle ABC and P any point on the in-circle, $\qquad \Sigma a . AP^2 = \Sigma a . AO^2 + 2r\Delta.$

8. If ABC be an equilateral triangle, and L a tangent to the in-circle, $\qquad\qquad\qquad \dfrac{1}{AL} + \dfrac{1}{BL} + \dfrac{1}{CL} = 0.$

[For $AL + BL + CL = 3r$, therefore on squaring $\Sigma AL^2 + 2\Sigma BL . CL = 9r^2$. Also $\Sigma AL^2 = 3r^2 + \frac{3}{2}R^2$, or since $R = 2r$, $\Sigma AL^2 = 9r^2$; hence $\Sigma BL . CL = 0$, therefore, etc.]

9. Perpendiculars are let fall from P on the sides of any polygon ABC ... and their feet joined ; prove that if the area of the in-

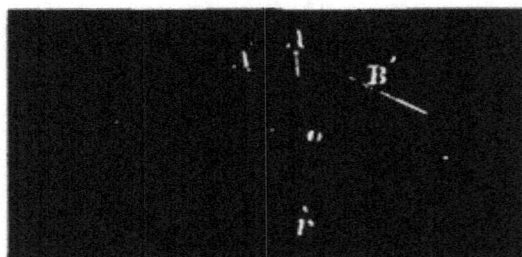

scribed figure $A'B'C'$ ... is constant, the locus of P is a circle concentric with the mean centre of A, B, C, ... for the multiples in $2A$, $\sin 2B$, $\sin 2C$, ...

[Let O be the middle point of AP. Then
$$2A'OB' = 2A'B'P - AA'B'P;$$
hence $\qquad 2\Sigma A'OB' = 2\Sigma PA'B' - \Sigma PAA'B',$
or $\qquad \tfrac{1}{4}.PA^2 \sin 2A = 2A'B'C' \ldots - ABC \ldots.$
Therefore $\Sigma \sin 2A . AP$ is constant ....

For a triangle the mean centre of A, B, C for multiples $\sin 2A$, $\sin 2B$, $\sin 2C$ is the circum-centre, showing Art. 23, Ex. 2, to be a particular case of this theorem. M'Vicker.]

56. Theorem.—*If ΣAB denote the sum of the mutual distances of a system of points $A, B, C \ldots$ from each other; to prove that $\Sigma(ab . AB)^2 = \Sigma(a) . \Sigma(a . AO^2)$.*

In Art. 55 if we suppose P to coincide with each point of the system successively we have the following relations :—

$$a . AA^2 + b . AB^2 + c . AC^2 + \ldots = \Sigma a . AO^2 + \Sigma(a) OA^2,$$
$$a . BA^2 + b . BB^2 + c . BC^2 + \ldots = \Sigma a . AO^2 + \Sigma(a) OB^2,$$
$$a . CA^2 + b . CB^2 + c . CC^2 + \ldots = \Sigma a . AO^2 + \Sigma(a) OC^2,$$
....

Multiplying these results by a, b, c ... respectively and adding $2\Sigma ab . AB^2 = \Sigma(a) . \Sigma a . AO^2 + \Sigma(a) . \Sigma a . AO^2$,
therefore $\qquad \Sigma ab . AB^2 = \Sigma(a) . \Sigma a . AO^2.$

Cor. 1. If the multiples are each equal to unity,
$$\Sigma AB^2 = n . \Sigma AO^2.$$

Cor. 2. The sum of the squares of all the lines joining the vertices of a regular polygon $= n^2 R^2$; where R is the radius of the circum-circle.

Cor. 3. For three points A, B, C, the sum of the squares of the sides of a triangle = *three times the sum of the squares of the lines joining the vertices to the centroid;* or *three times the sum of the squares of the sides = four times the sum of the squares of the medians.*

Cor. 4. If O be the in-centre and a, b, c the sides of a triangle ABC $\Sigma(ab.AB^2)=\Sigma(a).\Sigma a.AO^2$ reduces to (Art. 52, Ex. 1)
$$abc(a+b+c)=(a+b+c)\Sigma a.AO^2,$$
$$\Sigma a.AO^2=abc,$$
with analogous results for O_1, O_2, and O_3.

Cor. 5. The sum of the squares of the six lines joining the centres of the in- and ex-circles $=48R^2$.

Since the centre O of the circum-circle is (Art. 52, Ex. 9.) the mean centre of O_1, O_2, O_3, O_4,
$$\Sigma O_1 O_2 = 4\Sigma OO_1^2 = 4\{R^2-2Rr+\Sigma(R^2+2Rr_1)\}$$
$$= 16R^2+8R(r_1+r_2+r_3-r),$$
but $r_1+r_2+r_3-r=4R$; therefore, etc.*

EXAMPLES.

1. If S denote the symmedian point of a triangle,
$$a^2AS^2+b^2BS^2+c^2CS^2=\frac{3a^2b^2c^2}{a^2+b^2+c^2}. \qquad \text{(Art. 52, Ex. 4.)}$$

2. For the Brocard points Ω, Ω',

a°. $\qquad \dfrac{A\Omega^2}{b^2}+\dfrac{B\Omega^2}{c^2}+\dfrac{C\Omega^2}{a^2}=1.$

β°. $\qquad \dfrac{A\Omega'^2}{c^2}+\dfrac{B\Omega'^2}{a^2}+\dfrac{C\Omega'^2}{b^2}=1.$ $\qquad$ (Art. 52, Ex. 4.)

3. The distance OP of any point P from the in-centre of a triangle is given by the equation
$$\Sigma a.AP^2=abc+\Sigma(a).OP^2.$$
[Eliminating $\Sigma a.AO^2$ between the equations,
$$\Sigma a.AP^2=\Sigma a.AO^2+\Sigma(a).OP^2,$$
and $\qquad\qquad \Sigma(ab.AB^2)=\Sigma(a).\Sigma a.AO^2,$
the above result follows.]

---

* Otherwise thus :—Since O_1 is the orthocentre of $O_2O_3O_4$, if perpendiculars OX, OY, OZ be drawn to the sides from the circum-centre O of $O_2O_3O_4$, $O_1O_2=2OX$, $O_1O_3=2OY$, ... ; also $OO_2=2R$; hence
$$O_1O_4^2+O_2O_3^2=4(2R)^2=16R^2,$$
therefore $\qquad\qquad \Sigma O_2O_3^2=48R^2.$

4. If P coincides with the circum-centre, prove the following where D, D_1, D_2, D_3 are the distances of the circum-centre from the in- and ex-centres :—

$$D^2 = R^2 - 2Rr \; ; \; D_1{}^2 = R^2 + 2Rr_1, \text{ etc., etc.}$$

5. Prove that the distance δ of the symmedian point S from the circum-centre O of a triangle ABC is given by the equation

$$\delta^2 = R^2 - \frac{3a^2b^2c^2}{(a^2 + b^2 + c^2)^2}.$$

[For any point P (Art. 52, Ex. 4) $\Sigma a^2 AP^2 = \Sigma a^2 AS^2 + \Sigma(a^2)\delta^2$, letting P coincide with O; therefore

$$(a^2 + b^2 + c^2)R^2 = \frac{3a^2b^2c^2}{a^2 + b^2 + c^2} + (a^2 + b^2 + c^2)\delta^2 \; ;$$

hence *
$$\delta^2 = R^2 - \frac{3a^2b^2c^2}{(a^2 + b^2 + c^2)^2} \; ;$$

therefore, etc.]

6. The distances of Ω and Ω' from the circum-centre are given by the equations $O\Omega = O\Omega' = R\sqrt{1 - 4\sin^2\omega}$.

$$\left[\text{For } \frac{1}{a^2} + \frac{1}{b^2} + \frac{1}{c^2} = \frac{1}{4R^2}\Sigma \operatorname{cosec}^2 A = \frac{1}{4R^2}\operatorname{cosec}^2\omega.\right]$$

7. For the in-centre O_1 and the ex-centres O_2, O_3, O_4 prove the relations

$\alpha^\circ.$
$$\frac{O_1O_2{}^2}{r_1} + \frac{O_1O_3{}^2}{r_2} + \frac{O_1O_4{}^2}{r_3} = 8R.$$

$\beta^\circ.$
$$\frac{O_3O_4{}^2}{r_2r_3} + \frac{O_4O_2{}^2}{r_3r_1} + \frac{O_2O_3{}^2}{r_1r_2} = \frac{8R}{r}.$$

8. For any point P
$$(s-a)PO_1{}^2 + (s-b)PO_2{}^2 + (s-c)PO_3{}^2 - sPO^2 = 2abc.$$

9. Find the following expression for the square of the distance δ between the circum- and ortho-centre of a triangle ABC.
$$\delta^2 = R^2(1 - 8\cos A \cos B \cos C)$$
$$= \Sigma a^2(a^2 - b^2)(a^2 - c^2)/16\Delta^2.$$

[By the previous method, or more simply by finding the area of the pedal triangle of ABC, $(2\,\text{area} = R^2\sin 2A \sin 2B \sin 2C)$, and using Art. 23, Ex. 1, and reducing.]

---

* This expression is equivalent to
$$\delta^2 = R^2\sec^2\omega(1 - 4\sin^2\omega),$$
where ω is the Brocard angle.

Reciprocal Theorems.

57. Theorem. — *For any two points M and N, and systems of lines A, B, C, ... and multiples a, b, c, ... having given* $\Sigma a \cdot MA = 0$ *and* $\Sigma a \cdot NA = 0$ *to prove that*

$$\Sigma a \cdot LA = 0,$$

where L is any point on the line O connecting M and N.

For　　　$MN \cdot LA + NL \cdot MA + LM \cdot NA = 0.$

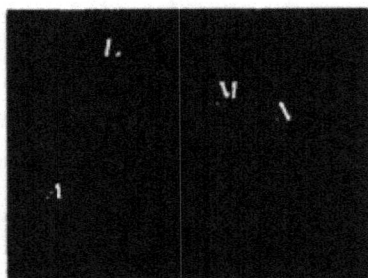

Similarly for the lines B and C,

$$MN \cdot LB + NL \cdot MB + LM \cdot NB = 0,$$
$$MN \cdot LC + NL \cdot MC + LM \cdot NC = 0.$$

Multiplying these equations respectively by $a, b, c, \ldots$ and adding, we get

$$MN\Sigma a \cdot LA + NL\Sigma a \cdot MA + LM\Sigma a \cdot NA = 0 \ldots\ldots(1)$$

hence if $\Sigma a \cdot MA$ and $\Sigma a \cdot NA$ each $= 0$, $\Sigma a \cdot LA = 0$, for any other point L on the line MN.

More generally: If $\Sigma a \cdot MA$ *and* $\Sigma a \cdot NA$ *are equal* $\Sigma a \cdot LA$ *has the same value.*

For, let $\Sigma a \cdot MA = \Sigma a \cdot NA = k$; substituting in (1)

$$MN\Sigma a \cdot LA + (NL + LM)k = 0;$$

dividing by $MN (= LN + ML)$ and transposing

$$\Sigma a \cdot LA = k.$$

Hence, the locus of a point L such that the sum of given multiples of the perpendiculars from it upon a system of lines A, B, C, ... is constant ($\Sigma a . LA = k$) is a right line.

Def. When the constant vanishes $\Sigma a . LA = 0$, the locus O is termed the *Central Axis* of the system of lines for the given system of multiples.

It is evident that the central axis is one of a system of parallel lines obtained by taking different values, $k_1, k_2, k_3, ...$ of k.

For if L in (1) lies on O then

$$NL\Sigma a . MA + LM\Sigma a . NA = 0 \dots\dots\dots\dots(2)$$

or $\quad\dfrac{\Sigma a . MA}{\Sigma a . NA} = \dfrac{ML}{NL} = \dfrac{MO}{NO} \quad$ (Euc. VI. 4.)

hence the values of the summation corresponding to any point is proportional to the distance of that point from the central line or axis.

Otherwise thus :—If M_1 and N_1 are the loci of M and N such that $\Sigma a . MA = k_1$ and $\Sigma a . NA = k_2$ and P if possible their point of intersection ; then since P is on both lines $\Sigma a . PA = k_1$ and $\Sigma a . PA = k_2$, which is absurd ; therefore, etc.

58. **Problem.**—*To find the Central Axis O of a given system of lines A, B, C, ... for a given system of multiples a, b, c, ...*

Take any three points P, Q, R, and calculate $\Sigma a . PA$, $\Sigma a . QA$, and $\Sigma a . RA$.

On QR find a point L such that

$$\frac{\Sigma a . QA}{\Sigma a . RA} = \frac{QL}{RL}.$$

L is by (2) on the required line; similarly obtaining points M and N on the other sides of the triangle P, Q, R, their line of connection is that required.

59. Let the multiples a, b, c ... denote segments of the given lines A, B, C... respectively; $a . LA$, $b . LB$, $c . LC$ ... are each twice the area of the triangle subtended by the corresponding segment at the point L; hence, *the locus of a point such that the sum of the areas subtended at it by any number of finite lines is constant, (k) is a right line;* and if different values be assumed for k the locus varies in position by moving parallel to itself.

60. **Theorem**.—*The locus of the mean centre O of the points of intersection A_1, B_1, C_1, D_1, of a variable line L, moving parallel to itself, with the sides of a given polygon is a right line.*

Let a, b, c, etc., be the given multiples and a, β, γ ... the angles at A_1, B_1, C_1 ... made by the variable line with the sides A, B, C ... of the given polygon.

By hyp.　　　　　　$\Sigma a . A_1 O = 0$,

but $A_1 O = OA/\sin a$; $B_1 O = OB/\sin \beta$; $C_1 O = OC/\sin \gamma$, etc., substituting these values,

$$a/\sin a . OA + b/\sin \beta . OB + C/\sin \gamma . OC + \text{etc.} = 0,$$

hence O describes a line, viz., the central axis of the system for the multiples $a \cosec a$, $b \cosec \beta$, $c \cosec \gamma$ ....

Def. This locus of the mean centre for the system of parallels, is termed a *Diameter of the Polygon* when the multiples $a = b = c = ... = 1$; a name suggested by the property to which the theorem is reducible when the polygon becomes a circle.

61. **Problem**.— *To find a point P such that for any systems of lines A, B, C ... and multiples a, b, c ...*
　　　　　　$\Sigma a . PA^2$ *is a minimum.*

Let any line L through P meet the sides of the polygon in A', B', C' ... at angles a, β, γ .... Then $\Sigma a . PA^2$

is a minimum when $\Sigma a \sin^2 a \cdot PA'^2$ is a minimum, that is when P is the mean centre of A', B', C' ... for the multiples $a \sin^2 a$, $b \sin^2 \beta$ .... As L varies parallel to itself the locus of P is a diameter. Let it meet the sides of the polygon in A_1, B_1, C_1 ...; the mean centre of these points for the multiples $a \sin^2 a$, $b \sin^2 \beta$ ... is obviously the point required.

EXAMPLES.

1. If a line is drawn through O the centre of an escribed circle to meet the sides in X and Y such that $CX = CY$; prove that $AY \cdot BX = (\frac{1}{2} XY)^2$; and conversely, if $AY \cdot BX = (\frac{1}{2} XY)^2$, AB is a tangent to the circle.

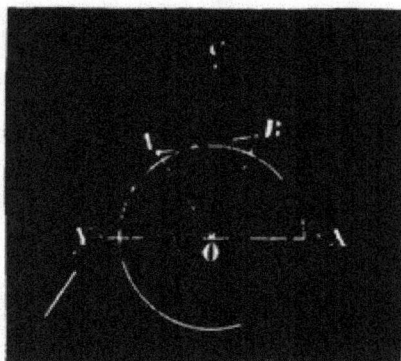

[The angles of the triangles BOX and AOY are as follows :—
$$BXO = 90 - \tfrac{1}{2}C, \quad OBX = 90 - \tfrac{1}{2}B, \text{ therefore } BOX = 90 - \tfrac{1}{2}A ;$$
$$AYO = 90 - \tfrac{1}{2}C, \quad OAY = 90 - \tfrac{1}{2}A, \text{ therefore } AOY = 90 - \tfrac{1}{2}B.$$
Hence they are similar ; therefore, etc.]

2. The diameters of an equilateral triangle envelope the in-circle.
[Suppose the multiples to be equal to unity, through B and C draw any two parallel lines terminated by the opposite sides of the triangle and trisect them in X and Y towards the vertices. Since X and Y are the mean centres of their intersections with the sides,

the line XY is a diameter. Draw parallels XX'', YY''' to the sides AB and AC respectively.

Then the triangles $XX'X''$ and $YY'Y''$ are similar, therefore

$$X'X'' . Y'Y'' = XX'' . YY'''.$$

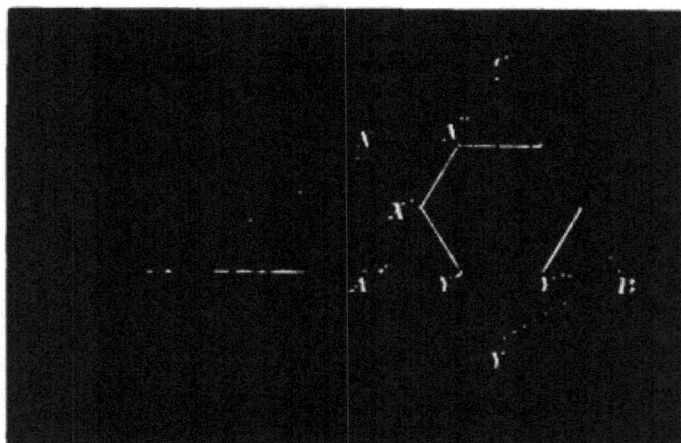

Again, the triangles CXX'' and BYY'' are similar, since the sides are parallel, therefore

$$XX'' . YY'' = CX'' . BY'' = (\tfrac{1}{2}X''Y'')^2,$$

therefore $\qquad X'X'' . Y'Y'' = (\tfrac{1}{2}X''Y'')^2;$

therefore, etc., by Ex. 1. M'Vicker.]

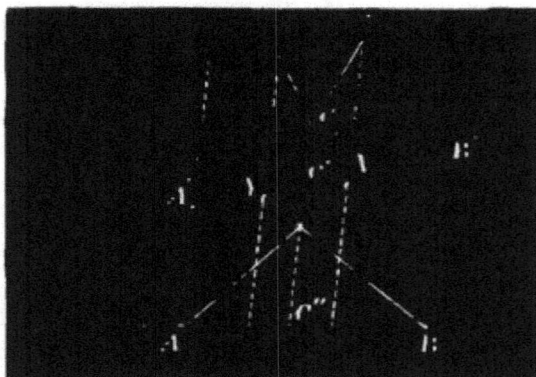

Otherwise thus :—Draw any system of parallels AA', BB', CC' terminated by the opposite sides and let A', B', C denote the mean centres of their points of intersection with the sides of ABC. Let the diameter $A'B'C'$ meet the sides in X, Y, Z; the parallels through X and Y are bisected at these points, hence AX and BY each bisect CC' and therefore meet at its middle point. Then from the complete quadrilateral $ABCXYZ$ the row $AC''BZ$ is harmonic, therefore AA', $C'C''$ and BB' are in harmonic progression, or

$$\frac{1}{AA'} + \frac{1}{BB'} = \frac{2}{C'C''} = \frac{1}{CC''}$$

but $\Sigma\dfrac{1}{AA'} = 0$ is the criterion for the tangent to the in-circle. See Art. 55, Ex. 8 ; therefore, etc.]

3. If a system of n points $A, B, C, \ldots, N$ be situated at equal distances on an arc of a circle O, r ; required to find the position of their mean centre.

[Through O draw a parallel L to the chord of the arc AN ; let the angle $AOL = \alpha$ and $AON = n\beta$. Then, if d be the distance of the mean centre from O, we have (Art. 53)

$$nd = R\{\sin\alpha + \sin\overline{\alpha + \beta} + \sin\overline{\alpha + 2\beta} + \ldots + \sin\overline{\alpha + n - 1}\beta\}$$
$$= \frac{\sin(\alpha + \tfrac{1}{2}\overline{n-1}\beta)\sin\tfrac{1}{2}n\beta}{\sin\tfrac{1}{2}\beta} ;$$

but $\alpha + \tfrac{1}{2}n\beta = \tfrac{1}{2}\pi$, therefore the above expression becomes, on reduction, $r\cot\tfrac{1}{2}\beta\sin\tfrac{1}{2}n\beta$.]

Note.—If the number of points on the arc is infinitely great, it follows, since β is indefinitely small, that

$$d = \frac{\text{chord} \times \text{radius}}{\text{length of arc}}.$$

CHAPTER V.

COLLINEAR POINTS AND CONCURRENT LINES.

62. Theorem.—*If a straight line be drawn cutting the sides of a triangle ABC in points X, Y, Z, to prove the relation*

$$\frac{BX}{CX}\cdot\frac{CY}{AY}\cdot\frac{AZ}{BZ}=1;$$

and conversely, *having given this relation to prove the points are collinear.* (Menelaus.)

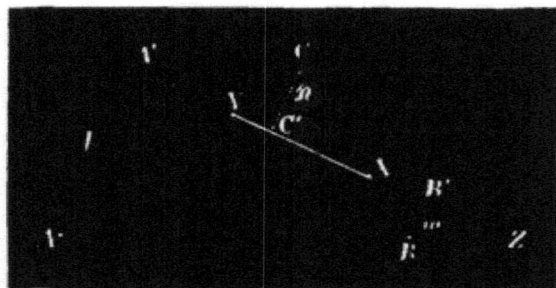

For denoting the perpendiculars from the vertices on the transversal by l, m, n; we have by similar pairs of triangles,

$$\frac{BX}{CX}=\frac{m}{n};\quad \frac{CY}{AY}=\frac{n}{l};\quad \frac{AZ}{BZ}=\frac{l}{m}.$$

Multiplying these equations* and reducing, the above result follows at once.

Conversely, if the line joining X and Y meet the base in Z' by the first part of the Proposition,

$$\frac{BX}{CX}\cdot\frac{CY}{AY}\cdot\frac{AZ'}{BZ'}=1,$$

but by hyp.

$$\frac{BX}{CX}\cdot\frac{CY}{AY}\cdot\frac{AZ}{BZ}=1,$$

hence

$$\frac{AZ}{BZ}=\frac{AZ'}{BZ'},$$

therefore Z and Z' coincide.

63. Theorem.—*If three lines AO, BO, CO be drawn from the vertices of a triangle ABC through any point O to meet the opposite sides in X, Y, Z; to prove the relation*

$$\frac{BX}{CX}\cdot\frac{CY}{AY}\cdot\frac{AZ}{BZ}=-1,†$$

and conversely, *if this relation be given the lines AX, BY, CZ are concurrent.* (Ceva.)

For the triangles BOC and COA on a common base are proportional to their altitudes, which are in the ratio BZ/AZ.

---

*The proof here given applies equally to the general proposition :—
Any right line meeting the sides of a polygon $ABCDEF$... in points X, Y, Z, U, V, W... gives the relation

$$\frac{AX}{BX}\cdot\frac{BY}{CY}\cdot\frac{CZ}{DZ}\cdot\frac{DU}{EU}\cdot\frac{EV}{FV}\cdot\frac{FW}{GW}\ldots=1.$$

† A line drawn across the sides of a triangle meets them either all externally, or two internally and one externally, *i.e.* the number of sides cut externally is always *odd*, and therefore the product of the ratios $\frac{BX}{CX}$, $\frac{CY}{AY}$, $\frac{AZ}{BZ}$ is positive. On the other hand, if three points on the sides connect concurrently with the opposite vertices, an *odd number is internal* and the product of the ratios is therefore *negative*.

Hence the following equations :—

$$\frac{BX}{CX}=\frac{AOB}{AOC}, \quad \frac{CY}{AY}=\frac{BOC}{BOA}, \quad \frac{AZ}{BZ}=\frac{COA}{COB};$$

on multiplying* and reducing, the above result is obtained.

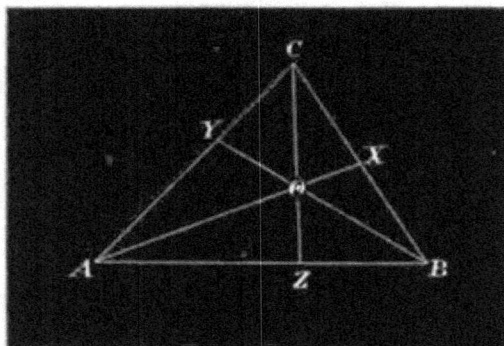

Conversely, let AX and BY meet in O.   Join CO and let it meet AB in Z'. Then by what has been proved

$$\frac{BX}{CX}\cdot\frac{CY}{AY}\cdot\frac{AZ'}{BZ'} = -1,$$

but by hyp. $$\frac{BX}{CX}\cdot\frac{CY}{AY}\cdot\frac{AZ}{BZ} = -1,$$

therefore the points Z and Z' coincide.

64. The relations of the previous Articles are equivalent to the two following :—

$$\frac{\sin BAX}{\sin CAX}\cdot\frac{\sin CBY}{\sin ABY}\cdot\frac{\sin ACZ}{\sin BCZ} = \pm 1.$$

For by the rule of sines $\dfrac{BX}{CX}=\dfrac{c\sin BAX}{b\sin CAX}$, with similar values for the remaining ratios, compounding and reducing, the above results are obtained.

---

* More generally, if the vertices of a polygon $ABCD...$ of any odd number of sides be joined to any point O and the lines produced to meet the opposite sides in X, Y, Z, U, V, W, it follows by similar reasoning that $\dfrac{AX}{BX}\cdot\dfrac{BY}{CY}\cdot\dfrac{CZ}{DZ}\cdot\dfrac{DU}{EU}... = -1.$

These formulae may be regarded as criteria of points on the sides of a triangle lying on a line and connecting concurrently with the opposite vertices.

We shall now apply them to the following remarkable particular cases:—

I. Let the points X, Y, Z be at infinity on the sides, thus $BX = CX$, $CY = AY$, and $AZ = BZ$; hence the criterion of Art. 62 is satisfied and it follows that *every three and therefore all points at infinity in the same plane may be regarded as lying on a line.**

II. Let AX, BY and CZ be any three parallel lines. Since

$$\frac{BX}{CX} \cdot \frac{CY}{AY} \cdot \frac{AZ}{BZ} = -1,$$

every three, and therefore all, parallel lines are concurrent.

Of these properties Townsend says: "Paradoxical as these conclusions appear when first stated, all doubt of their legitimacy has been long set at rest by the number and variety of the considerations tending to verify and confirm them."—*Modern Geometry,* Vol. I., Art. 136.

III. When $AC = BC$, and O is a point on the circle touching the equal sides at A and B.

By Euc. III. 32, $\angle BAO = \angle CBO$; $\angle ABO = \angle CAO$.

Substituting in the above equation, and

$$\frac{\sin ACO}{\sin BCO} = \frac{\sin^2 ABO}{\sin^2 BAO} = \frac{AO^2}{BO^2}.$$

---

* This conception of elements situated at an infinite distance is due to Desargues. About the year 1640 he showed that parallel straight lines meet at an infinitely distant point; and that parallel planes may be regarded as intersecting in the line at infinity. More recently the celebrated Poncelet proved that all points at infinity may be considered to lie in a plane.

Similarly, if CO meet the circle again in O',

$$\frac{\sin ACO}{\sin BCO} = \frac{AO'^2}{BO'^2}.$$

Hence:—*A variable chord OO' of a circle passing through a fixed point C divides harmonically the arc AB, intercepted by the tangents CA and CB.*

Also, since AB is divided harmonically at O and O', OO' is divided harmonically by AB; hence the variable pairs of tangents at O and O' intersect on the fixed line AB.

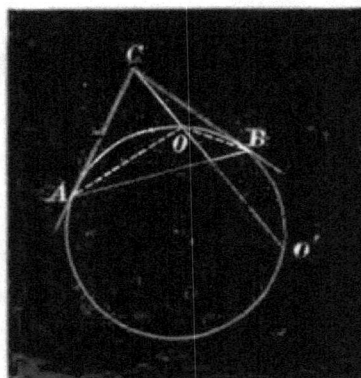

IV. Describe a circle about AOB, and let it meet the lines AC, BC, CO again in $A'B'O'$.

Then, for the point O,

$$\frac{\sin BAO}{\sin ABO} \cdot \frac{\sin CBO}{\sin CAO} = \frac{\sin BCO}{\sin ACO};$$

but $CBO = CO'B$ and $CAO = CO'A'$. (Euc. III. 22.)

Substituting these values and reducing by rule of sines,

$$\frac{OB}{OA} \cdot \frac{OB'}{OA'} = \frac{\sin BCO}{\sin ACO} \quad\cdots\cdots\cdots\cdots(1)$$

Similarly, for O',

$$\frac{O'B}{O'A} \cdot \frac{O'B'}{O'A'} = \frac{\sin BCO}{\sin ACO} \quad\cdots\cdots\cdots\cdots(2)$$

Equating these values,

$$\frac{AO}{BO} \div \frac{AO'}{BO'} = \frac{A'O'}{B'O'} \div \frac{A'O}{B'O} * \quad \dots\dots\dots\dots\dots(3)$$

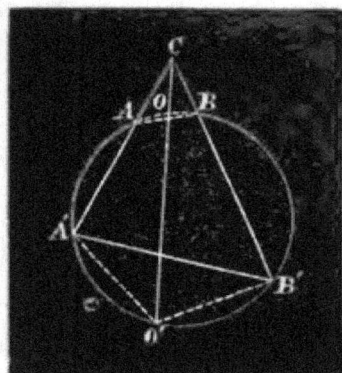

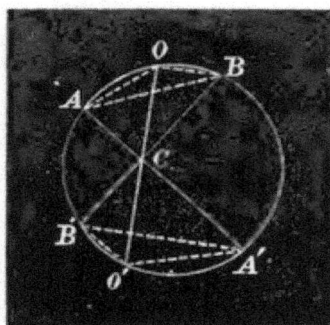

Hence:—*If two arcs of a circle AB and A'B' are divided in O and O' so as to fulfil the relation (3), AA', BB' and OO' are concurrent.*

EXAMPLES.

1. The internal bisectors of the angles of a triangle are concurrent.

2. Any two external and the internal bisector of the remaining angle are concurrent.

3. The lines joining the vertices ($a°$) to the points of contact of the in-circle ($\beta°$) to the internal points of contact of the ex-circles, are concurrent.

[The centres of perspective are named respectively† *point de Gergonne* and *point de Nagel* of the triangle.]

---

* The function $\dfrac{AO}{BO} \div \dfrac{AO'}{BO'}$ is termed the Anharmonic Ratio of the points A, B, O, O'; and (3) may be expressed thus :—"If the arcs AB and $A'B'$ are divided *equi-anharmonically* in O and O', the lines AA', BB' and OO' are concurrent; and conversely."

† *Educational Times*, July, 1890.

4. The perpendiculars of a triangle are concurrent.

5. The tangents to the circum-circle at A, B, C meet the opposite sides collinearly.

6. If a circle meet the sides of a triangle in X, X', Y, Y', Z, Z' such that either triad X, Y, Z is collinear or connects concurrently with the opposite vertices; a similar relation exists amongst the remaining points X', Y', Z'.

7. If three points are collinear, their *isotomic conjugates* with respect to the sides are collinear.

7a. If they connect concurrently with the vertices, their *isogonal conjugates* with respect to the angles also connect concurrently.

8. For any triangle ABC and transversal XYZ; if any point O is joined to the six points

$$\frac{\sin BOX}{\sin COX} \cdot \frac{\sin COY}{\sin AOY} \cdot \frac{\sin AOZ}{\sin BOZ} = 1.*$$

[For $\dfrac{BX}{CX} = \dfrac{BO \sin BOX}{CO \sin COX}$, with similar values for $\dfrac{CY}{AY}$ and $\dfrac{AZ}{BZ}$; therefore, etc. ...]

9. If the sides of a triangle and any three concurrent lines

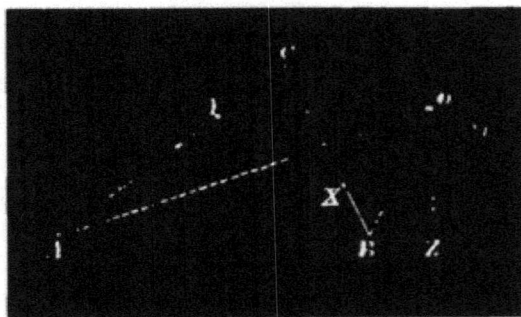

---

* Examples 8 and 9 will be afterwards enunciated as follows:

8°. The lines joining any point to the six vertices of a quadrilateral form a pencil of rays in Involution.

9°. Any line drawn across the sides and diagonals of a quadrilateral is cut in Involution.

through its vertices are cut by a transversal in six points X and X', Y and Y', Z and Z'; (BC in X, AO in X' ...)

$$\frac{YX'}{ZX'} \cdot \frac{ZY'}{XY'} \cdot \frac{XZ'}{YZ'} = 1,*$$

and conversely.

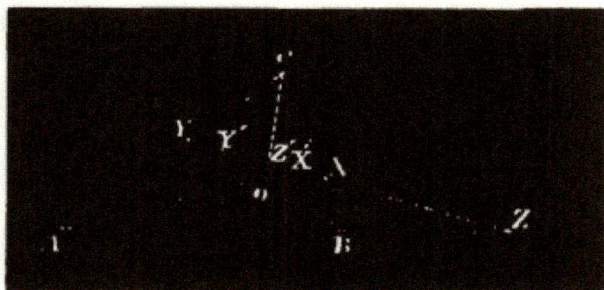

[For $\dfrac{\sin BAO}{\sin CAO} = \dfrac{ZX'}{YX'}$, with similar values for $\dfrac{\sin CBO}{\sin ABO}$ and $\dfrac{\sin ACO}{\sin BCO}$; therefore, etc.]

10. If AX, BY, CZ are concurrent, the intersections of YZ and BC (X'), ZX and CA (Y'), XY and AB (Z') are collinear.

[For $\dfrac{BX'}{CX'} \cdot \dfrac{CY}{AY} \cdot \dfrac{AZ}{BZ} = 1$. Compounding this with two similar equations involving Y' and Z' and reducing, we have

$$\frac{BX'}{CX'} \cdot \frac{CY'}{AY'} \cdot \frac{AZ'}{BZ'} = 1.]$$

11. Given two points A and B on a circle MNP, on the same side of the diameter MN; find a point P on the other side such that the intersections X and Y of AP and BP respectively with MN may be equidistant from the centre.

[Let AB and MN meet in Z; then it is easily proved that $PX^2/PY^2 = BZ/AZ$; hence the species of the triangle PXY is known; therefore, etc.]

12. Draw two circles in contact each touching a given line at a given point and having their radii in a given ratio.

---

* Will be afterwards seen to be an *Equation of Involution* of the pencil.

13. If lines be drawn from the vertices of ABC to a point Ω such that $\Omega BC = \Omega CA = \Omega AB = \theta$, prove that θ is given by the equation
$$\cot \theta = \cot A + \cot B + \cot C.$$
[For $\sin^3 \theta = \sin(A - \theta) \sin(B - \theta) \sin(C - \theta)$; etc.   Cf. Art. 28.]

14. In the general case if the lines in Ex. 13 making equal angles (a) with the sides are not concurrent, they form a triangle $A'B'C'$ similar to ABC and the ratio of similitude is equal to
$$\cos a - \sin a(\cot A + \cot B + \cot C) : 1.$$

Defs. The *Centres of Perspective* of two lines AB and $A'B'$ are the points of intersection of the pairs of lines AB', $A'B$ and AA', BB' joining their extremities.

Two triangles are said to be in perspective when the lines joining corresponding vertices meet in a point. This point is called the *Centre of Perspective* of the triangles.

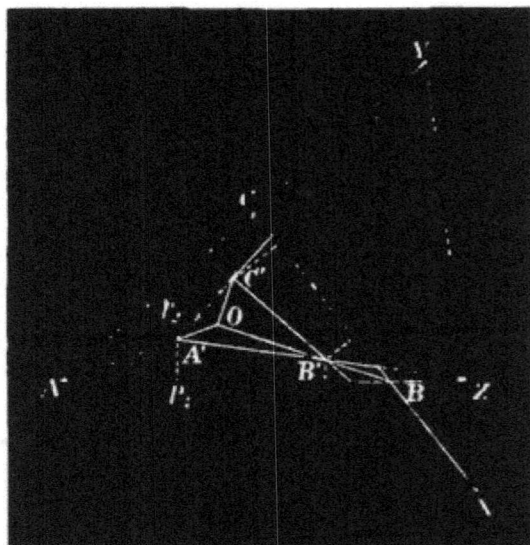

65. Criterion of Perspective of Triangles. Theorem.

If the perpendiculars from the vertices of a triangle $A'B'C'$ on the sides of another ABC be denoted by p_1, p_2, p_3;

q_1, q_2, q_3; r_1, r_2, r_3 (*i.e. from A' on BCp_1, A' on CAp_2, and so on*), *the two are in perspective if*

$$\frac{q_1}{r_1} \cdot \frac{r_2}{p_2} \cdot \frac{p_3}{q_3} = 1; \ and \ conversely.$$

For let AA' meet BC in X'. Then

$$\sin BAX'/\sin CAX' = p_3/p_2,$$

with similar values for r_2/r_1, and q_1/q_3; multiplying these equations together, therefore, etc., by Art. 64, which also proves the converse* proposition.

66. Theorem.—*If the vertices of two triangles connect concurrently, their pairs of corresponding sides intersect collinearly (BC and $B'C'$ in X, etc. ...).*

For, by similar triangles,

$$\frac{q_1}{r_1} = \frac{B'X}{C'X}, \ \frac{r_2}{p_2} = \frac{C'Y}{A'Y}, \ and \ \frac{p_3}{q_3} = \frac{A'Z}{B'Z}.$$

Multiplying, we have

$$\frac{B'X}{C'X} \cdot \frac{C'Y}{A'Y} \cdot \frac{A'Z}{B'Z} = \frac{q_1}{r_1} \cdot \frac{r_2}{p_2} \cdot \frac{p_3}{q_3} = 1, \ therefore, \ etc.$$

Def. The line of collinearity is termed the *Axis of Perspective* or *Homology*† of the triangles.

EXAMPLES.

1. Any triangle escribed to a circle is in perspective with that formed by joining the points of contact of its sides.

[The centre of perspective is the symmedian point of the inscribed triangle.]

---

*Or thus :—Let O be the centre of perspective of the triangles and a, β, γ the perpendiculars from it on the sides of ABC; since $\beta/\gamma = p_2/p_3$, $\gamma/a = q_3/q_1$, and $a/\beta = r_1/r_2$; multiply and reduce ; therefore, etc.

† The term Homology is due to Poncelet who first studied the properties of homological figures in space, *v. Traité des propriétés projectives des figures* (1822).

2. If three triangles ABC, $A_1B_1C_1$, $A_2B_2C_2$ have a common axis of perspective XYZ, their centres of perspective when taken two and two are collinear.

[For the triangles (fig. of Ex. 3) BB_1B_2 and CC_1C_2 are in perspective, their centre being at X; similarly Y is the centre of perspective of CC_1C_2, AA_1A_2 and Z of AA_1A_2 and BB_1B_2. Hence the corresponding sides of these pairs of triangles intersect in collinear points. But these points (*e.g.* AA_1, BB_1) are the centres of perspective of the given triangles in pairs; therefore, etc.]

3. If three triangles ABC, $A_1B_1C_1$, $A_2B_2C_2$ have a common centre of perspective, their axes are concurrent.

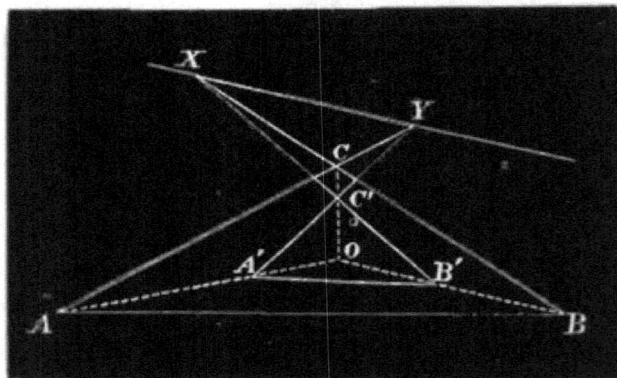

[Consider the three triangles whose sides are respectively the directions BC, B_1C_1, B_2C_2; CA, C_1A_1, C_2A_2; AB, A_1B_1, A_2B_2.

It is manifest they are in pairs in perspective, the axis of the first pair being CC_1; and XY is a line joining corresponding vertices.

Thus the axis of perspective XY of any two and therefore of every two of the given triangles passes through the centre of perspective of the conjugate triad.]

NOTE.—It will be noticed that the common centre O of the three given triangles is the point of concurrence of the axes AA_1, BB_1, CC_1 of the conjugate triad, and the common centre of the conjugate triad taken in pairs is the point of concurrence of the axes of the given triangles.

4. Brocard's first triangle is in perspective in three ways with ABC.

[The Brocard points are evidently two centres of perspective (Art. 28); also the lines AA', BB', CC' are concurrent, for p_2/p_3 found by aid of the property of Art. 28, Ex. 2, to be c^3/b^3; therefore, etc.

The three centres of perspective are the mean centres of the vertices ABC for multiples proportional to (Art. 52)

$$\frac{1}{a^2}, \frac{1}{b^2}, \frac{1}{c^2}; \frac{1}{b^2}, \frac{1}{c^2}, \frac{1}{a^2}; \frac{1}{c^2}, \frac{1}{a^2}, \frac{1}{b^2}]$$

5. If Ω, Ω', Ω'' denote the three centres of perspective of ABC and its first Brocard triangle $A'B'C'$, to prove that the corresponding vertices of their three median triangles lie on three right lines. (Stoll.)

[For $A'B'C'$ and ABC have a common centroid G (Art. 53, Ex. 6). But $\Omega\Omega'\Omega''$ has the same centroid; for its vertices are the mean centres of A, B, C for multiples proportional to $\frac{1}{b^2}, \frac{1}{c^2}, \frac{1}{a^2}; \frac{1}{c^2}, \frac{1}{a^2}, \frac{1}{b^2};$ $\frac{1}{a^2}, \frac{1}{b^2}, \frac{1}{c^2};$ therefore (Art. 53, Cor. 3) the mean centre of Ω, Ω', Ω'' is that for A, B, C for multiples each $= 1/a^2 + 1/b^2 + 1/c^2$. Now let L, L', L'' be the middle points of the corresponding sides of the three triangles such that $GA = 2GL$, $GA' = 2GL'$, and $G\Omega'' = 2GL''$: since A, A', Ω'' are collinear; L, L', L'' are also collinear, and the two lines of collinearity parallel.]

where X and X' are the points of intersection of BC with $C'A'$ and $A'B'$, etc. ; *and conversely.*

Using the previous notation, we have by similar triangles $\dfrac{BX}{CX} = \dfrac{q_2}{r_2}, \dfrac{BX'}{CX'} = \dfrac{q_3}{r_3}$, etc.

Hence $\dfrac{BX \cdot BX'}{CX \cdot CX'} = \dfrac{q_2 q_3}{r_2 r_3}$;

therefore the left side of the above equation becomes

$$\frac{q_2 q_3}{r_2 r_3} \cdot \frac{r_3 r_1}{p_3 p_1} \cdot \frac{p_1 p_2}{q_1 q_2},$$

which is equal to, on reduction,

$$\frac{r_1}{q_1} \cdot \frac{p_2}{r_2} \cdot \frac{q_3}{p_3} ; \text{ therefore, etc.}\quad \text{(Art. 65.)}$$

Cor. 1. **Pascal's Theorem.**—If $XX'YY'ZZ'$ be any cyclic hexagon, then (Euc. III. 36)

$$AY \cdot AY' = AZ \cdot AZ' ; \quad BZ \cdot BZ' = BX \cdot BX', \text{ etc.}$$

Hence :—*The two triangles formed by the two triads of alternate sides of any cyclic hexagon are in perspective ;* or, *the opposite sides of a cyclic hexagon meet in three collinear points.*

The centre and axis of perspective of any two triangles in perspective are called the *Pascal* * *Point* and *Line* of the hexagon $XX'YY'ZZ'$, which is termed a *Pascal Hexagon.*

Cor. 2. If X, X' ; Y, Y' ; Z, Z' coincide in pairs on the circle, the sides of the hexagon become the tangents to the circle at X, Y, Z, and the chords of contact YZ, ZX and XY ; the Pascal point is therefore the symmedian point of the triangle XYZ. (Art. 66, Ex. 1.)

---

* When only sixteen years old, Pascal discovered this property of the *mystic hexagram. Essai sur les Coniques,* Pascal, 1640.

Cor. 3.

$$\frac{\sin BA'X \sin CA'X}{\sin BA'X' \sin CA'X'} \cdot \frac{\sin CB'Y \sin AB'Y}{\sin CB'Y' \sin AB'Y'} \cdot \frac{\sin AC'Z \sin BC'Z}{\sin AC'Z' \sin BC'Z'}$$
$$= 1.$$

[For $\quad \dfrac{\sin BA'X}{\sin BA'X'} = \dfrac{q_2}{q_3}; \quad \dfrac{\sin CA'X}{\sin CA'X'} = \dfrac{r_2}{r_3},$ etc. ;

hence the above expression is equivalent to

$$\frac{q_2 r_2}{q_3 r_3} \cdot \frac{r_3 p_3}{r_1 p_1} \cdot \frac{p_1 q_1}{p_2 q_2} = \frac{q_1}{r_1} \cdot \frac{r_2}{p_2} \cdot \frac{p_3}{q_3} = 1.]$$

Cor. 4. **Brianchon's Theorem.**—Let $AC'BA'CB'$ be an escribed hexagon and x, y, z the intercepts made by the circle on the sides of the triangle $A'B'C'$; since

$$\frac{\sin BA'X \sin CA'X}{\sin BA'X' \sin CA'X'} = \frac{y^2}{z^2} *$$

with two other similar equations, Cor. 3 in this particular

---

*The property on which this depends is as follows :—*If from the point of intersection C of two tangents CA, CB to a circle a secant of length x is drawn dividing the angle ACB into segments a and β; then $\sin a \sin \beta \propto x^2$.*

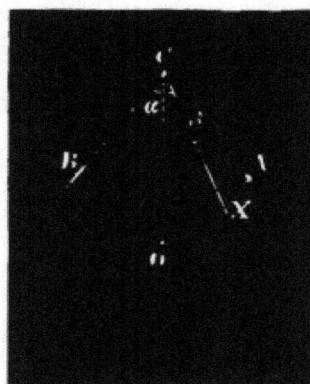

For if O be the centre of the circle and OX a perpendicular to the secant, we have

$$\sin a \sin \beta = \sin^2\tfrac{1}{2}(a+\beta) - \sin^2\tfrac{1}{2}(a-\beta) = r^2/OC^2 - OX^2/OC^2 = x^2/4OC^2 ;$$

therefore, etc.

case reduces to :—*The lines connecting the opposite ver-*
tices of an escribed hexagon are concurrent; or, the
two triangles formed by joining the alternate vertices of
an escribed hexagon are in perspective.

The centre and axis of perspective of the triangles are
termed the *Brianchon* * *Point* and *Line* of the hexagon
$AC'BA'CB'$, which for the same reason is called a
Brianchon Hexagon.

COR. 5. If two of the sides AF and EF of an escribed
hexagon coincide, the vertex F is the point of contact of
the tangent AE (Art. 6); *hence for an escribed pentagon*
$ABCDE$, *if the lines AD and BE meet in O, the points*
C, O, F *are collinear* (cf. Art. 63, foot-note).

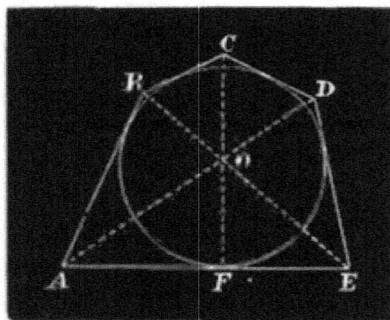

COR. 6. If two pairs of sides BC, CD and AF, EF
coincide, the hexagon reduces to a quadrilateral $ABDE$;
hence the diagonals AD and BE meet on CF; similarly
they meet on $C'F'$; therefore *the internal diagonals of*
an escribed quadrilateral and of the corresponding in-
scribed meet in a point.

---

* Published by Brianchon in the year 1806, and derived by him from
Pascal's Theorem by the process of reciprocation with respect to the
circle.   (See Art. 80, 2°.)

COR. 7. Consider the cyclic hexagon $FFC'CCF'$.

Its Pascal line is the line of collinearity of the three points (1) FF, CC; (2) FC', CF'; (3) FF', CC': but the line

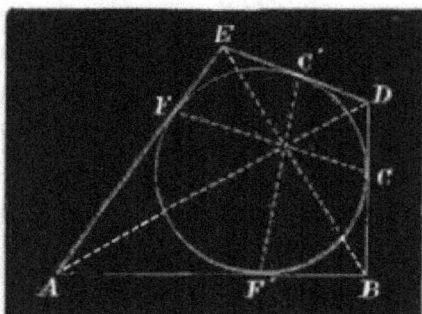

joining (2) and (3) is the third diagonal of the inscribed quadrilateral $CFC'F'$ and (1) is the intersection of the tangents at C and F, and therefore one extremity of the third diagonal of the escribed quadrilateral; hence :—*the third diagonals of any inscribed and corresponding escribed quadrilaterals coincide.*

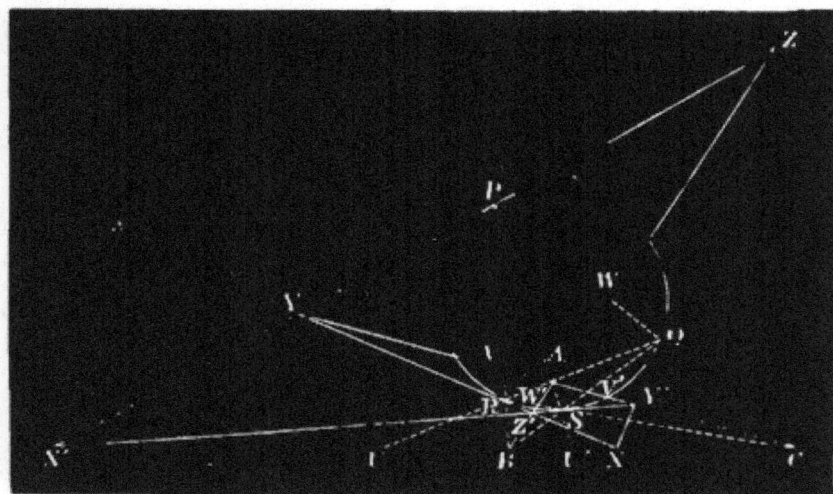

COR. 8. Let $PQRS$ be any cyclic quadrilateral; and let $XX'YY'ZZ'$, the corresponding escribed quadrilateral,

be regarded as a Brianchon hexagon $ZPX'Z'RX$ whose two pairs of coincident sides are the tangents from Y. Then the lines ZZ', PR, XX' are concurrent at the Brianchon point B; similarly, if the pairs of coincident sides are the tangents from Y', we have ZZ', QS, XX' concurrent, *i.e.* the pairs of opposite connectors PR and QS of the inscribed quadrilateral and ZZ' and XX' of the corresponding escribed cointersect. We see therefore from Cors. 7 and 8 that *any pair of opposite connectors of an inscribed quadrilateral and the corresponding pair for the quadrilateral escribed at its vertices are concurrent.* The three points of concurrence on the figure are A, B, C.

The points U, V, W, U', V', W' lie in triads on four lines.

EXAMPLES.

1. Three pairs of tangents are drawn from the vertices of a triangle to any circle to meet the opposite sides in points XX', YY', ZZ'; show that if X, Y, Z are collinear, X', Y', Z' are also collinear.

[Apply Cor. 4.]

2. ABC is a triangle inscribed in and in perspective with $A'B'C'$; the tangents from ABC to the in-circle of $A'B'C'$ meet the opposite sides in three collinear points X, Y, Z (BC in X, etc.).

[Let the axis of perspective of the two triangles be $X'Y'Z'$, therefore by Cor. 4 we have $\left(\dfrac{BX . BX'}{CX . CX'}\right)(\ldots)(\ldots)=1$; therefore, etc., by Ex. 1.]

3. If points XX', YY', ZZ' be taken on the sides of a triangle such that
$$\frac{BX}{CX} . \frac{BX'}{CX'} . \frac{CY}{AY} . \frac{CY'}{AY'} . \frac{AZ}{BZ} . \frac{AZ'}{BZ'}=1,$$
they are the vertices of a Pascal hexagon.

4. The lines joining each pair of points to the opposite vertex (AX and AX', etc.) of the triangle determine a Brianchon hexagon.

5. (*a*°) Any two transversals XYZ, $X'Y'Z'$ determine on the sides the vertices of a Pascal hexagon.

(*β*°) Two triads of points on the sides which connect concurrently with the opposite vertices determine a Pascal hexagon.

(*γ*°) A transversal XYZ and three points X', Y', Z' which connect concurrently with the opposite vertices determine a Brianchon hexagon.

6. A hexagon is inscribed in a circle; prove that the continued products of the perpendiculars from any point on the Pascal line on the alternate sides are equal ($xyz = x'y'z'$).

[Let $AB'CA'BC'$ be the hexagon whose pairs of opposite sides BC', $B'C$; CA', $C'A$; AB', $A'B$ meet in points X, Y, Z respectively and the Pascal line L (XYZ) at angles a, a', β, β', γ, γ'; then

$$\frac{BL \cdot CL}{B'L \cdot CL} = \frac{BX \cdot C'X \sin^2 a}{B'X \cdot CX \sin^2 a'} = \frac{\sin^2 a}{\sin^2 a'}. \qquad \text{(Euc. III. 36)}$$

Similarly, $\dfrac{C'L \cdot AL}{CL \cdot A'L} = \dfrac{\sin^2 \beta}{\sin^2 \beta'}$ and $\dfrac{AL \cdot B'L}{A'L \cdot BL} = \dfrac{\sin^2 \gamma}{\sin^2 \gamma'}$.

Multiplying these equations and reducing,

$\sin^2 a \sin^2 \beta \sin^2 \gamma = \sin^2 a' \sin^2 \beta' \sin^2 \gamma'$; therefore, etc.]

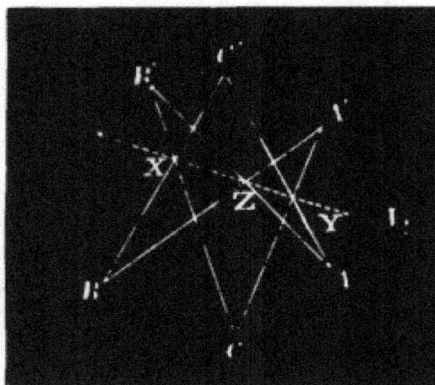

7. From the middle points L, M, N of the sides of a triangle tangents are drawn to the in-circle; show that these tangents form a triangle ($A'B'C'$) in perspective with that (PQR) obtained by joining the points of contact of the in- or ex-circles with the sides, and the centre of perspective is the median point of ABC.

[For since the sides of ABC with any two of the tangents form an escribed pentagon, *e.g*, $BCMNA'$, by Cor. 5, the lines BM, CN, $A'P$ are concurrent; that is, $A'P$ passes through the centroid (BM, CN). Similarly for $B'Q$, $C''R$; therefore, etc.]

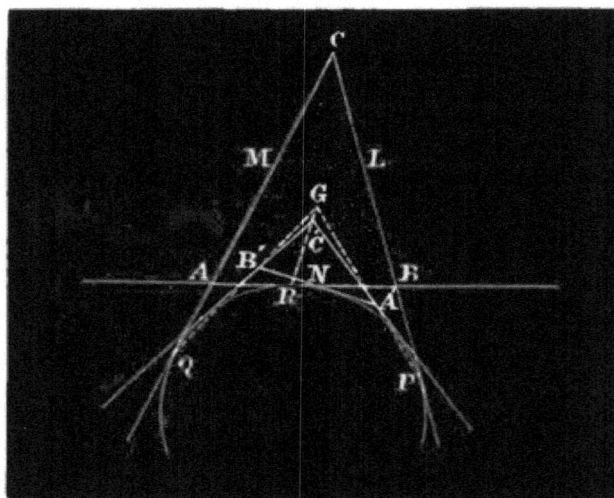

Note.—If LMN is *any* inscribed triangle in perspective with ABC, the above reasoning applies to prove that $A'B'C'$ and PQR have the *same* centre of perspective.

8. If two triangles ABC and $A'B'C'$ are in perspective, $A'BC$, $AB'C'$; $AB'C$, $A'BC'$; ABC', $A'B'C$ are also in perspective.

9. If AA', BB', CC' denote the lengths of three lines whose directions are concurrent, their six centres of perspective (of BB' and CC', X and X', etc.) taken in pairs lie in triads on four lines.

[For they are the axes of perspective of the triangles in Ex. 8.]

10. If X, Y, Z are on the sides of a triangle and fulfil the relation
$$\Sigma(BX^2 - CX^2) = 0,$$
the perpendiculars through them to the sides are concurrent; and conversely.

11. If two triangles are such that the perpendiculars from the vertices of either upon the sides of the other are concurrent, then conversely the perpendiculars from the vertices of the latter upon the sides of the former are concurrent.

[By Ex. 10.]

12. State the particular cases of the Theorem of Ex. 11 for a given triangle taken with the ($a°$) pedal, ($\beta°$) median, ($\gamma°$) triangles formed by joining the points of contact with the sides of the in- or ex-circles.

13. If XYZ be a transversal to a triangle ABC, X', Y', Z' the harmonic conjugates of X, Y, Z, with respect to the sides ; prove that

1°. The triads of points $Y'Z'X$, $Z'X'Y$, $X'Y'Z$ are collinear.

2°. $X'Y'Z'$, $X'YZ$, $Y'ZX$, $Z'XY$ connect concurrently with the opposite vertices.

14. The middle points of the segments XX', YY', ZZ' are collinear.

[For they are the middle points of the diagonals of a complete quadrilateral by Ex. 3. For another proof *v.* Art. 91.]

15. The perpendiculars from the vertices of a triangle ABC on the sides of $A'B'C'$, its first Brocard triangle, are concurrent on the circum-circle. (Tarry's Point.)

[By the theorem of Ex. 11.]

16. The perpendiculars from the middle points of the sides of $A'B'C'$ on the sides of ABC are concurrent. (Cf. Ex. 15.)

17. The Simson line of Tarry's point is perpendicular to OK, the line joining the circum-centre to the symmedian point.

18. In the figure of Art. 28 show that

$$OA' : OB' : OC' = \cos(A + \omega) : \cos(B + \omega) : \cos(C + \omega) ;$$

and deduce the formula for the Brocard angle,

$$\sin A \cos(A + \omega) + \sin B \cos(B + \omega) + \sin C \cos(C + \omega) = 0.$$

Note on Tarry's Point.—It will appear obvious that the diameter of the circum-circle containing Tarry's point is related to the triangle ABC in the same manner as OK is to $A'B'C'$; and that the circum- and Brocard circles are divided similarly by these *corresponding* diameters. Also, if a, β, γ denote the perpendiculars from Tarry's point on the sides of ABC,

$$a : \beta : \gamma = \sec(A + \omega) : \sec(B + \omega) : \sec(C + \omega).$$

A point of interest may be here noticed. From Art. 28, Ex. 18 (note) it is evident that the centres O_1, O_2, O_3 of Neuberg's circles with respect to the sides of ABC are the vertices of similar isosceles triangles described on a, b, c respectively, whose equal base angles are $\frac{1}{2}\pi - \omega$. Therefore, if T denote Tarry's point, it easily follows that AT, AO_1; BT, BO_2; CT, CO_3 divide the angles of ABC isogonally. *But the isogonal conjugate of a point on the circum-circle is at infinity; hence the lines AO_1, BO_2, CO_3 are parallel.*

HARMONIC PROPERTIES OF THE QUADRILATERAL.

68. Theorem.—*In any complete quadrilateral each of the diagonals XX', YY', ZZ' is divided harmonically by the other two.*

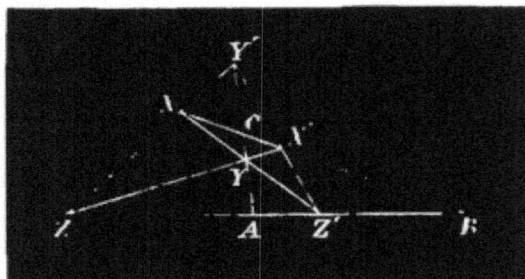

Consider the triangle $ZZ'Y'$ and transversal BXX',

$$\frac{Z'X}{Y'X'} \cdot \frac{Y'X}{ZX} = \frac{Z'B}{ZB} \quad\dots\dots\dots\dots\dots(1)$$

And since YY', YZ, YZ' are three concurrent lines through its vertices, we have

$$\frac{Z'X'}{Y'X'} \cdot \frac{Y'X}{ZX} = -\frac{Z'A}{ZA} \quad\dots\dots\dots\dots\dots(2)$$

Equating these results, we have $ZA/Z'A = -ZB/Z'B$.

Hence the row of points $ZZ'AB$ is harmonic.

Similarly, $BCXX'$ and $CAYY'$ are harmonic.

COR. 1. The angles of the triangle ABC, formed by the diagonals (*the diagonal triangle*) are divided harmonically by the pairs of lines AX, AX'; BY, BY'; CZ, CZ'.

COR. 2. If two lines be given in magnitude and position (ZZ' and XX') their two centres of perspective (Y

and Y') joined to their point of intersection (B) form a harmonic pencil. They also divide the line joining their centres of perspective (in A and C) harmonically.

Problem.—*To determine the number of polygons which can be formed from n points.*

Each point joined to the remaining $n-1$ points gives $n-1$ lines. Taking any one of these lines as the first side of the polygon we have similarly $n-2$ selections for the second side, $n-3$ for the third side, and so on. Therefore we have $(n-1)(n-2)$ selections for the first two sides, $(n-1)(n-2)(n-3)$ for the first three sides, etc.; hence we have finally $\lfloor n-1$ equal to twice the number of polygons, since any sequence of sides when reversed gives the same polygon.

Thus four points may be joined in three ways as in figure.

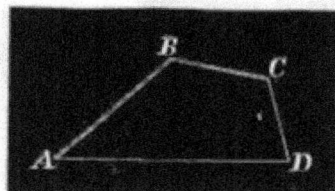

1.

2.

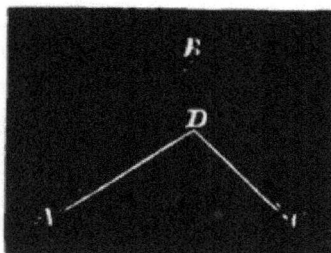

3.

Fig. 1 is called a *Convex*, Fig. 2 an *Intersecting*, and Fig. 3 a *Re-entrant* Polygon.

By application of the general formula to the hexagon we find that six points in general determine a system of sixty hexagons.

EXAMPLES.

1. The conditions that the quadrilaterals in the figures are escribed are :—

      1°. $BC + AD = AB + CD$.
      2°. $BC \sim AD = AB \sim CD$.
      3°. $BC \sim AD = AB \sim CD$.

[Since tangents from any point to a circle are equal.]

.2. To prove that the quadrilateral whose angles and perimeter are given is of maximum area when it is escribed to a circle. (Hermite.)

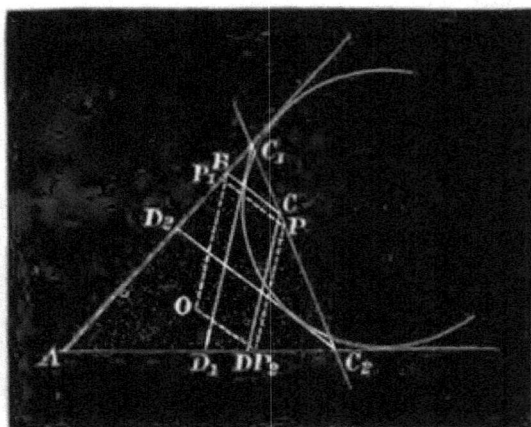

[Let two of the sides AB and AD be fixed in position and the remaining two vary. It is easy to see that *the locus of C is a line.* Suppose C_1 and C_2 to be the positions of C on the fixed lines and C_1D_1, C_2D_2 parallels to the fixed directions CD and CB.

The perimeters of the triangles AC_1D_1 and AC_2D_2 are each equal to the perimeter of the quadrilateral $ABCD$; the ex-circle of AC_1D_1 is the ex-circle of AD_2C_2 and $D_2C_2 - D_2C_1 = D_1C_1 - D_1C_2$.

Now, for any point P and parallels PP_1 and PP_2 by similar triangles $PC_1/C_1C_2 = (PP_1 - P_1C_1)/(D_2C_2 - D_2C_1)$
and $PC_2/C_1C_2 = (PP_2 - P_2C_2)/(D_1C_1 - D_1C_2)$;
adding these equations, we get

      $PP_1 + PP_2 - P_1C_1 - P_2C_2 = D_2C_2 - D_2C_1$;

to each side add $AC_1 + AC_2$, and

      $AP_1 + P_1P + PP_2 + P_2A = AD_2 + D_2C_2 + C_2A =$ given perimeter.

Regarding P and C as consecutive points on the locus, the area of the quadrilateral is a maximum when $BCPP_1 = CDPP_2$, i.e. BD is parallel to C_1C_2. Hence the parallels BO and DO to CD and BC respectively form with AB and AD a re-entrant escribed quadrilateral, and therefore $AB + BO = AD + DO$, or (Euc. I. 34) $AB + CD = AD + BC$; therefore, etc.]

It may at once be inferred that *the maximum polygon of any order, of given angles and perimeter, is escribed to a circle.*

3. If three common tangents D, E, F to three circles A, B, C taken two and two are concurrent; prove that the conjugate triad D', E', F' are also concurrent.*

4. Let the lines joining the middle points of the three pairs of opposite connectors, BC and AD, etc., of four points A, B, C, D be λ, μ, ν; prove by means of the following evident formulae,

$$4\lambda^2 = \delta^2 + \delta'^2 + 2\delta\delta'\cos\widehat{\delta\delta'} = a^2 + c^2 + 2ac\cos\widehat{ac}, \quad\ldots\ldots\ldots\ldots(1)$$
$$4\mu^2 = b^2 + d^2 - 2bd\cos\widehat{bd} = a^2 + c^2 - 2ac\cos\widehat{ac}, \quad\ldots\ldots\ldots\ldots(2)$$
$$4\nu^2 = \delta^2 + \delta'^2 - 2\delta\delta'\cos\widehat{\delta\delta'} = b^2 + d^2 + 2bd\cos\widehat{bd}, \quad\ldots\ldots\ldots\ldots(3)$$

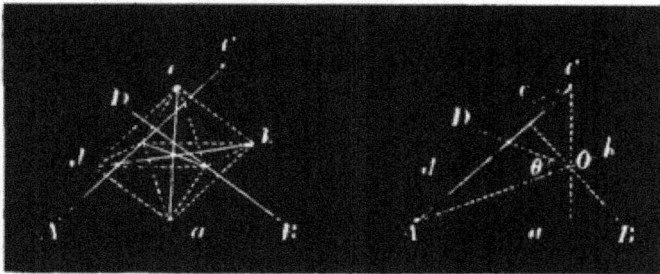

the relations,

1°. $2(\mu^2 + \nu^2) = c^2 + a^2$; $2(\nu^2 + \lambda^2) = b^2 + d^2$; $2(\lambda^2 + \mu^2) = \delta^2 + \delta'^2$:

2°. $4(\lambda^2 + \mu^2 + \nu^2) = a^2 + b^2 + c^2 + d^2 + \delta^2 + \delta'^2$:

3°. $4\lambda^2 = b^2 + d^2 - c^2 - a^2 + \delta^2 + \delta'^2$:

with similar expressions for μ and ν:

4°. $\mu^2 - \nu^2 = ac\cos\widehat{ac}$; $\nu^2 - \lambda^2 = -bd\cos\widehat{bd}$; $\lambda^2 - \mu^2 = -\delta\delta'\cos\widehat{\delta\delta'}$;
$2(\mu^2 - \nu^2) = \delta^2 + \delta'^2 - b^2 - b'^2$, etc.; $\Sigma ac\cos\widehat{ac} = 0$.

---

* Catalan's *Théorèmes et Problèmes de Géométrie Élémentaire*, pp. 53, 54 (1879).

5. 4(area of quadrilateral) $= (b^2 + d^2 - c^2 - a^2)\tan \widehat{\delta\delta'}$.

5a. Hence, or otherwise construct a quadrilateral, having given its four sides and area.

6. To find the cosine of the angle between any pair of opposite connectors.

[Equate the values of λ^2, μ^2, ν^2 in 3° with those of (1), (2), (3).

7. If any point D be joined to the vertices of a triangle ABC; the area of the triangle formed by joining the orthocentres of BCD, CDA, DAB is equal to ABC.

[Let O_1, O_2, O_3 denote the orthocentres. DO_1, DO_2, DO_3 are equal to $a \cot A$, $b \cot B$, $c \cot C$ respectively, and are mutually inclined at angles A, B, C; therefore, etc.]

8. If the vertices of a quadrilateral $ABCD$ be joined to the orthocentres O, O_1, O_2, O_3 of the four triangles formed by their vertices taken in triads ; to prove that

$$O \cdot ABCD = O_1 \cdot ABCD = O_2 \cdot ABCD = O_3 \cdot ABCD.$$

[Let the angle $AOD = \theta$. Taking any of the anharmonic ratios of the pencil $O \cdot ABCD$ and reducing, we obtain

$$\frac{\sin\theta \sin B}{\sin(B+\theta)\sin A} = \frac{b}{a}\frac{\sin\theta}{\sin(B+\theta)} = \frac{bd\sin OAD}{ac\sin OCD} = \frac{bd\cos\widehat{bd}}{ac\cos\widehat{ac}} = \frac{\nu^2 - \lambda^2}{\nu^2 - \mu^2}$$

(Ex. 4, 4°). It follows generally that the six anharmonic ratios of the pencil $O \cdot ABCD$ are $\dfrac{\lambda^2 - \mu^2}{\lambda^2 - \nu^2}, \dfrac{\mu^2 - \nu^2}{\mu^2 - \lambda^2}, \dfrac{\nu^2 - \lambda^2}{\nu^2 - \mu^2}$ and their reciprocals. Similarly for the remaining pencils $O_1 \cdot ABCD$, etc. Russell.]

Note on Pascal and Brianchon's Hexagons.

When two triangles ABC and $A'B'C'$ are in perspective, the lines AA', BB', CC' are concurrent ; therefore A and A', B and B', C and C' may be regarded as the opposite vertices of a Brianchon hexagon, and the centre of perspective of the two triangles is the Brianchon point of the hexagon.

But in this case we have three other pairs of triangles in perspective, viz., BCA' and $B'C'A$, CAB' and $C'A'B$, ABC' and $A'B'C$. Hence with the vertices of two triangles in perspective we can form four Brianchon hexagons having the same Brianchon point, the opposite vertices of the hexagons being in each case corresponding vertices of the two triangles.

Again, if the non-corresponding sides of the triangles intersect as in figure in points X and X', Y and Y'', Z and Z', and the corresponding sides in UVW, UVW is the axis of perspective.

But in this case we have three other pairs of triangles in perspective to the same axis, viz., those obtained by interchanging a pair of corresponding sides, *e.g.*, if L, M, N and L', M', N' denote the sides of the given triangles, it is obvious that the triangles LMN' and $L'M'N$, MNL' and $M'N'L$, NLM' and $N'L'M$ have the same axis of perspective; hence with the sides of two triangles in perspective we may form four Pascal hexagons having a common Pascal line, *i.e.*, the axis of perspective of the triangles, the corresponding sides of the triangles being in each case opposite sides of the hexagons.

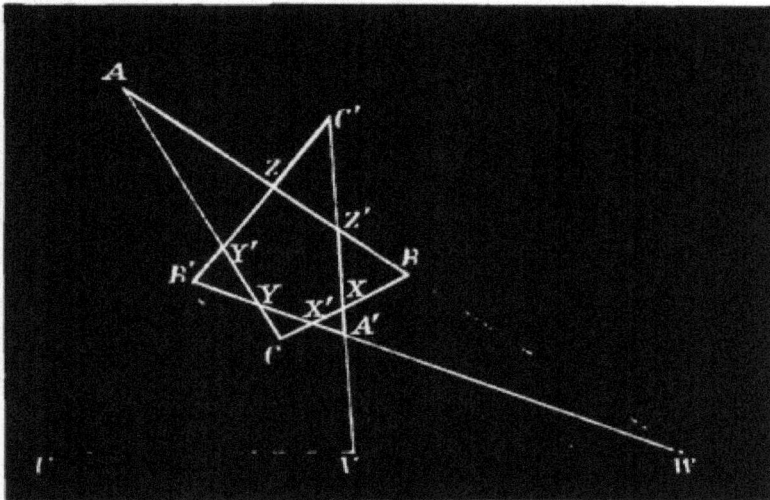

In the accompanying figure the legs of the angles whose vertices are at U, V and W intersect again in twelve points, viz.,

$$X,\ X',\ Y,\ Y'',\ Z,\ Z',\ A,\ A',\ B,\ B',\ C,\ C'',$$

and these we have seen may be grouped in four different ways into two groups of six $(XX'YY''ZZ')$, and $AA'BB'CC''$ determining Pascal and Brianchon hexagons respectively; also that the alternate sides

XX' and YY') of the Pascal hexagon intersect (in C) in six points, which form a Brianchon hexagon.

Again, since sixty Pascal hexagons may be formed from the points $XX'YY'ZZ'$, and YY' and ZZ' meet in A, and YX' and $Z'X$ in A', by taking these lines as pairs of opposite sides of one of the hexagons ($YY'XZ'ZX'$), AA' is its Pascal line ; similarly BB' and CC' are Pascal lines of the hexagons $XX'YZ'ZY'$ and $XX'ZY''YZ'$ respectively ; but AA', BB' and CC' are concurrent, hence *the sixty Pascal lines pass in threes through twenty points.*

Similarly it may be shewn that of the sixty *Brianchon hexagons* formed by the conjugate hexad of points $AA'BB'CC'$, their Brianchon points lie in triads on twenty lines. And either property involves the other as will be seen by reciprocation with respect to a circle.

CHAPTER VI.

INVERSE POINTS WITH RESPECT TO A CIRCLE.

Def. Two points P and Q are inverse with respect to a circle when the line PQ passes through the centre O and $OP \cdot OQ =$ the square of the radius of the circle.

For the circle of unit radius $OP \cdot OQ = 1$ or OP is the inverse, or reciprocal, of OQ.

69. It appears from the definition (1°) That inverse points are in the same direction from the centre when the circle is real and in opposite directions when the radius is imaginary, that is when it is of the form $R\sqrt{-1}$. (2°) They coincide on the circle; and when the radius is not real the inverse Q of a point P at a distance OP from the centre is given by the equation $OP \cdot OQ = -R^2$. (3°) When either coincides with the centre the other is at infinity.

70. **Theorem.**—*If a line AB be divided internally and externally in P and Q in the same ratio, P and Q are inverse points with respect to the circle on AB as diameter; also A and B are inverse points with respect to the circle on PQ.*

For if M be the middle point of AB, by hyp.,

$$\frac{AP}{BP} = \frac{AQ}{BQ}, \text{ hence } \frac{AM + MP}{BM - MP} = \frac{AM + MQ}{QM - MB}.$$

139

by taking the sum to difference on each side we have $\dfrac{AM+BM}{2MP}=\dfrac{2MQ}{AM+BM}$; therefore $MP \cdot MQ = MA^2$.

A similar proof applies to show that
$$NA \quad NB = NP^2 = NA^2,$$
where N is the middle point of PQ

71. Since $\qquad MP \cdot MQ = MN^2 - PN^2,$ $\qquad$ (Euc. II. 6)
therefore (Art. 70) $\quad AM^2 = MN^2 - PN^2,$
or transposing, $\qquad MN^2 = AM^2 + PN^2$

Hence *for any two segments AB and PQ placed to divide each other harmonically, the square of the distance (MN) between their middle points = the sum of the squares of half the segments.*

EXAMPLES.

1. The distances of the points of contact of the in- and ex-circles of a triangle with the sides measured from any vertex on either of the sides passing through it are s, $s-a$, $s-b$, $s-c$.

2. If M denote the middle point of the base (c) of a triangle, Q the intersection with the base of the fourth common tangent to the ex-circles O_1 and O_2, P the foot of the perpendicular from the vertex on the base, $\quad MP \cdot MQ = \left(\dfrac{a+b}{2}\right)^2.$

[For O_1O_2 is divided harmonically in C and Q, project O_1, O_2, and C on base and apply Art. 70].

3. Show also that the rectangle under the distances of the middle point of the base from the feet of the perpendicular and internal bisector of vertical angle = square on half the difference of sides.

72. **Theorem.**—*The distances of any point X on a circle from a pair of inverse points have a constant ratio.*

Since $OQ : OX = OX : OP$; the two triangles OQX and OXP are similar (Euc. VI. 6), and (Euc. VI. 4) $\dfrac{PX^2}{QX^2} = \dfrac{PO^2}{OX^2} = \dfrac{OX^2}{OQ^2}$;

therefore $\dfrac{PX^2}{QX^2} = \dfrac{OP}{OQ}$,

or *the squares of the distances of a variable point (X) on a circle from a pair of inverse points (P, Q) are as the distances of these points from the centre.*

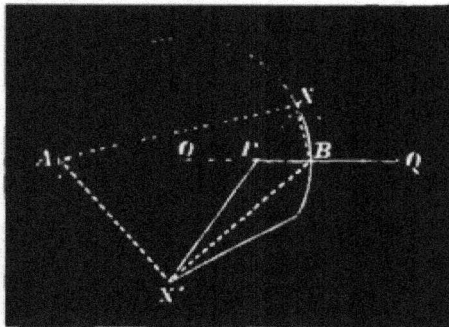

COR. 1. Let X coincide with each extremity of the diameter AB containing the points, then

$$\frac{PX^2}{QX^2} = \frac{PA^2}{QA^2} = \frac{PB^2}{QB^2} = \frac{OP}{OQ}$$

COR. 2. Given a triangle (PQX), the base (PQ), and ratio of sides, the locus of the vertex is a circle (ABX) with respect to which the extremities of the base are inverse points.

COR. 3. If the ratio of sides in Cor. 2 = 1, the locus is a line bisecting the base at right angles, therefore *the reflexion of a point is its inverse with respect to the line.*

COR. 4. From Cor. 1. AX and BX are the bisectors of the angle PXQ.

COR. 5. If PX be produced to meet the circle again in X', A and B are the centres of the in- and ex-circles of the triangle QXX'. (By Cor. 4.)

Cor. 6. The line PQ containing a pair of inverse points bisects the angle (XQX') which any chord through either (P) subtends at the other.

Cor. 7. The quadrilateral $OQXX'$ is cyclic.

[For $OXX' = PQX$, but $OXX' = OX'X$; therefore, etc. Euc. III. 21.]

Cor. 8. For any other pair of inverse points P', Q' on the diameter AB; the angles PXP' and QXQ are equal or supplemental according as the pairs of points are taken in the same or opposite directions from the centre.

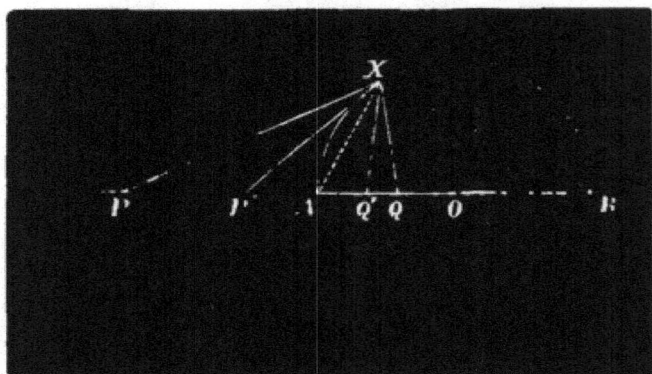

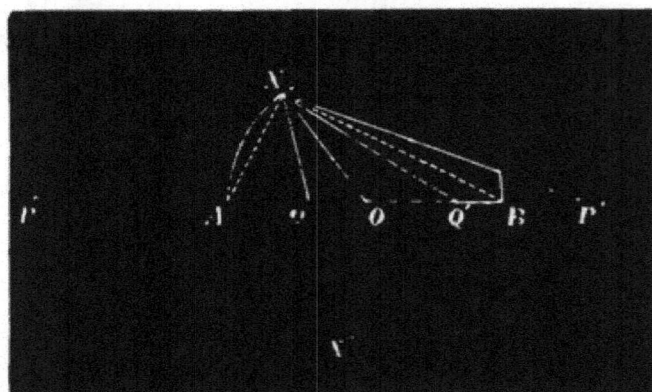

[The angles PXQ and $P'XQ'$ have in either case the same bisectors AX and BX.]

Examples.

1. Any circle passing through a pair of inverse points P and Q with respect to a given one cuts the latter orthogonally.

[From the definition of inverse points and Euc. III. 37.]

2. To find two points P and Q which shall be inverse with respect to two given circles.

[The circle passing through any point and its inverses with respect to each of the given circles meets their line of centres in the points required.]

3. The line L bisecting PQ at right angles is such that the tangents from any point O on it to either of the circles in Ex. 2 are equal to OP or OQ.

[For the circle with O as centre and $OP=OQ$ as radius meets the given circles orthogonally ; * therefore, etc.]

4. Any two pairs of inverse points are concyclic.

5. Any chord XY of a circle passing through P is divided harmonically by P and the perpendicular to PQ through Q.

[For the angle XQY is bisected internally and externally by the lines at right angles.]

6. The radical axes L, M, N of three circles taken in pairs are concurrent.

[For the point (L, M) of intersection of any two is the centre of the circle cutting the three given ones orthogonally.]

Def. This point of concurrence O is the *Radical Centre* of the circles, and is such that for any three secants XX', YY', ZZ' drawn through it to the circles respectively

$$OX . OX' = OY . OY' = OZ . OZ'.$$

The common value of these rectangles is called the *Radical Product* of the circles, and is equal to the square of the tangents to them when O is outside the circles. (See Art. 23, Ex. 11, footnote.)

---

* Hence the locus of a point from which tangents to two circles are equal is a right line, viz., the axis of reflexion of their common pair of inverse points. It is termed the *Radical Axis* of the circles, and is their chord of intersection, *real or imaginary.*

7. The radical axis of two intersecting circles is their chord of intersection ; hence show that the common chords of three circles taken in pairs are concurrent.

8. Describe a circle meeting three given circles at right angles.

9. For any triangle ABC find a point O such that
$$OA : OB : OC = \text{given ratios.}$$

10. For any four collinear points A, B, C, D find the loci of points (1°) such that the angles AOB and COD are equal, (2°) BOC is supplement of AOD.

11. For any six collinear points taken in the order $ABCC'B'A'$ find O such that the angles BOC, COA, AOB are respectively equal to $B'OC'$, $C'OA'$, $A'OB'$.

[By Ex. 10.]

12. The four sides of an escribed quadrilateral $ABCD$ being given in magnitude and AB in position ; find the locus of the centre O of the in-circle.

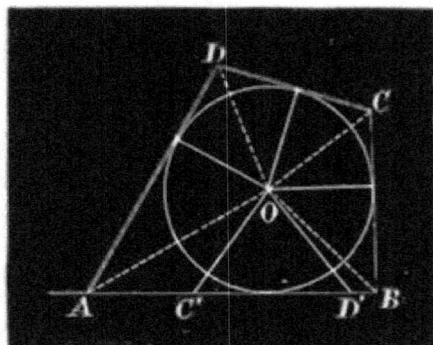

[Make $AD' = AD$ and $BC' = BC$. Since OA, OB, OC, OD are the bisectors of the angles of the quadrilateral, it is easy to see that $AOB + COD = \pi$. Again the triangles AOD and AOD' are equal in every respect (Euc. I. 4) ; hence $\angle ADO = AD'O$; similarly $\angle BCO = BC'O$; therefore by addition it follows that $\angle C'OD' = COD$ or $AOB + C'OD' = \pi$, and the required locus is a circle having A, B and C', D' pairs of inverse points. Dilworth.]

13. The centres of perspective P and Q of any two parallel chords AA' and BB' of a circle are inverse points with respect to the circle, and the circle touching the chords at their middle points.

[For we have $PA = PA'$, $PB = PB'$, $QA = QA'$ and $QB = QB'$; hence $PA/QA = PB/QB =$ etc. ; therefore, etc.

The second part follows, since MN is divided harmonically by P and Q. Art. 70.]

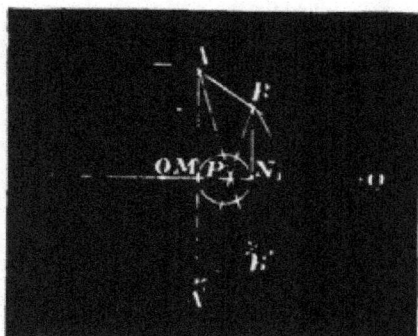

13a. To what does the theorem reduce when AA' and BB' coincide ?

14. For *any* two pairs of inverse points P, Q and P', Q' prove that

$$\frac{PP' . PQ'}{QP' . QQ'} = \frac{OP}{OQ} \quad \left(= \frac{PA^2}{QA^2} = \frac{PB^2}{BQ^2} \right).$$

[$PP'QQ'$ is a cyclic quadrilateral (Ex. 4); hence the triangles OPP' and OQQ' are similar ; so also are OPQ' and $OP'Q$; therefore, etc. (Euc. VI. 4). Otherwise if p and q denote the perpendiculars from P and Q on $OP'Q'$, we have

$$PP' . PQ' = p . D, \text{ and } QP' . QQ' = q . D ;$$

hence $\qquad \dfrac{PP' . PQ'}{QP' . QQ'} = \dfrac{p}{q} = \dfrac{OP}{OQ}.$]

15. If P, Q, R be any three collinear points on the diagonal triangle of a quadrilateral ; their harmonic conjugates $P'Q'R'$ with respect to the *diagonals* XX', YY', ZZ' are also collinear.

[For XX' is divided harmonically in B and C (Art. 68) and P and P; hence, by Ex. 14,

$$\frac{BP . BP'}{CP . CP'} = \frac{BX^2}{CX^2} = \frac{BL}{CL} \text{ (where } LX = LX').$$

Similarly $\qquad \dfrac{CQ . CQ'}{AQ . AQ'} = \dfrac{CY^2}{AY^2} = \dfrac{CM}{AM}$ (where $MY = MY'$): etc.

κ

Multiplying, we have

$$\frac{BP}{CP}\cdot\frac{CQ}{AQ}\cdot\frac{AR}{BR}\cdot\frac{BP'}{CP'}\cdot\frac{CQ'}{AQ'}\cdot\frac{AR'}{BR'}=1=\frac{BL}{CL}\cdot\frac{CM}{AM}\cdot\frac{AN}{BN};^*$$

but P, Q, R are in a line ;† therefore, etc.]

16. To what does Ex. 15 reduce when the line PQR is at infinity?

17. The angles subtended by the diagonals of a complete quadrilateral at any point O have a common angle of harmonic section, real or imaginary.

[O is the point of intersection of the lines PQR and $P'Q'R'$ in Ex. 16; therefore, etc.]

18. The circles on the diagonals of a complete quadrilateral pass through two points, real or imaginary.

[In Ex. 17, if two of the angles XOX', YOY' are right; ZOZ' must also be right,‡ since it is divided harmonically by PQR and $P'Q'R'$.]

19. Any transversal to the pencil in Ex. 17 is cut in six points which, taken in pairs, have a common segment of harmonic section.

20. To what does Ex. 17 reduce when O is at infinity?

21. If the sides of a triangle ABC are divided harmonically in XX', YY', ZZ'; if X, Y, Z are collinear, the middle points L, M, N of these segments are collinear.

22. If perpendiculars be let fall on the sides of a triangle from a pair of inverse points O and O' and their feet joined; the triangles PQR and $P'Q'R'$ thus formed are similar and their areas are as the distances of O and O' from the circum-centre.

[For $\qquad QR = AO \sin A$, and $Q'R' = AO' \sin A$,

therefore $\qquad QR/Q'R' = AO/AO'$;

similarly $\qquad RP/R'P' = BO/BO'$, etc. $\qquad\qquad$ Art. 72.]

---

* Hence also *the middle points L, M, N of the diagonals of a complete quadrilateral are collinear.*

† PQR and $P'Q'R'$ are termed *Conjugate Lines* of the quadrilateral.

‡ Generally, For a number of angles at a common vertex having a common angle of harmonic section if any two are right, all the others are also right.

23. Through a point P in the diameter of a semi-circle draw a chord AB such that the area of the quadrilateral $ABA'B'$, where $A'B'$ is the projection of AB on the diameter, may be a maximum.

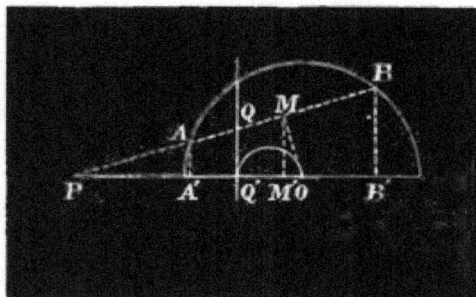

[Let Q' be the inverse of P with respect to the circle; draw QQ' at right angles to $A'B'$. Project M the middle point of AB on $A'B'$ and let X be the intersection of MM' with the semi-circle on $Q'O$. Then area S of quadrilateral $ABA'B' = A'B' . MM'$, hence

$$S^2 = 4MM'^2 . A'M'^2 = 4OM' . PM' . M'P . M'Q, \text{ by Art 70,}$$
$$= 4PM'^2 . OM' . M'Q = 4PM'^2 . M'X^2 ;$$

or S is equal to the area of the maximum rectangle that can be inscribed in a given circle, one of whose sides is parallel to a given line. Art. 14, Ex. 2.]

24. Six perpendiculars are drawn from the inverse of the intersection of the diagonals of a cyclic quadrilateral to the sides and diagonals. Show

1°. The feet of those to the sides are collinear.

2°. The line of collinearity bisects at right angles the line joining the feet of perpendiculars on the diagonals.
[By method of Ex. 22.]

25. If XX'; YY'; ZZ' denote the feet of the bisectors of the angles of a triangle ABC, show that the pedal triangles of two points O and O' inverse to any of the circles on these segments as diameters, with respect to ABC, are *inversely* similar. (Neuberg.)

[Let O and O' be inverse with respect to $ZZ'C$,* PQR and $Q'P'R'$ their pedal triangles respectively. M the middle point of ZZ'.

Then
$$\frac{PQ^2}{PQ'^2}=\frac{MO}{MO'},$$

and
$$\frac{R'P}{R'Q'}=\frac{AO'\sin A}{BO'\sin B}=\frac{BO\sin B}{AO\sin A}\text{ (by Ex. 14)}=\frac{RP}{RQ};$$

also the angles R and R' are equal; therefore etc.

Note.—If O is on the circle $ZZ'C$ the pedal triangle is isosceles, similarly if it is the point of intersection of the circles $ZZ'C$ and $YY'B$ it is isosceles in a double aspect, *i.e. equilateral.*

Hence we may infer that *the circles AXX', BYY', and CZZ' pass through two points O and O' which are inverse* (Ex. 22) *with respect to the circum-circle of ABC and whose pedal triangles with respect to ABC are equilateral.*]

---

* *Le cercle d'Apollonius* du triangle ABC par rapport à AB. V. *Educ. Times*, Dec., 1890.

CHAPTER VII.

POLES AND POLARS WITH RESPECT TO A CIRCLE.

SECTION I.

CONJUGATE POINTS, POLAR CIRCLE.

73. Def. The perpendicular to the line joining a pair of inverse points passing through either is the *Polar* of the other with respect to the circle. In the figure of Art. 74 C and Z are inverse points; and C and the line AB are termed *Pole and Polar* with respect to the circle.

Any point A or B on the polar is the *Conjugate* of C, hence the polar of a point is the locus of its conjugates.

Again, since the circle on BC as diameter passes through Z and therefore cuts the given one orthogonally :— 1°. *The circle described on the line joining two conjugate points cuts the given circle orthogonally.* 2°. *The distance between two conjugate points is equal to twice the length of the tangent to the circle from the middle point of the line connecting them.*

74. Theorem.—*For any two conjugate points B and C, to prove that each lies on the polar of the other with respect to the circle.*

Suppose the polar of C to be AB, we require to prove that the polar B passes through C. Join AO, draw a

perpendicular to it CX. Then evidently (Euc. III. 36)
$OA . OX = OC . OZ = r^2$; hence CX is the polar of A.
Thus *as the point B moves along the line AB its polar*

(a)

*turns around, or envelopes, C. At Z therefore the polar is
the chord of contact of tangents through that point to the
circle.*

EXAMPLES.

1. The extremities of any diameter of a circle which cuts a given
one orthogonally are conjugate points with respect to the latter.

2. If a variable chord AB of a circle pass through a fixed point
P; the locus of the intersection of tangents to the circle at A and
B is a line.
     [The polar of P with respect to the circle.]

3. The diameter AB of a circle is the polar of a point at infinity
in a direction perpendicular to AB.

4. The locus of a point which has a common conjugate with
respect to three circles is their common orthogonal circle.

75. Theorem.—*If A and B be any two points and L and M their polars with respect to a circle, the point LM is the pole of the line AB.*

For *LM* is conjugate to both *A* and *B*, hence the line joining *A* and *B* is its polar (Art. 73), or "the *line of connexion of any two points is the polar of the point of intersection of the polars of the points.*" Townsend.

76. More generally for three points *A*, *B*, *C* and their polars *L*, *M*, *N*, denoting the points *MN*, *NL*, *LM* by *A'*, *B'*, *C'* respectively; we see as above that *A'*, *B'*, *C'* are the poles of *BC*, *CA*, *AB*; hence, *for any two triangles if the vertices of either are the poles of the corresponding sides of the other;* then, reciprocally, *the vertices of the latter are the poles of the corresponding sides of the former.*

Def. Such triangles are said to be *Reciprocal Polars* with respect to the circle.

77. In the particular case when *ABC* and *A'B'C'* coincide, the triangle is *Self-Reciprocal* with respect to the circle. It is manifest, since each vertex is the pole of the opposite side, every two of its vertices are conjugate points; and the triangle is therefore termed *Self-Conjugate* with respect to the circle.

Its centre *O* coincides with the orthocentre *O* of *ABC* and the square of its radius (ρ) is given by

$$\rho^2 = OA \cdot OX = OB \cdot OY = OC \cdot OZ,$$

where *X*, *Y*, *Z* are the feet of the perpendiculars of the triangle.

This circle is called the *Polar Circle* of the triangle.

NOTE.—In order that the polar circle may be real, the pairs of points *A* and *X*, *B* and *Y*, *C* and *Z*, which are inverse with respect to it, must lie in the same direction from its centre *O*. It is therefore real when the triangle is obtuse angled, and imaginary for acute angled triangles.

78. *Expressions for the Radius (ρ) of the Polar Circle.*

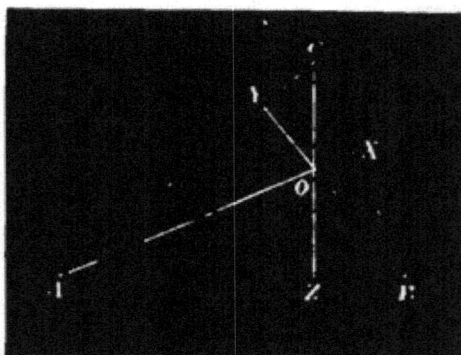

(β)

Let O be the ortho-centre of ABC, then it appears that A, B, C are the ortho-centres of BOC, COA, and AOB respectively. For this reason the four points A, B, C, O are said to form an *Orthocentric System.*

Also the circum-circles of the four triangles BOC, COA, AOB, and ABC are equal.

Hence since a and AO, chords of equal circles, subtend complementary angles at the circumferences,

$$a^2 + AO^2 = b^2 + BO^2 = c^2 + CO^2 = d^2, \ldots \ldots \ldots (1)$$

also (fig. β) $a^2 = BO^2 + CO^2 + 2CO \cdot OZ$,   (Euc. II. 13)

therefore by substitution from (1)

$$a^2 = 2d^2 - b^2 - c^2 + 2CO \cdot OZ,$$

or $-CO \cdot OZ = d^2 - \tfrac{1}{2}(a^2 + b^2 + c^2) = \rho^2 \ldots \ldots \ldots (2)$

This formula is equivalent to $\rho^2 = d^2 \cos A \cos B \cos C$, by reduction or independently, as follows:—

$$-\rho^2 = OC \cdot OZ = OC \cdot \frac{OA \cdot OB}{d} = d^2 \cos A \cos B \cos C, \ldots (3)$$

since a chord is equal to the diameter of the circle into the sine of the angle it subtends.

Examples.

1. The four polar circles of the triangles BOC, COA, AOB and ABC are mutually orthogonal.*

[Let their radii be ρ_a, ρ_b, ρ_c, ρ. Since their centres are at A, B, C, O, by Euc. II. 2,

$$AB^2 = AB . AZ + AB . BZ = \rho_a^2 + \rho_b^2 ;$$

therefore, etc.]

2. B and C, C and A, A and B are pairs of conjugate points with respect to the polar circles of BOC, COA and AOB respectively.

3. The square of the distance BC between any two conjugate points is equal to the sum of the squares of the tangents drawn from them to the circle.

[By Ex. 1 the tangents from B and C to the circle ρ_a are the radii of ρ_b and ρ_c, but $BC^2 = \rho_b^2 + \rho_c^2$; therefore, etc.]

4. Prove that $AZ . BZ = t^2$, where t is the tangent to the polar circle from Z, the *Polar Centre*† of AB; and conversely.

[By similar triangles ACZ and OBZ, $AC : CZ = OZ : BZ$, etc.]

5. Conjugate points A and B with respect to any chord MN are conjugate with respect to the circle.

[For the polar centre Z of AB is the middle point of MN; but (hyp.) $ZA . ZB = ZM^2 = -ZM . ZN$ or the square of the imaginary tangent from Z to the circle; therefore, etc., by Ex. 4.]

6. If a number of circles have a common orthogonal circle, the extremities of any diameter of the latter are conjugates with respect to the entire system.

7. On a given line find two points which shall be conjugates to each of two given circles.

[The middle point of the required segment is such that the tangents from it to the circles are equal; therefore, etc., by Art. 72, Ex. 3.]

---

* Hence :—If four circles are mutually orthogonal, their centres form an orthocentric system and one of the circles is imaginary.

† Z the foot of the perpendicular from the centre on AB is also called the *Middle Point* of the line. (Cf. Euc. III. 3.)

8. On a given circle O find two points A and B which shall be conjugates to each of the circles C, r_1 ; D, r_2.

[The middle point M of the required chord is on the radical axis L of the given circles (Art. 72, Ex. 3). Let $2t$ be the length of AB; then $\quad CM^2 = t^2 + r_1^2 = r_1^2 + AM^2 = r_1^2 + r_2^2 - OM^2$; hence $CM^2 + OM^2$ is known, and the triangle COM is completely determined ; therefore, etc.]

9. Place a chord of given length in a circle so that its extremities may be conjugates with respect to another.

[See Ex. 8.]

10. If a right line AB meet either (C, r) of two circles in conjugate points (A, B) with respect to the other ; then reciprocally it meets the latter (C', r') in conjugate points $(A'$ and $B')$ with respect to the former.

[For by Ex. 5 AB divides $A'B'$ harmonically, hence $A'B'$ divides AB harmonically ; therefore, etc.]

11. Find the locus of the middle points M and N of the chords AB and $A'B'$ in Ex. 10.

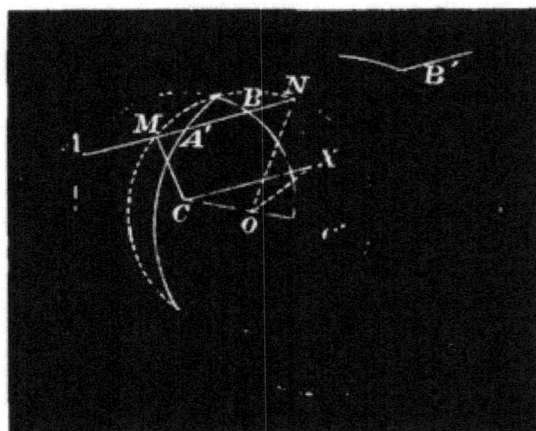

$$[CM^2 + C'M^2 = CN^2 + C'N^2 = CM^2 + C'N^2 + MN^2$$
$$= CM^2 + C'N^2 + MB^2 + A'N^2 \qquad \text{(Art. 71.)}$$
$$= r^2 + r'^2 = \text{const.} ;$$

hence the required locus is *a circle whose centre O is at the middle point of CC' and the square of whose radius is equal to* $\frac{1}{2}(r^2 + r'^2) - \delta^2$, where $2\delta = CC'$. It evidently passes through the intersections of the given circles.]

12. Show that $CM . C'N = $ const.

[Draw CX at right angles to $C'N$. Join OX. Since $OC'X$ is an isosceles triangle and N a point in the base produced,

$$CM . C'N = C'N . NX = ON^2 - OX^2 = ON^2 - OC'^2$$
$$= \tfrac{1}{2}(r^2 + r'^2 - 4\delta^2) = rr'\cos\theta,$$

where θ is the angle between the given circles ; therefore, etc.]

13. Any circle described around the polar centre of a triangle ABC meets the corresponding sides of the median triangle in A', B', C' such that $AA' = BB' = CC'$.

14. A tangent is drawn from the polar centre to the circumcircle, and from the point of contact a tangent is drawn to the polar circle, show that the angle between these lines is 45°.

15. Draw through P a line cutting each of two given circles in conjugate points with respect to the other.

[By Exs. 10 and 11.]

16. Draw a line cutting each of two circles X and Y in conjugate points with respect to a third (Z).

[Let the required line meet Z in the points A and B. The middle point M of AB is the intersection of two known circles passing through the intersections of Z and X and Z and Y (Ex. 11), and is thus determined ; therefore, etc.]

SECTION II.

79. **Salmon's Theorem.**—*The distances of any two points A and B from the centre O of a circle are proportional to the distances AM and BL of each from the polar of the other.*

Draw AB' and BA' perpendiculars to OB and OA respectively.

Then $OA.OL=OB.OM=r^2$, and since $AA'BB'$ is a cyclic quadrilateral, $OA.OA'=OB.OB'$; therefore

$$\frac{OA}{OB}=\frac{OB'}{OA'}=\frac{OM}{OL}=\frac{OM-OB'}{OL-OA'}=\frac{B'M}{A'L}=\frac{AM}{BL};$$

therefore, etc. By alternation $OA/AM=OB/BL$.

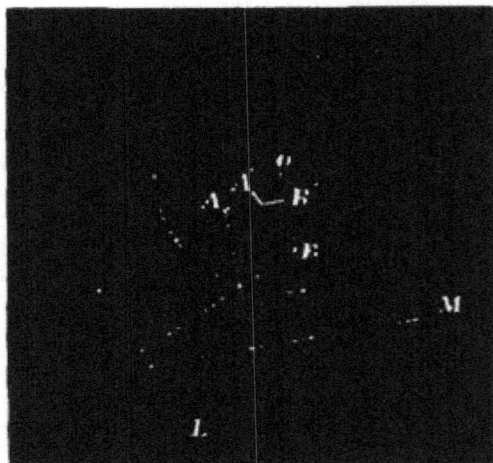

COR. 1. If M is a fixed line and OA/AM a constant ratio, B is a fixed point and the envelope of L is a circle; or, *the pole of a variable tangent to a circle with respect to another given circle is such that its distance from the centre of the latter bears a fixed ratio to the distance from a fixed line.*

COR. 2. If A and B are both on the circle (O, r); $OA=OB$, and therefore $AM=BL$; or, the points of contact of tangents to a circle are equidistant from the tangents as is otherwise evident (Euc. I. 26).

COR. 3. Let B and its polar M vary and the different positions be denoted by $B_1, B_2, B_3, ..., M_1, M_2, M_3, ...$; then since

$$\frac{OA}{AM}=\frac{OB}{BL},\ \frac{OA}{AM_1}=\frac{OB_1}{B_1L},\ \frac{OA}{AM_2}=\frac{OB_2}{B_2L},\ \text{etc.};$$

by multiplication of ratios, we have

$$\frac{OA^n}{AM . AM_1 . AM_2 ...} = \frac{OB . OB_1 . OB_2 ...}{BL . B_1L . B_2L ...};$$

or, *the product of the distances of a point (A) from any number of lines (M) is to the product of the distances of their poles (B) from the polar (L) of the point as the nth power of the distance of the point from the centre is to the product of the distances of the poles from the centre.*

Cor. 4. If M, M_1, M_2 in Cor. 3 form an inscribed polygon, B, B_1, B_2, ... are the vertices of the corresponding escribed one; hence the product of the distances of any point from the sides of an in-polygon is to the product of the distances of the vertices of the corresponding ex-polygon from the polar of the point as the nth power of the distance of the latter from the centre is to the product of the distances of the vertices of the ex-polygon from the centre.

Cor. 5. The rectangle under the distances of the extremities of any chord from a tangent is equal to the square of the distance of its point of contact from the chord.

Examples.

1. The opposite vertices of an escribed quadrilateral are AA', BB', CC'; to prove that

$$OA . OA' : OB . OB' : OC . OC' = AX . A'X : BX . B'X : CX . C'X,$$

where X is a tangent to the circle at any point P.

[Let the corresponding pairs of sides of the in-quadrilateral be L, L'; M, M'; N, N'; then since

$$\frac{OA}{AX} = \frac{OP}{PL} \text{ and } \frac{OA'}{A'X} = \frac{OP}{PL'},$$

multiplying these equations, $\dfrac{OA . OA'}{AX . A'X} = \dfrac{OP^2}{PL . PL'};$

but $PL . PL' = PM . PM' = PN . PN'$; therefore, etc.]

2. If a, β, γ denote the perpendiculars from any point on the circum-circle on the sides of an in-triangle,

$$\beta\gamma \sin A + \gamma a \sin B + a\beta \sin C = 0$$

or

$$\frac{a}{a} + \frac{b}{\beta} + \frac{c}{\gamma} = 0.$$

3. If λ, μ, ν be the perpendiculars from the vertices of any triangle upon a variable tangent to the in-circle,

$$\frac{\cot\tfrac{1}{2}A}{\lambda} + \frac{\cot\tfrac{1}{2}B}{\mu} + \frac{\cot\tfrac{1}{2}C}{\nu} = 0.$$

[Let A', B', C', P be the points of contact with the sides and any tangent, then $\dfrac{OA}{\lambda} = \dfrac{r}{a'}$, where a' is the perpendicular from P on $B'C'$.

Hence
$$\Sigma \frac{OA \cdot B'C'}{\lambda} = r\Sigma \frac{B'C'}{a'} = 0 ; *$$ (Ex. 2)

but $OA \cdot B'C' = 2r^2 \cot\tfrac{1}{2}A$; substituting, we have

$$\Sigma \cot\tfrac{1}{2}A/\lambda = 0.$$

A particular case of this has been noticed in Art. 55, Ex. 8.]

4. If the perpendiculars from the vertices on any tangent to the circum-circle of a triangle be λ, μ, ν ; to prove that

$$a\sqrt{}/\lambda + b\sqrt{}/\mu + c\sqrt{}/\nu = 0.$$

[If P be the point of contact of the tangent to the circle, by Ptolemy's Theorem,

$$a \cdot AP + b \cdot BP + c \cdot CP = 0,$$

but $AP^2 = 2r\lambda$, etc., hence $\Sigma a\sqrt{}/\lambda = 0$.]

5. For any point P on the in-circle whose distances from the sides are a, β, γ ; to prove that

$$\cos\tfrac{1}{2}A\sqrt{}/a + \cos\tfrac{1}{2}B\sqrt{}/\beta + \cos\tfrac{1}{2}C\sqrt{}/\gamma = 0.$$

[Let λ', μ', ν' be the distances of the points of contact A', B', C of the sides of ABC from the tangent at P; a', β', γ' the distances of P from the sides of $A'B'C'$.

By Ex. 4, $\Sigma a'\sqrt{}/\lambda' = 0$ or $\Sigma \dfrac{a'}{\sqrt{\mu'\nu'}} = 0$,

but $\sqrt{\mu'\nu'} = a' = \sqrt{\beta\gamma}$;   (Art. 79, Cor. 5.)

---

* The angles of $A'B'C'$ are respectively $90 - \tfrac{1}{2}A$, $90 - \tfrac{1}{2}B$, $90 - \tfrac{1}{2}C$; therefore $a' : b' : c' = \cos\tfrac{1}{2}A : \cos\tfrac{1}{2}B : \cos\tfrac{1}{2}C.$

hence, on substituting, since $a' = 2r \cos \frac{1}{2}A$,

$$\Sigma a' \sqrt{\lambda'} = 0 = \cos \tfrac{1}{2} A \sqrt{a}, \text{ therefore, etc.}]$$

Note.—The equations in Exs. 2 and 5 are known in Analytical Geometry to be those of the circum- and in-circles respectively, the given triangle ABC being taken as the triangle of reference. The expressions in Exs. 3 and 4 are the *Tangential Equations of the In- and Circum-Circles.*

6. If two triangles ABC, $A'B'C'$ are reciprocal polars, they are in perspective.

[Let the perpendiculars from $A'B'C'$ on the sides of ABC be p_1, p_2, p_3; q_1, q_2, q_3; r_1, r_2, r_3 respectively; then, by Salmon's Theorem,

$$\frac{OB'}{OC} = \frac{q_3}{r_2}; \quad \frac{OC'}{OA'} = \frac{r_1}{p_3}; \quad \frac{OA'}{OB'} = \frac{p_2}{q_1};$$

multiplying these equations we have

$$\frac{p_2}{p_3} \cdot \frac{q_3}{q_1} \cdot \frac{r_1}{r_2} = 1; \text{ therefore, etc.} \quad (\text{Art. 65.})]$$

7. A triangle inscribed in a circle is in perspective with the corresponding escribed one.

[By Ex. 6.]

8. Any two triangles may be so placed that the vertices of either are the poles of the sides of the other with respect to a circle.

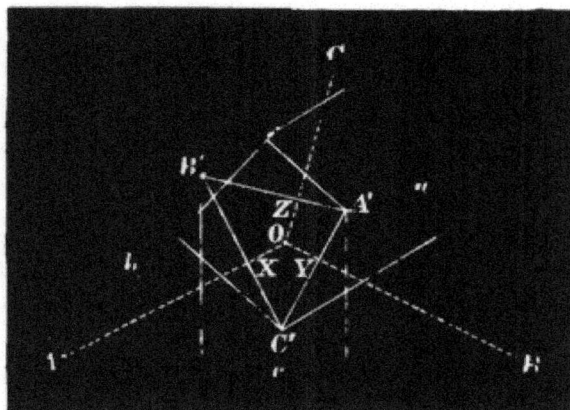

[At the centre O of the required circle the sides of each triangle subtend angles *similar* to those of the other triangle. Find points satisfying these conditions with respect to each triangle and place the latter with the points coincident and AO at right angles to $B'C'$; then OB and OC will be at right angles to $C'A'$ and $A'B'$. Again, since the perpendiculars from ABC on the sides of $A'B'C'$ are concurrent, those from $A'B'C'$ on the sides of ABC are also concurrent; it follows obviously that OA', OB', OC' are perpendicular to the sides of ABC; and

$$OA . OX = OB . OY = \ldots = OA' . OX' = \ldots = \rho^2.$$

9. To find the radius ρ of the circle in Ex. 8.

$$\left[\frac{\text{area } B'OC'}{\text{area } ABC} = \frac{OB'.OC'}{bc} = \frac{\rho_4}{bc.OY'.OZ'} = \frac{\rho_4}{4COA.AOB}.\right.$$

Similarly, $\dfrac{C'OA'}{ABC} = \dfrac{\rho^4}{4} . \dfrac{1}{AOB.BOC}$, etc.

Adding these results, we have

$$\frac{A'B'C'}{ABC} = \frac{\rho^4}{4} . \Sigma \frac{1}{BOC.COA} = \frac{\rho^4}{4} . \frac{ABC}{BOC.COA.AOB},$$

or $$\rho^4 = \frac{4 BOC . COA . AOB . A'B'C'}{(ABC)^2}.\left.\right]$$

10. The area of the reciprocal polar $A'B'C'$ of a given triangle with respect to a circle is given by the equation of Ex 9.

11. The minimum value of $A'B'C'$ is obtained when the centre O coincides with the centroid of ABC; and $= \dfrac{27\rho^4}{4ABC}$.

[In this case $BOC = COA = AOB$. Art. 14, Ex. 5.]

12. The reciprocal polar of the median triangle with respect to the in-circle or ex-circles of the given one is equal to ABC.

13. The reciprocal polar triangle may be of any species.

[Species depends on the position of the centre O.]

14. In Ex. 8 O is one or other of two fixed points.

[One of them is obviously within both triangles and the sides of each subtend at it angles equal to the supplements of the angles of the other.

The other is the common intersection of the circles described *externally* on the sides of ABC containing angles equal to $\pi - A'$, $\pi - B'$, $\pi - C'$. On making the figure it will be observed that these circles, intersecting in pairs at the vertices of the triangle, can only

meet again in *one* point ; hence, if a *point O be reflected with respect
to the three sides of a triangle, the circles* BCO_1, CAO_2, ABO_3 *meet in
a point.**

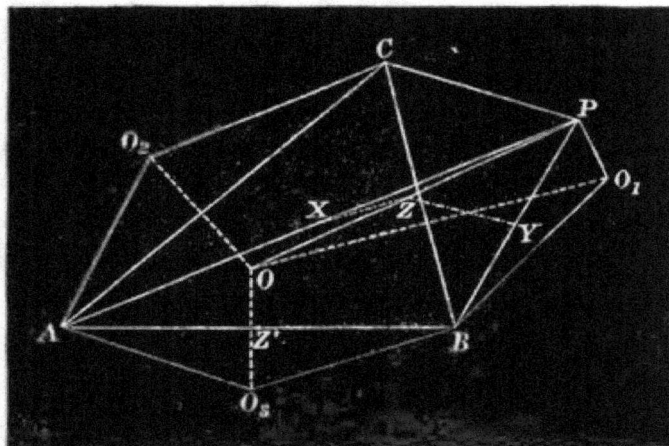

15. If the triangles ABC and $A'B'C'$ are similar the second centre
is any point on the circum-circle of ABC ; also if P be joined to
$A, B,$ and O and X, Y, Z be the middle points of these lines and Z'
the middle point of AB ; $XYZZ'$ is a cyclic quadrilateral for

$$\angle XZY = AOB \text{ and } XZ'Y = APB = \pi - AOB ;$$

hence $$XZY + XZ'Y = \pi ;$$

therefore Z the middle point of OP is on the nine-points-circle of
ABP. Similarly it is on the nine-points-circles of the triangles
with BC and AC as bases and P as vertex. Hence for any four
points $A, B, C, P,$ the nine-points-circles of three of the triangles
formed by them are concurrent. It is therefore obvious that all
four *nine-points-circles of the four triangles BCP, CAP, ABP, ABC
are concurrent.*†

16. A triangle reciprocates into a similar one from either of the
Brocard points as origin. (Art. 27.)

---

* The points O and P are *reciprocally* related to the triangle ABC.
For it will be seen that, if P be reflected with respect to the sides, the
circles BCP_1, CAP_2, ABP_3 will meet in O. It follows thence that the
nine points circles of the triangles BCO, CAO and ABO also pass
through this point of concurrence.

† Van de Berg, *Mathesis*, t. 2, p. 141.

SECTION III.

RECIPROCATION.

80. If ABC ... be any polygon and $A'B'C'$ ... another derived from it by taking the poles A', B', C', ... of the sides BC, CA, AB, etc., with respect to any circle, then we have seen (Art. 76) that the vertices A, B, C, etc., of the former are the poles of the sides of the latter, and the two polygons are said to be *Reciprocal Polars* with respect to the circle. The process of deriving $A'B'C'$... is termed *Reciprocation*, and the circle, radius, and centre are the *Circle, Radius,* and *Centre,* or *Origin of Reciprocation.*

More generally, if ABC ... be any curve to which tangents T_1, T_2, T_3, ... are drawn at the points A, B, C, ..., the locus of their poles is the *Reciprocal Polar Curve* of ABC ... with respect to the circle. If the tangents at A and B are indefinitely near, their poles A', B' are also indefinitely near on the reciprocal curve; but the point T_1T_2 is (Art. 76) the pole of the line $A'B'$; hence in the limit the point A is the pole of the tangent at A'. The point A and tangent at A' are said to *correspond.* Thus, of two polar reciprocal curves any tangent to either corresponds to a point on the other, and each point of contact and the corresponding tangent are pole and polar with respect to the circle.

The following fundamental properties of two Reciprocal figures will appear obvious :—

1°. The line joining any two points of either is the

polar of the intersection of the corresponding lines of the other.

2°. Concurrent lines reciprocate into collinear points.

3°. The angle subtended by any two points of one at the origin is equal to the angle between the corresponding lines of the other.

4°. For any two figures X and Y and their reciprocals X' and Y', the points of intersection of X and Y correspond to the common tangents to X' and Y'; in other words, a common tangent to two curves corresponds to a point of intersection of their reciprocals.

5°. If X and Y touch, their reciprocals X' and Y' also touch, and each point of contact is the pole of the common tangent at the other.

6°. Since two circles have four common tangents, real or imaginary, they reciprocate into curves which intersect in four points. (By 4°.)

7°. Any point connected with X and the tangents through it to the curve correspond to a line and its points of intersection with the reciprocal curve X'.

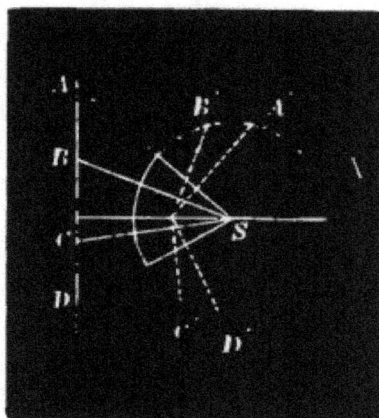

8°. The reciprocal of a circle is a *curve of the second degree,* i.e. one which meets every line in two points, real or imaginary. (By 7°.)

9°. The pencils determined by any four collinear points A, B, C, D at the origin S and the corresponding lines A', B', C', D' are similar.

[For the corresponding rays of pencils are at right angles.]

10°. Harmonic rows of points reciprocate into harmonic pencils of rays ; and in the particular case when one point D of the row A, B, C, D coincides with the origin S ; SA', SB', SC' are in arithmetical progression.

11°. Parallel lines reciprocate into points collinear with the origin.

12°. A point and its polar reciprocate into a line and its pole with respect to the reciprocal curve.   (Cf. 7°.)

Reciprocation of the Circle.

81. Let the origin S be outside the circle (O, r) ; $OS = \delta$; L the polar of O with respect to the Circle of Reciprocation, and P the pole of any tangent to the circle at Z.

For the two points O and P we have, by Salmon's Theorem, $\quad \dfrac{SP}{PL} = \dfrac{SO}{OZ} = \dfrac{\delta}{r} = \text{const.} = e \text{ (say)}.$

The locus of P given by the equation $SP/PL = e$ is a *Conic Section*, of which S is termed a *Focus*, L a *Directrix*, and e the *Eccentricity*.   (See Art. 79, Cor. 1.)

When $e > 1$, the conic is called a Hyperbola,

    „ $e = 1$,   „   „   Parabola,

    „ $e < 1$,   „   „   Ellipse.

Thus the reciprocal polar of a circle is a hyperbola, parabola, or ellipse, according as the origin is outside, upon, or within the circle.

In the particular case when the origin coincides with the centre of the given circle, the reciprocal curve is a concentric circle.

Since the tangents to a circle are real and distinct from any points outside it, and reciprocate from S as

origin to two points at infinity; their points of contact
X and Y reciprocate into two real tangents to the conic,
meeting in C the correspondent of XY, whose points of
contact are at infinity.

These lines are termed the *Asymptotes* of the hyperbola. They are *imaginary* for the ellipse, though they
intersect in a real point, and *coincident* with the line at
infinity for the parabola.

The tangents A' and B' at the extremities of the
diameter OS correspond to points A and B called the
Vertices of the conic; also since the distances of S from
A', XY, B', are in H.P., SA, SC, SB their reciprocals are
in A.P.; hence C is the middle point of the segment AB,
and it is obviously the point at which the asymptotes
intersect.*

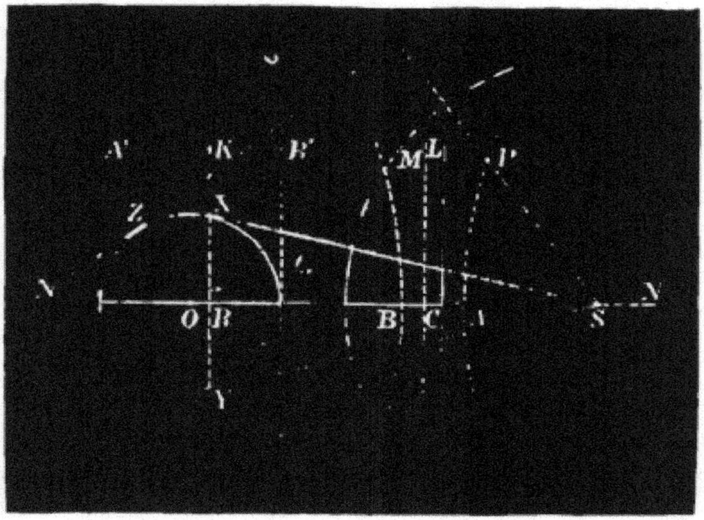

---

* When the origin is outside the circle its polar divides the circumference into two parts which are respectively concave and convex to it.

These portions reciprocate into two distinct curves convex and concave to the origin as shown in the figure, and both branches reach to

Also since SA', SO and SB' are in A.P., their reciprocals SA, SL, SB respectively are in H.P.

The tangents from any point K, on XY, to the circle, with XY and KS form an harmonic pencil (Art. 78, Ex. 5); hence by reciprocation any line through C meets the conic in an harmonic row of points, one of which, corresponding to the ray KS, is at infinity. Thus every chord of the conic through C is bisected. On account of this property C is termed the *Centre* of the curve.

Again, the tangents to the circle from any point on the perpendicular through S to RS and the lines joining that point to R and S form an harmonic pencil; hence by reciprocation any line parallel to OS meets the conic in an harmonic row of points, one of which, corresponding to the ray through S, is at infinity; another, that corresponding to the ray through R, is on M the perpendicular through C to OS. It is therefore manifest that the conic is symmetrically situated with respect to this line. It is moreover symmetrical with respect to ON. These rectangular lines OM, ON through the centre C are termed the *Axes* of the curve.

---

infinity. If, however, we assume in general that consecutive tangents to the circle reciprocate into consecutive points on the conic, by taking two tangents indefinitely near, one on the convex and the other on the concave part of the circle, we are led to the conclusions that the points at infinity on the opposite branches of the curves are indefinitely near, that the asymptotes are tangents at the points of coincidence, and that the hyperbola is a continuous curve.

EXAMPLES.

1. A circle, any point and its polar with respect to the circle, *e.g.*

A conic, a line and its pole with respect to the conic.

Circle, centre and line at infinity.

Conic, directrix and focus.

Circle, origin and polar of origin.

Conic, line at infinity and centre of conic.

Circle and inscribed polygon.

Conic and escribed polygon.

Circle (or conic) and self conjugate triangle.*

Conic and self conjugate triangle.

2. The opposite sides of a cyclic hexagon meet in three collinear points. (Pascal.)

The opposite vertices of an escribed polygon connect by three concurrent lines. (Brianchon.)

This result follows when the circle described about the hexagon is taken as the circle of reciprocation.

In general, from any origin, the theorem of Pascal with respect to a circle reciprocates into Brianchon's property for a conic.

3. Four points on a circle subtend at a variable point on it equianharmonic pencils.

Four fixed tangents to a circle meet a variable tangent to it in equianharmonic rows ;

hence, generally from any origin, the property of Euc. III. 21 becomes :—A variable tangent to a conic meets four fixed tangents in rows of points which are equianharmonic ; and reciprocally four fixed points on a conic subtend equianharmonic rows at a variable fifth point on it.

And again it follows conversely that, if two points connect equianharmonically with four others, all six lie on a conic ; hence :— Any two of the hexad of points connect equianharmonically with the remaining four. This system is sometimes called an *Equianharmonic Hexagon*. (Townsend, *Mod. Geom.* vol. II. p. 168.)

4. Concentric Circles.

Conics having same focus (origin) and directrix.

---

* If the origin is taken at one of the vertices of the triangle the corresponding side of the reciprocal triangle is therefore at infinity, and its other two sides are diameters (*conjugate*) of the conic. See Exs. S, 9.

5. Circles having a common pair of inverse points (from either point as origin).

Conics having a common focus and centre.

From the symmetry of the conic we infer that such a system has a second common focus ; hence :—*Coaxal Circles reciprocate from either of their common pair of inverse points into a system of Confocal Conics.*

6. Euc. III. 35, 36.

The rectangle under the distances of either focus from a pair of parallel tangents is constant ;

hence from symmetry we infer that the rectangle under the distances of the foci from any tangent is constant ; and conversely, the envelope of a variable line, the product of whose distances from two fixed points is constant, is a conic having the fixed points for foci.

7. A chord of a circle which subtends a right angle at the origin envelopes a conic.

The locus of the intersection of rectangular tangents to a conic is a circle.

(*Director Circle.*)

8. A variable chord of a circle passing through a fixed origin is divided harmonically by the point and its polar.

The variable chord of contact of two parallel tangents passes through and is bisected at the centre of the conic.

Def. The diameter of a conic parallel to a tangent is said to be *Conjugate* to that which passes through its point of contact.

9. Conjugate points with respect to a circle (from the pole of line joining them as origin).

Conjugate diameters of a conic.

10. If a variable point P moves on a line through the origin, S its polar passes through Q the pole of the line with respect to the circle ; and the tangents from P and the lines PQ and PS form an harmonic pencil.

If a variable chord of a conic moves parallel to a fixed direction, the harmonic conjugates of the points on it at infinity (*i.e.* the middle points) are collinear ;

hence *the locus of the middle points of any system of parallel chords is a line.*

11. Conjugate points coincide on the circle.

Each asymptote is its own conjugate.

12. The rectangle under their distances from the *middle* of the line joining them is constant.

The product of the tangents of the angles made by a pair of conjugate diameters with either axis of the conic is constant.

13. Euc. III. 21, 22.

The angles subtended at a focus by either pair of opposite sides of an escribed quadrilateral are equal or supplemental.

14. The locus of intersection of tangents containing a given angle is a concentric circle.

The envelope of a chord which subtends a constant angle at the focus is a conic having the same focus and directrix.

Their chord of contact envelopes a concentric circle.

The locus of the point of intersection of the tangents at the extremities is another conic having same focus and directrix.

15. If the vertex of an angle of given magnitude is on a circle, its variable chord of intersection envelopes a concentric circle.

If two points are taken on a fixed tangent so as to subtend a constant angle at the focus, the locus of the intersection of the tangents through them is a conic having same focus and directrix.

16. If the angle is right, the chord envelopes the centre (from vertex as origin).

The locus of intersection of rectangular tangents to a parabola is the directrix.

17. The perpendiculars of a triangle are concurrent.

The diagonals of a complete quadrilateral each subtend a right angle at a certain point ;

or the circles on the diagonals are concurrent.

It follows, because their centres lie on a line, that they pass

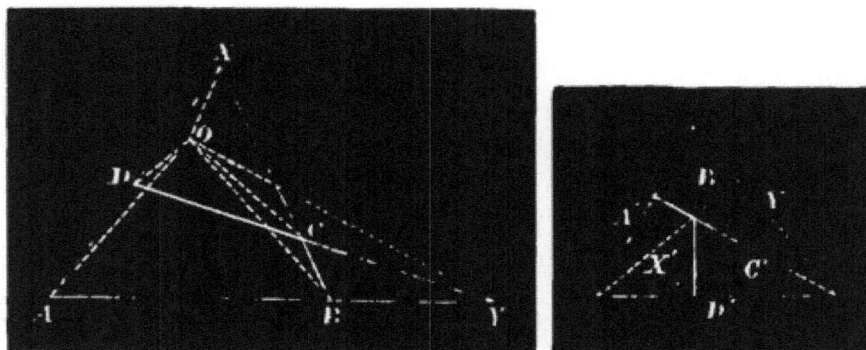

through a second point, the reflexion of the first with respect to
the line, *i.e.*, they are coaxal.

18. Having given the base
and ratio of sides of a triangle,
the locus of the vertex is a
circle to which the extremities
of the base are inverse points
(origin at either).

The line joining the centre of
a conic to the foot of the perpendicular from focus on any
tangent is constant.

The locus of the foot of the perpendicular is called the *Auxiliary
Circle* of the conic. The circle and conic evidently touch at the
extremities of the major axis.

Since the centre of a parabola is at infinity, its auxiliary circle
degenerates into the tangent at the vertex.

19. Common tangents to two
circles subtend right angles at
either common inverse point.

Confocal conics cut at right
angles.

20. The feet of the perpendiculars from any point on a
circle on the sides of an inscribed
triangle are collinear.

The perpendiculars through
the vertices of a triangle,
escribed to a parabola, to the
lines joining them to the focus
are concurrent ;

in other words, the circum-circle of a triangle described about a
parabola passes through the focus (cf. Ex. 18). We infer that the
circum-circles of the four triangles formed by four tangents (that is
any four lines whatever) meet in a point.

It follows also, since any point (origin) on the circum-circle and the orthocentre are equidistant from the Simson line of the point, that *the locus of the orthocentre of a variable triangle escribed to a parabola is the directrix.*

21. Having given base and vertical angle, the locus of the vertex of the triangle is a circle. (Euc. III. 21.)

If the extremities of a variable line, which subtends a constant angle at a fixed point, move on two fixed lines, it envelopes a conic to which these lines are tangents.

It therefore cuts them equianharmonically.

22. Since *inverse* segments subtend similar angles at any point on the circle, the segments of a line drawn across two circles subtend similar angles at either common inverse point.

The pairs of tangents to confocal conics from any point are equally inclined.

23. All circles meet in two imaginary points on the line at infinity.

Confocal conics have pairs of imaginary common tangents passing through the foci.

24. The polars of a point with respect to a system of coaxal circles are concurrent.

The poles of a line with respect to a system of confocal conics are collinear.

25. The two points in Ex. 24 are in perpendicular directions from either common inverse point.

The locus of the poles is a line perpendicular to the given one.

26. The sum of the squares of the segments of two rectangular chords of a circle is constant.

The sum of the squares of the reciprocals of the distances of the foci from two rectangular tangents is constant ;

hence if p_1, p_2, π_1, π_2 denote the distances of the foci from the tangents $\Sigma 1/p_1^2 =$ constant.

27. In Ex. 26, if the square of the radius of reciprocation is the power of the point with respect to the circle.

$p_1^2 + p_2^2 + \pi_1^2 + \pi_2^2 =$ constant ; or the locus of the intersection of rectangular tangents is a *concentric* circle (*Director Circle*).

28. From the properties of the conic, rectangular tangents, director circle, centre and line at infinity.

A variable chord of a conic which subtends a right angle at any point envelopes a conic ; and the focus and directrix of the envelope are pole and polar with respect to the given conic.

If the point is on the given conic *the envelope reduces to a point* [*] on the perpendicular to the tangent passing through its point of contact. (*The Normal.*)

29. The base *BC* of a triangle *ABC* inscribed in a circle is fixed and the origin taken at its pole. Applying the formula of Art. 79, Ex. 10, we have the area of the reciprocal triangle constant, hence :— the *area cut off by any tangent with the asymptotes is constant.* And conversely, *given the vertical angle in position and area of a triangle, the envelope of the base is a conic ; and the sides are divided equianharmonically by the extremities of the base.*

30. Show by reciprocating from a vertex of a self conjugate triangle with respect to a circle that

$a°$. The sum of the squares of any two conjugate diameters of an ellipse is constant.

$\beta°$. The difference of the squares of any two conjugate diameters of a hyperbola is constant.

31. Find by the methods of Art. 79, Exs. 3 and 4, the tangential equations of a conic circumscribed or inscribed to the triangle of reference.

---

[*] This is proved independently as follows : If two right lines are drawn at right angles through a fixed point and intercept a variable segment *AB* on a fixed tangent to a circle ; the locus of the intersection of tangents through *A* and *B* is a line.

For it is a locus that can only meet the given tangent in one point ; therefore, etc., by reciprocation.

CHAPTER VIII.

Section I.

Coaxal Circles.

82. Definitions.—The *Radical Axis* L of two circles A, r_1 and B, r_2 is the line perpendicular to AB and dividing it so that $AL^2 \sim BL^2 = r_1^2 \sim r_2^2$. Cf. Art. 72, Ex. 3.

It follows from the definition that L is the common chord of the circles when they intersect, and we may generalize this statement by regarding the radical axis as their chord of intersection real or imaginary.

Thus all circles having a common radical axis pass through two real or two imaginary points.

Such a group is termed a *Coaxal System*.

83. It has been seen, Art. 72, Ex. 3, that a variable circle cutting two given ones orthogonally passes through two fixed points, viz., their common pair of inverse points; this orthogonal system is therefore coaxal; and from their mutual relations the two groups are said to be *Conjugate Coaxal Systems*. It is obvious that if either set possesses real points of intersection, the other does not; also the common points of one set are the common pair of inverse points with respect to the other Art. 72, Ex. 1.

Since the line of centres AB bisects the common chord

MN it is the axis of reflexion of each common point with respect to the other.

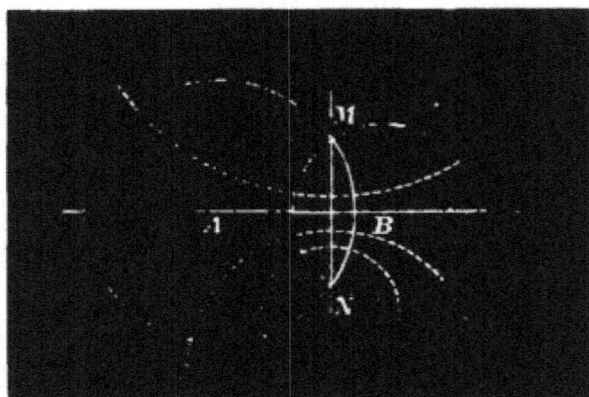

NOTE.—If two circles are concentric their radical axis is the line at infinity ; therefore a system of concentric circles passes through two imaginary points at infinity.

These are called the *Circular Points*.

If the circles touch, their radical axis is the common tangent at the point of contact.

If the circles reduce to points, the radical axis of two points is their axis of reflexion.

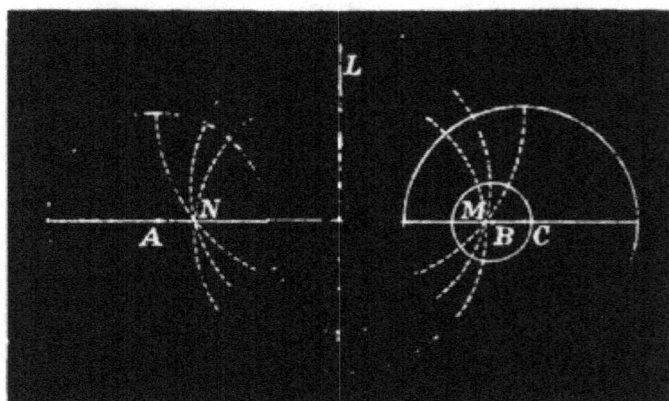

84. Let A, r_1; B, r_2; C, r_3 ... denote the circles of a coaxal system. Then, since

$$AL^2 - BL^2 = r_1^2 - r_2^2,\ AL^2 - CL^2 = r_1^2 - r_3^2,\ \text{etc.,}$$

we have by transposing

$$AL^2 - r_1^2 = BL^2 - r_2^2 = CL^2 - r_3^2 = \ldots = \pm k^2 \ldots \ldots (1)$$

The common value of these quantities ($\pm k^2$) is the *Modulus* of the system. It is *positive* for a non-intersecting system and *negative* for the intersecting or common point species.

85. It follows from Art. 84 (1) that the position of the centre C of any circle of given radius of a coaxal system is determined, and conversely. In the former case

$$CL^2 = AL^2 - r_1^2 + r_3^2 = \text{a known quantity.}$$

Two values of CL equal in magnitude but of opposite signs are thus found. Hence the reflexion of every circle of the system with respect to the radical axis is also a circle of the system. The radical axis is therefore the line around which the entire group is symmetrically disposed.

86. The radical axes of three circles taken in pairs are concurrent (Art. 72, Ex. 6). In the particular case when their centres are collinear the axes are parallel, and the point of concurrence (*Radical Centre*) is at infinity. If the circles are coaxal the radical axes coincide and the tangents from any point on this line to the three circles are therefore equal.

Conversely, if three circles whose centres are collinear have a radical centre not at infinity they form a coaxal system.

87. *Limiting Values of the Radius* given by the equation $AL^2 - r_1^2 = \text{const.}$

Since $AL^2 - r_1^2$ is constant, AL and r_1 increase and diminish in value together; or according as the centre approaches to or recedes from the radical axis, the radius diminishes or increases.

It follows in the limit when C is at infinity that the circle loses its curvature, and a portion of it coincides with the radical axis. The remainder being at infinity is the line at infinity; hence *we regard the line at infinity, and the radical axis, together as forming the circle of the system whose radius is infinitely great.*[*]

Again, since $AL^2 - r_1^2 = CL^2 - r_3^2$ if $r_3 = 0$,

$$CL^2 = AL^2 - r_1^2 \dots\dots\dots\dots\dots\dots(1)$$

The two values of CL in this equation determine therefore the positions of the centres of the circles of infinitely small radii. These are the *Points* or evanescent *Circles* of the group, and are termed the *Limiting Points.*

By (1) $r_1^2 = AL^2 - CL^2 = (AL - CL)(AL + CL)$
$$= AC \cdot AC',$$

where C' is the reflexion of C with respect to the radical axis; therefore the limiting points are the common pair of inverse points of the coaxal system. (Cf. Art. 72, Ex. 1.) Hence the radical axis of a circle and point is the axis of reflexion of the point and its inverse with respect to the circle.

88. **Theorems.**—I. *The radical axis of a coaxal system is the locus of a point the tangents from which to the circles are equal.*

Let the tangents from P be t_1 and t_2.

---

[*] Since two circles meet on their radical axis, we infer that any two circles pass through two imaginary points on the line at infinity. Also, because every two circles intersect on this line, therefore *all circles* pass through the same two imaginary points, *i.e. the Circular Points at Infinity.*

Then $t_1{}^2 = PA^2 - r_1{}^2,\ t_2{}^2 = PB^2 - r_2{}^2$;
hence, by subtraction,

$$t_1{}^2 - t_2{}^2 = PA^2 - PB^2 - (r_1{}^2 - r_2{}^2) = 0; \quad \text{(Art. 82)}$$

therefore, etc.

II. More generally, *The difference of the squares of the tangents* $(t_1{}^2 \sim t_2{}^2)$ *from any point* P *to two circles* = *twice the rectangle under the distance between their centres and the distance of* P *from their radical axis; or*

$$t_1{}^2 - t_2{}^2 = 2AB \cdot PL.$$

For, draw PP' perpendicular to AB and take M the middle point of AB.

Then $t_1{}^2 = AP^2 - r_1{}^2$, and $t_2{}^2 = BP^2 - r_2{}^2$;

hence $\quad t_1{}^2 - t_2{}^2 = AP^2 - BP^2 - (r_1{}^2 - r_2{}^2)$

$$= AP'^2 - BP'^2 - (AL^2 - BL^2) \quad \text{(Euc. I. 47)}$$

$$= 2AB \cdot P'M + 2AB \cdot ML; \quad \text{(Euc. II. 5 or 6)}$$

therefore $\quad t_1{}^2 - t_2{}^2 = 2AB \cdot PL.$

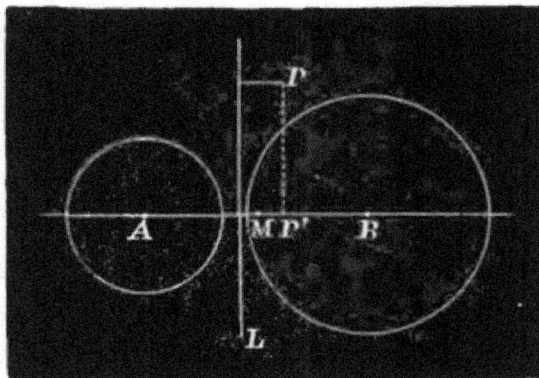

COR. 1. If P be any point on one of the circles (B, r_2),

$$t_2 = 0, \text{ and } t_1{}^2 = 2AB \cdot PL, \text{ or } t_1{}^2 \propto PL;$$

or, *if the square of the tangent from a variable point to a given circle varies as its distance from a fixed line, the locus of the point is a circle coaxal with the given circle and line.*

COR. 2. More generally, if C be the centre of a circle coaxal with A and B passing through P, t_1 and t_2 the tangents from P, we have, by Cor. 1,

$$t_1^2 = 2AC \cdot PL \quad (1) \quad \text{and} \quad t_2^2 = 2BC \cdot PL \quad (2);$$

dividing (1) by (2), we have

$$\frac{t_1^2}{t_2^2} = \frac{AC}{BC}; \quad \dots\dots\dots\dots\dots\dots\dots(3)$$

hence *the locus of point such that the ratio of the tangents drawn from it to two circles is constant is a coaxal circle whose centre is determined by (3).*

COR. 3. *The common tangents to two circles each subtend right angles at the limiting points.*

For, if M be a limiting point, XY one of the common tangents, and L its intersection with the radical axis, $LX = LY = LM$; therefore, etc.

COR. 4. If a variable chord XY of a circle be divided at P such that $PX \cdot PY \propto PM^2$, where M is a fixed point; the locus of P is a circle coaxal with the given circle and point.

The line PM is the tangent from P to the limiting point M; therefore, etc.

EXAMPLES.

1. If a variable chord (AB) of a circle (O, r) subtend a right angle at a fixed point (M), the loci—

   $a°$. of its middle point N;

   $\beta°$. of N' the foot of the perpendicular on it from M;

   $\gamma°$. of the pole P of AB

are circles each coaxal with the given point and circle.

[To prove $\alpha°$ and $\beta°$; we have

$$\frac{NM^2}{NA \cdot NB} = \frac{N'M^2}{N'A \cdot N'B} = -1 ;$$

hence N and N' lie on the same circle coaxal with M and O, r, whose centre bisects internally the interval OM, by Cors. 2 and 4.

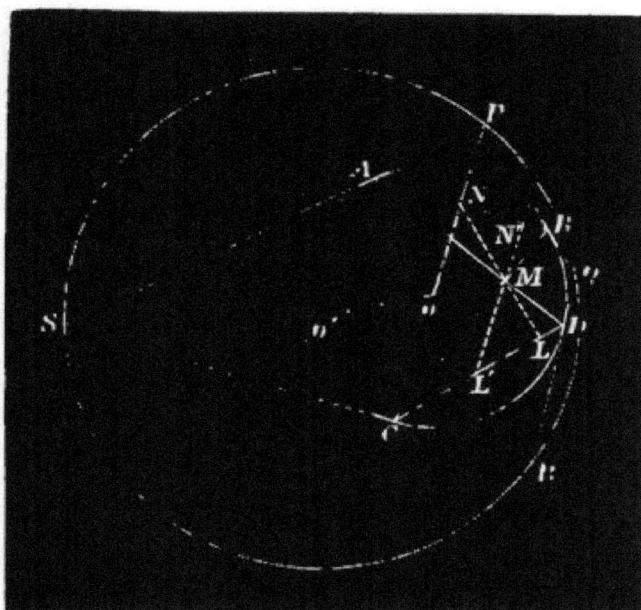

To prove $\gamma°$. Since N describes a circle, its inverse P describes a circle coaxal with O, r and the locus of N. For the locus of P is a circle ; and it is coaxal with the other two, because the three circles have a common pair of points real or imaginary.]

2. The orthocentre of a triangle is the radical centre of the circles described on the sides as diameters ; and the common value (Art. 77) of the rectangles under the segments of the perpendiculars is the radical product of the point with respect to the circles.

3. The middle points of the four common tangents to two circles the collinear.

[Each point of bisection is on the radical axis.]

4. Find the radical centre and product of the ex-circles of a given triangle.

[The middle point of the base is the middle point of the common tangent to the two circles which touch the base externally ; therefore the line through it parallel to the internal bisector of the vertical angle, *i.e.* at right angles to their line of centres, is their radical axis. Similarly for each of the remaining pairs. Hence the radical centre is the in-centre of the median triangle ; and, generally, the ex-centres of the median triangle are the radical centres of the three triads of circles formed by taking the in-circle and two ex-circles of the original triangle.

For the values of the radical products, see Art. 48, Ex. 1.]

5. The circum-centre of a triangle is the radical centre of any three coaxal systems which have B and C, C and A, A and B for limiting points.

6. The extremities of any two secants to two given circles which intersect on their radical axis are concyclic.

7. Any circle P, R cutting two circles A, r_1 ; B, r_2 at angles α and β meets the radical axis at an angle θ given by the equation

$$\cos \theta = \frac{r_1 \cos \alpha - r_2 \cos \beta}{AB}.$$

[Denote the secants by PXX' and PYY'. Applying the formula $t_1{}^2 - t_2{}^2 = 2AB \cdot PL$, we have

$$2AB \cdot PL = R(R + XX') - R(R + YY')$$
$$= R(XX' - YY') = 2R(r_1 \cos \alpha - r_2 \cos \beta) ;$$

hence $$\frac{PL}{R} = \frac{r_1 \cos \alpha - r_2 \cos \beta}{AB}.$$

But $PL/R =$ the cosine of the angle in the segment of P, R made by the intercept on the radical axis ; therefore, etc.]

8. The axis of perspective of ABC and its pedal triangle is the radical axis of the circum- and nine-points-circles.

[By Art. 88, I. and Euc. III. 36.]

8*a*. The line joining the orthocentre and circum-centre is at right angles to the axis of perspective of ABC and the pedal triangle.

[It is the line of centres of the circum- and nine-points-circles.]

9. Two circles touch at M and a chord AB of either touches the other at P; prove that PM is a bisector of the angle AMB.

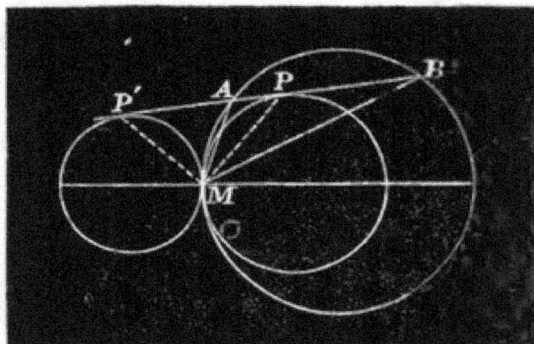

[By Art. 88, Cor. 2, $AP/AM = BP/BM$.]

10. For any cyclic quadrilateral whose diagonals intersect in M; prove that, if the bisectors of the angles between the diagonals meet the four sides in X, Y, X', Y',

$$AL . BL . CL . DL = XL . YL . X'L . Y'L,$$

where L is the radical axis of the circle and point.

11. If L, M, N denote the radical axes of three pairs of circles X and A, X and B, X and C, and L', M', N' the radical axes of Y and A, Y and B, Y and C; to prove that the two triangles LMN and $L'M'N'$ are in perspective; and that the centre of perspective is the radical centre of A, B and C; and their axis of perspective the radical axis of X and Y.

[For MN is a point on the radical axis of B and C (Art. 72, Ex. 6); similarly $M'N'$ a vertex of the triangle $L'M'N'$ is on the same line; therefore, etc.]

12. If three lines AX, BY, CZ be drawn from the vertices of a triangle to the opposite sides; the radical centre of the circles on these lines as diameters is the orthocentre and their common orthogonal circle the polar circle of the triangle.

[The perpendiculars of the triangle are respectively chords of these circles; therefore, etc. Art. 77.]

13. For any three circles A, B, C and three others taken with them such that B, C, X; C, A, Y; A, B, Z form three coaxal systems; to prove that, 1°, the system of six circles have the same radical

centre and product ; and, 2°, if the centres of X, Y, Z are collinear, these circles are coaxal.

[In 1° the radical centre and product is obviously that of the circles A, B, C; 2° follows at once since, if the circles be not coaxal, their radical centre is at infinity.   Art. 86.]

14. Two coaxal systems have a common circle ; find the locus of the points of contact of the circles which touch.

[Let L and L', the radical axes of the systems, meet at P, and T be one point of contact. The common tangent at T passes through P, and PT is the radius of the common orthogonal circle of the two systems, which is therefore the required locus.]

15. The radical axis of any two circles bisects the distance between the polars of the centre of each circle with respect to the other.

*16. Three circles are described each touching two sides of a triangle and the circum-circle internally in points L, M, and N ; to prove that the triangles ABC and LMN are in perspective.

[Let one of the circles touch the sides a and b in the points P and Q and the circum-circle in N. Then N being a limiting point of the two circles $AQ^2/AN^2 = BP^2/BN^2 = (R-\rho)/R$, where ρ is the radius of the inner circle ; but $AQ = b - CQ = b - ab/s$, Art. 6, Ex. 3 ; similarly, $BP = a - ab/s$; substituting these values and reducing we get $\dfrac{AN}{BN} = \dfrac{s-a}{a} \Big/ \dfrac{s-b}{b}$. Also, $AN/BN = $ the ratio of the perpendiculars from N on the sides b and a respectively. (Euc. III. 22.)

Similarly, the ratios of the perpendiculars from L and M on the corresponding pairs of sides of ABC are $\dfrac{s-b}{b} \Big/ \dfrac{s-c}{c}$ and $\dfrac{s-c}{c} \Big/ \dfrac{s-a}{a}$; therefore, etc., by Art. 65.]

*17. If circles are described as in Ex. 16 touching the circum-circle externally in points L', M', N', the triangles ABC and $L'M'N'$ are in perspective.

*18. The centres of perspective in Exs. 16 and 17 are respectively the isogonal conjugates of the centres of perspective of ABC and

---

* Professor de Longchamps, *Educ. Times*, July, 1890.

the triangle formed by joining the internal points of contact of the escribed circles with the sides (*point de Nagel*) ; and of ABC and the triangle formed by joining the points of contact of the in-circle with the sides (*point de Gorgonne*).

[Make use of the property given in Art. 64, Ex. 3.]

19. The nine-points-circle of a triangle touches the in- and three ex-circles.

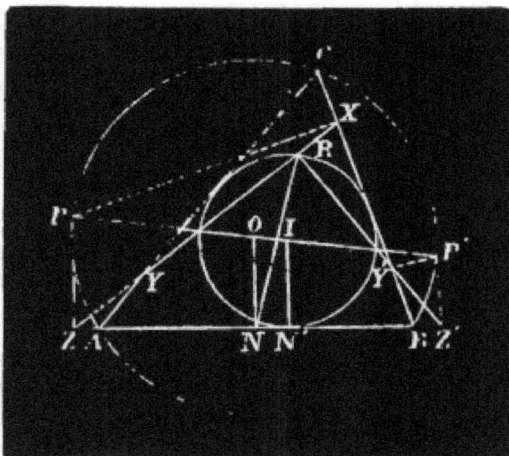

[Let ABC be the triangle, O and I the centres of the circum- and in-circles, PP' the common diameter, XYZ and $X'Y'Z'$ the Simson lines of P and P', R their point of intersection, L, M, N the middle points of the sides, L', M', N' the points of contact of the in-circle.

Since $OP = OP''$, $NZ = NZ'$. But the Simson lines of two points diametrically opposite meet at R at right angles on the nine points circle ; therefore $NZ = NZ' = NR$. Again, $OP/OI = NZ/NN'$ $= NR/NN'$; therefore $NR/NN' = MR/MM' = LR/LL'$; hence it follows that R is a limiting point of the in-circle and the circum-circle of the triangle LMN. See Art. 83 Note. This elegant proof of the well-known property is due to M'Cay.]

20. A variable circle O, ρ touches two circles A, r_1 ; B, r_2 ; prove that the polar M of its centre with respect to either (A, r) envelopes a fixed circle.

[Since it touches the two circles, it cuts their radical axis L at a constant angle (Art. 88, Ex. 7), or $\rho/OL = $ const. Draw a parallel

L' to L such that $\rho/OL=r_1/LL'$, then each of these ratios $=AO/OL'$. Let P be the pole of L' with respect to A, r_1; by Salmon's Theorem, we have $AO/OL'=AP/PM$, therefore PM is constant, and the envelope of M is the circle described with P as centre and PM as radius.]

Note.—If four positions O_1, O_2, O_3, O_4 of the centre and their corresponding polars M_1, M_2, M_3, M_4 are taken; since the anharmonic ratios made by the four tangents on any variable one M is constant, therefore (Art. 80, 9°), the envelope circle reciprocates into a curve of such a nature that the anharmonic ratios of the pencils joining four fixed points on it to a variable fifth are equal. This we have seen Art. 81, Ex. 3, to be a conic section; and the ratio AO/OL' is the eccentricity, A the focus, and L' the directrix of the conic.

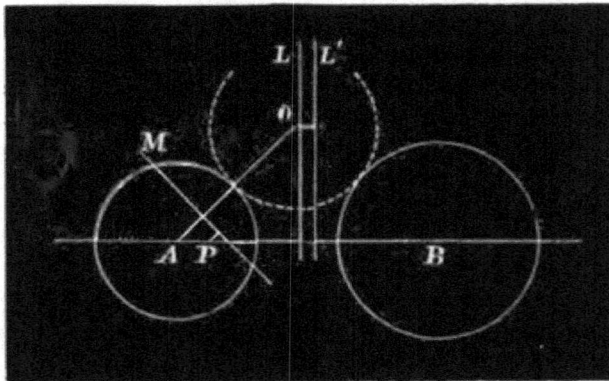

89. **Theorem.**—*A straight line is drawn to meet two circles A, r_1; B, r_2 in points X, X' and Y, Y' respectively, to prove that the tangents at these points intersect in four points P, Q, R, S which lie on a circle coaxal with the given ones.*

Let α and β be the angles of intersection of the line with the circles. Then

$$\sin\alpha/\sin\beta=PY'/PX=QY/QX'=RY/RX=SY'/SX';$$

therefore, since the ratios of the tangents $(t_1 : t_2)$ from each of the points P, Q, R, S to the given circles are

equal, they lie on a coaxal circle, whose centre C is given
by the relation $\qquad \dfrac{AC}{BC}=\dfrac{\sin^2\beta}{\sin^2\alpha}=\dfrac{t_1^2}{t_2^2}.$ $\qquad$ (Art. 88, Cor. 2)

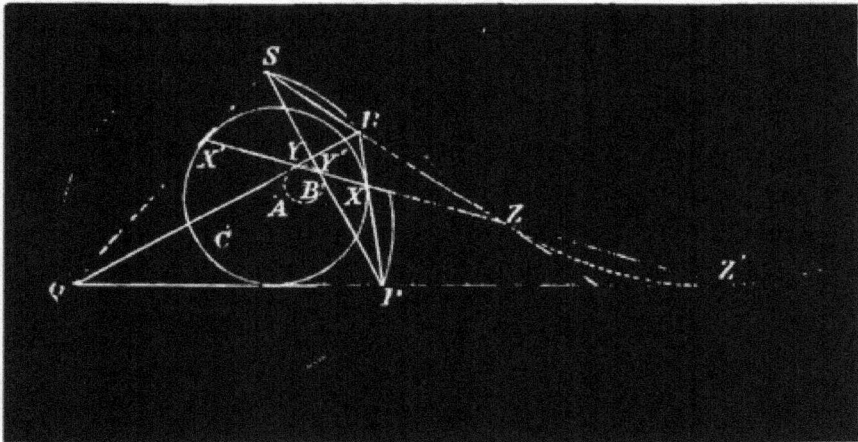

Cor. 1. Since $\sin\alpha = XX'/2r_1$ and $\sin\beta = YY''/2r_2$, we
have by division

$$t_1/t_2 = \sin\beta/\sin\alpha = YY''/XX' \div r_2/r_1 ; \ldots\ldots\ldots(1)$$

therefore, *if the intercepts made by two fixed circles on a
variable line are in a constant ratio* (XX'/YY'), *the
tangents at the points of intersection meet on a fixed
circle coaxal with the given ones.*

Cor. 2. If the intercepts in Cor. 1 have the ratio of the
radii $t_1 = t_2$, $\alpha = \beta$, C is at infinity, and the locus of the
intersection of the tangents is the radical axis.

Cor. 3. If the intercepts are in the sub-duplicate ratio
of the radii $XX'^2/YY'^2 = r_1/r_2$, then

$$t_1^2/t_2^2 = r_1/r_2 = AC/BC,\text{*}$$

---

* The two points C_1 and C_2 satisfying this relation are easily seen to
be the points of intersection of the direct and transverse common
tangents to the two circles and are called their *Centres of Similitude.*
The corresponding coaxal circles are the *External and Internal Circles
of Anti-similitude of the two given ones.*

hence the circle coaxal with two given ones whose centre divides the distance between their centres in the ratio of the radii is the locus of a point, the tangents from which to the given circles are in the sub-duplicate ratio of the radii.

COR. 4. If the intercepts are equal, $XX' = YY'$, the tangents are in the ratio of the radii and the locus of their intersection is called the *Circle of Similitude* of the given ones; its centre C is given by the equation

$$AC/BC = r_1^2/r_2^2, \dots\dots\dots\dots\dots\text{(Cor. 1.) (1)}$$

COR. 5. Since AB is divided internally and externally in C_2 and C_1 such that $\dfrac{AC_1}{BC_1} = \dfrac{AC_2}{BC_2} = \dfrac{r_1}{r_2}$ and again in C, by Cor. 4, such that $\dfrac{AC}{BC} = \dfrac{r_1^2}{r_2^2}$, it follows (Art. 70) that C is the middle point of the segment C_1C_2 and that the circle of similitude is the circle on it as diameter.

COR. 6. If the line $XX'YY'$ passes through the intersections (QS, PR and PS, QR) of opposite connectors of the quadrilateral; when PQ and RS are parallel; the circles A and B reduce to points and are therefore the limiting points of the system; *i.e. the common pair of inverse points of the circum-circle of the trapezium PQRS and that touching the parallel sides at Z and Z'.* (Art. 72, Ex. 13.)

EXAMPLES.

1. Any line meeting a pair of opposite sides of a cyclic quadrilateral at equal angles makes equal angles with each of the remaining pairs (Euc. III. 21, 22); intersects them in points XX', YY', ZZ' such that the circles touching the pairs of opposite con-

nectors at these points are coaxal with the given one ; and one of them lies on the side of the radical axis opposite to the other two.

2. A variable quadrilateral inscribed in a circle moves so that a pair of opposites envelope a circle, then each of the remaining pairs of opposites always touch circles coaxal with the given ones.

3. A variable triangle *ABC* is inscribed in a circle of a coaxal system, and two of its sides each envelope a circle of the system ; to prove that the third side *AC* envelopes another.

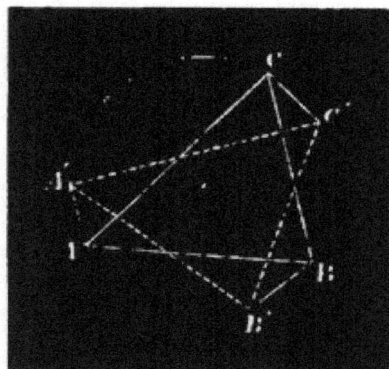

[Let *A'B'C'* be any other position of the given triangle. Then *ABA'B'* is a cyclic quadrilateral, and one pair of opposites *AB* and *A'B'* touch a given circle, therefore *AA'* and *BB'* touch one circle of the system.

Similarly *BB'* and *CC'* touch one circle of the system. But *BB'* can touch only one circle of the group on either side of the radical axis, Art. 92, Ex. 6 ; hence *AA'*, *BB'*, *CC'* touch the same circle. Now consider the quadrilateral *AA'CC'* ; it is obvious by Ex. 2 that *AC* and *A'C'* touch one circle ; therefore the envelope of *AC* is a coaxal circle.*]

4. **Poncelet's Theorem.**—If a variable polygon inscribed in a circle of a coaxal system moves so that all the sides but one touch fixed circles of the system, the last side also touches in every position a fixed circle of the system.

[By Ex. 3.]

---

* Dr. Hart, *Quarterly Journal*, Vol. II. p. 143.

5. The problem " to describe a polygon having all its vertices on a given circle and all its sides touching another " is either impossible or indeterminate.

[Let all the circles in **Ex.** 4 touching all the sides but one of the polygon coincide ; it follows therefore that if the last side touches this circle in one position it touches it in every position.]

6. To find the relation connecting the radii r_1 and r_2 of two circles and the distance δ between their centres so that a quadrilateral may be inscribed to one and circumscribed to the other. (Art. 88, Ex. 1.)

[By **Ex.** 5, when this is possible the position of the quadrilateral is indeterminate. Assuming it to have the position of symmetry, *i.e.*, with a pair of opposite vertices at the extremities of the common diameter, and θ the angle between any side and this diameter. By right-angled triangles we have the relations

$$\frac{r_1}{r_2 - \delta} = \sin \theta \text{ and } \frac{r_1}{r_2 + \delta} = \cos \theta$$

squaring and adding these results

$$\frac{1}{(r_2 - \delta)^2} + \frac{1}{(r_2 + \delta)^2} = \frac{1}{r_1^2}.]$$

7. If $A, r_1,\ B, r_2,\ C, r_3$ be three coaxal circles such that a variable quadrilateral whose pairs of opposite sides envelope A and B is inscribed in C, prove that

$$\frac{r_2^2}{(r_3 - \delta_1)^2} + \frac{r_1^2}{(r_3 + \delta_2)^2} = 1$$

where δ_1 and δ_2 denote the distances AC and BC.

[By the method of **Ex.** 6.]

8. If a variable line L meet two circles $Ar_1,\ Br_2$ so that the chords intercepted, $2c$ and $2c'$ are in a constant ratio κ ; to show that two points $A',\ B'$ may be found on the line AB to satisfy the relation

$$A'L \cdot B'L = \text{const.}$$

[For $c^2 = r_1^2 - AL^2, \quad c'^2 = r_2^2 - BL^2,$

hence $r_1^2 - AL^2 = \kappa^2(r_2^2 - BL^2),$

or $(AL + \kappa BL)(AL - \kappa BL) = \text{const.},$

but $AL + \kappa BL = (1 + \kappa)A'L,$

and
$$AL - \kappa BL = (1 - \kappa)B'L,$$

where A' and B' divide the line AB internally and externally in the ratio $\kappa : 1$.]

Note.—The variable line in the present article is thus seen to envelope a conic of which the points A' and B' are the foci.

90. We have seen, Art. 86, that in general three circles have but one common orthogonal circle, and in the particular case when more than one can be drawn the three form a coaxal system.

This property is sometimes of use in determining whether circles are coaxal, and may be regarded as a criterion of coaxality. The following illustrations are due to Walker.

91. Let ABC be a triangle and XYZ any transversal to its sides. Join AX, BY, CZ. These lines are drawn from the vertices of each of the four triangles AYZ, BZX, CXY, ABC, and terminated by the opposite sides; therefore, Art. 88, Ex. 12, the orthocentres of the four triangles are each the radical centre of the circles described on AX, BY, CZ as diameters.

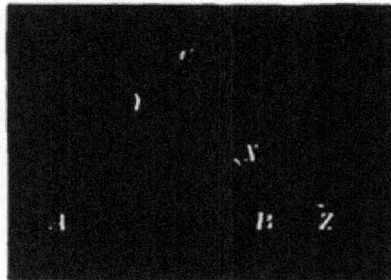

Hence we have the following theorems:—

1°. The orthocentres of the four triangles formed by any four lines are collinear.

$2°$. The middle points of the diagonals AX, BY, CZ of a complete quadrilateral are collinear.

$3°$. The line of collinearity of the orthocentres is at right angles to the line in $2°$, called the *Diagonal Line of the Quadrilateral*.

$4°$. The circles on the three diagonals as diameters are coaxal.

$5°$. The polar circles of the four triangles belong to the conjugate coaxal system.

EXAMPLES.

1. A, B, C, D are the vertices of a convex quadrilateral taken in order; A_e, B_e, C_e, D_e and A_i, B_i, C_i, D_i the external and internal bisectors of the angles; prove that

$a°$. The sixteen centres of the circles touching the sides of the four triangles formed by taking the sides of the quadrilateral in triads, lie in fours on these bisectors.

$\beta°$. The following groups of quadrilaterals are cyclic :—

$$\left.\begin{array}{l} A_e \ B_i \ C_i \ D_e \\ A_i \ B_e \ C_e \ D_i \end{array}\right\}(a) \qquad \left.\begin{array}{l} A_i \ B_i \ C_e \ D_e \\ A_e \ B_e \ C_i \ D_i \end{array}\right\}(c).$$

$$\left.\begin{array}{l} A_i \ B_e \ C_i \ D_e \\ A_e \ B_i \ C_e \ D_i \end{array}\right\}(b) \qquad \left.\begin{array}{l} A_e \ B_e \ C_e \ D_e \\ A_i \ B_i \ C_i \ D_i \end{array}\right\}(d)$$

$\gamma°$. Groups (a) and (c) are coaxal, and groups (b) and (d) conjugately coaxal.

[These properties are proved by employing Euc. III. 32 to show that any circle of either group is cut orthogonally by any circle of the other group. Russell.]

2^*. A, B, C, D are four points on a circle. Omitting each point in turn we have four triangles; prove that the sixteen centres of the circles touching the sides of these triangles lie in fours on four parallel lines, and also in fours on four lines each perpendicular

---

to the former set ; and that the two sets of lines are parallel to the bisectors of the angle between AC and BD. (M'Cay.)

3. ABC is a triangle, AA' a diameter of the circum-circle and H the orthocentre ; show that A' and H are equidistant from the base BC ; and hence deduce the theorem "the Simson line of any point is equidistant from the point and orthocentre of the triangle."

SECTION II.

ADDITIONAL CRITERIA OF COAXAL CIRCLES.

92. I. *Relation connecting the distances between the centres and the radii of three circles of a coaxal system.*

Let the circles be denoted by A, r_1; B, r_2; C, r_3.

Then for any point P on the radical axis, we have

$$BC . AP^2 + CA . BP^2 + AB . CP^2 = -BC . CA . AB ;$$

hence if t be the length of the tangent from P to the circles, since $AP^2 = r_1^2 + t^2$, etc., by substituting in this equation and reducing,

$$BC . r_1^2 + CA . r_2^2 + AB . r_3^2 = -BC . CA . AB, \dots \dots (1)$$

a result from which the radius r_3 of any circle of the system may be found when the position of its centre is known ; and conversely.

COR. 1. If $r_3 = 0$, C is a limiting point (Art. 87), by letting $AC = x$ in (1) we obtain a quadratic in x, the last term of which is r_1^2. Hence the *product of the distances of the limiting points from the centre of any circle of the system = the square of its radius.* Cf. Art. 87.

COR. 2. If $r_2 = r_3 = 0$, the criterion reduces to

$$AB . AC = r_1^2.$$

EXAMPLES.

1. If t_1, t_2, t_3 denote the tangents from any point P to three circles of a coaxal system; to prove that
$$BC \cdot t_1^2 + CA \cdot t_2^2 + AB \cdot t_3^2 = 0.$$
[For $\quad BC \cdot AP^2 + CA \cdot BP^2 + AB \cdot CP^2 = -BC \cdot CA \cdot AB,\ldots\ldots\ldots(1)$
and $\qquad BC \cdot r_1^2 + CA \cdot r_2^2 + AB \cdot r_3^2 = -BC \cdot CA \cdot AB\ldots\ldots\ldots(2)$
Subtracting (2) from (1); therefore, etc.]

2. Deduce as a particular case of Ex. 1 the theorem :—The locus of a point, the tangents from which to two given circles are in a constant ratio, is a coaxal circle.
$$[\text{Let } t_3 = 0.]$$

3. Explain the formula of Ex. 1 when $t_2 = t_3 = 0$.

4. Find the locus of a point P if the product of the tangents from it to two circles bears a constant ratio to the square of the tangent to any circle coaxal with them $(kt_1t_2 = t_3^2)$.

[In Ex. 1, substituting the given condition, the equation reduces to the form $(t_1 - mt_2)(t_1 - nt_2) = 0$; hence P describes two coaxal circles, since the ratio of the tangents t_1 and $t_2 = m$, or n.]

5. If the common tangent ZZ' to two circles meet a coaxal circle in the points A and B; to prove that MZ and MZ' are the bisectors of the angles subtended by the chord AB at either limiting point.

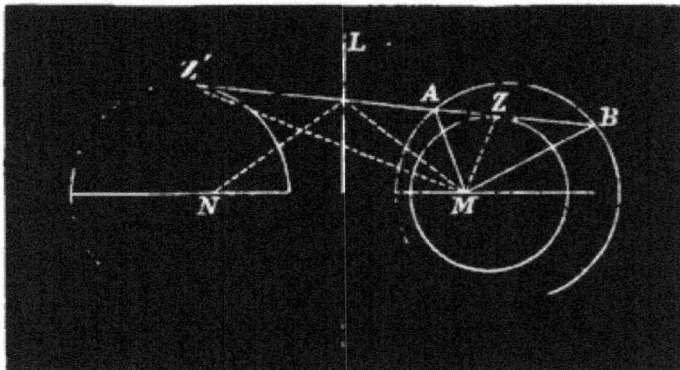

[For AZ, AM and BZ, BM being pairs of tangents drawn from two points A and B on the same circle to two circles of the system, it follows that $AZ/AM = BZ/BM$, by alternation $AM/BM = AZ/BZ$, and for a similar reason $= AZ'/BZ'$; therefore, etc.]

6. To describe two circles of a coaxal system touching a given line.

[In Ex. 5 divide the line AB internally and externally in Z and Z' in the given ratio AM/BM; therefore Z and Z' are the required points of contact. It will be noticed that the circles lie one on each side of the radical axis.]

7. A triangle ABC is inscribed in a circle of a coaxal system; prove that the points of contact X, X', Y, Y', Z, Z' of the three pairs of circles of the system which touch the sides BC, CA, and AB respectively,

$a°$. Lie three and three on four lines,

$\beta°$. Connect with the opposite vertices by six lines, passing three and three through four points.

[Apply the relations in Ex. 5 to the three sides; therefore, etc. Arts. 62 and 63.]

8. Apply the criterion of the Article to show that the nine-points-, circum- and polar circles are coaxal.

9. If points B and D are taken on any two circles whose centres are O and O' and joined to the limiting point M such that BMD is a right angle, the locus of the intersection of tangents at B and D to the circles is a coaxal circle.

Whence,
$$\frac{MB \cdot MC}{MA \cdot MD} = \frac{MO}{MO'} \quad\dots\dots\dots\dots\dots(1)$$

But since $BMD = 90°, \ AMC = 90°$ (Art. 72, Cor. 8),

and therefore $BMC + AMD = 180°$;

hence $\dfrac{BC}{AD} = \dfrac{MB \cdot MC}{MA \cdot MD} = \dfrac{MO}{MO'}$,

by (1), a constant quantity ; therefore, etc. (cf. Art. 89, Ex. 8).]

10. A quadrilateral *PQRS* is inscribed to one circle and escribed to another at the points *A*, *B*, *C*, *D*; prove that its position is *indeterminate*, and the diagonals *PR* and *QS*, *BC* and *AD* of the two cyclic quadrilaterals intersect (the latter at right angles) at the limiting point *M*.

[By Art. 89, Ex. 6. See also Art. 88, Ex. 1, and Art. 67, Cor. 6.]

11. Construct a quadrilateral in a given circle symmetrical with respect to a given diameter and circumscribed to a circle having its centre at a fixed point on the diameter.

[Find the radius of the second circle by Art. 89, Ex. 6.]

93. II. *A variable circle cuts three others of a coaxal system at angles a, β, γ, to prove the relation*

$$BC \cdot r_1 \cos a + CA \cdot r_2 \cos \beta + AB \cdot r_3 \cos \gamma = 0.$$

Let P, ρ be the variable circle meeting the given ones at the points R, S, T respectively; join PR, PS, PT, and produce the lines to meet the circles again in R', S', T'.

By Art. 92, Ex. 1, $BC \cdot t_1^2 + CA \cdot t_2^2 + AB \cdot t_3^2 = 0$, but $t_1^2 = PR \cdot PR' = \rho(\rho + RR') = \rho(\rho + 2r_1 \cos a)$, with similar values for t_2 and t_3. Substituting these values in the equation and reducing, we obtain the required result.

Cor. 1. If two of the circles are cut orthogonally, every circle of the system is cut orthogonally. For if $a = \beta = 90°$, two terms of the equation vanish, therefore $AB \cdot r_3 \cos \gamma = 0$ or $\gamma = 90°$.

Cor. 2. If the variable circle touch two of the given ones, it cuts the circle C, r_3 coaxal with them at an angle determined by the equation $AB.r_3\cos\gamma = \pm BC.r \pm CA.r_2$; like signs being taken when the contacts are similar and unlike signs when the contacts are dissimilar. The four possible values arising from the selections of sign on the right side of the equation give the values of γ corresponding to each assigned species of contact.

COR. 3. In Cor. 2, if $\cos\gamma = 0$, the centres C of the particular circles of the system which are cut at right angles are given by the relation

$$BC.r_1 \pm CA.r_2 = 0,$$

or $$AC/BC = \pm r_1/r_2.$$

Hence, the variable circle having similar contacts with two given circles cuts at right angles the coaxal circle whose centre is their external centre of similitude; and, if the contacts are dissimilar, the coaxal circle whose centre is the internal centre of similitude.

COR. 4. If $a = \pm\beta$ and $\gamma = 90$, the equation reduces to $AC/BC = \pm r_1/r_2$, as in Cor. 3. Hence, *the variable circle cutting two others at equal or supplemental angles cuts at right angles their external or internal circle of antisimilitude respectively.*

COR. 5. Let the radius of the variable circle be infinite; hence (Cor. 3) all lines cutting two circles at equal or supplemental angles are diameters of their external or internal circles of antisimilitude.

EXAMPLES.

1. To describe a circle cutting any three circles A, r_1; B, r_2; C, r_3 at given angles a, β, γ.

[The required circle cutting B, r_2; C, r_3 at given angles, therefore touches a known circle coaxal with them by Cor. 2; similarly for each of the remaining pairs of the given circles; hence the problem reduces to "*describe a circle touching three given circles with assigned contacts.*" There .are in consequence eight solutions. These are given in a subsequent chapter.]

2. Show that Ex. 1 cannot be reduced to describing a circle cutting three given circles orthogonally.

[For let X be the circle coaxal with B and C which is cut orthogonally by the required circle, and constructed by putting $\gamma = 90$ in the relation of the present Article; similarly let Y coaxal with C and A, and Z coaxal with A and B, be circles cut orthogonally by it. Their centres, being found by the relations

$$\frac{BX}{CX} = \frac{r_3 \cos \gamma}{r_2 \cos \beta}, \quad \frac{CY}{AY} = \frac{r_1 \cos \alpha}{r_3 \cos \gamma}, \quad \frac{AZ}{BZ} = \frac{r_2 \cos \beta}{r_1 \cos \alpha},$$

are collinear, Art. 62, and their common orthogonal circle therefore indeterminate.]

3. A variable circle P, ρ touches two others A, r_1; B, r_2; show that the square of the common tangent t, to it and any third circle C, r_3 coaxal with them, varies as its radius ($t^2 \propto \rho$).

[By Cor. 2 it cuts C, r_3 at a constant angle γ. But (Art. 4 (1)) $4 \sin^2 \frac{1}{2}\gamma = t^2 / \rho \cdot r_3$; therefore, etc. In the particular case when C, r_3 is a limiting point we have the theorem :—"*if a variable circle touch two fixed circles, its radius is in a constant ratio to the square of the tangent to it from either of the limiting points.*" Also, "*the ratio of the tangents from the limiting points is constant.*"]

4. A variable circle cuts two fixed circles at angles α and β, tangents are drawn from its centre to the circles, and tangents t_1 and t_2 from the points of contact to the variable circle; prove that

$$t_1^2 / t_2^2 = r_1 \cos \alpha / r_2 \cos \beta,$$

and deduce the properties of Ex. 3 as particular cases (Preston). See *Spherical Trigonometry*, Art. 159, Ex. 15.

5. Find the locus of the centre of a circle cutting any three circles at equal or supplemental angles.

[By Cor. 4.]

6. The vertex and base of a triangle are fixed in position and the vertical angle given in magnitude; find the envelope of the circumcircle.

SECTION III.

CIRCLE OF SIMILITUDE.

94. Let A, r_1 ; B, r_2 be any two circles, Z and Z' the points of section of AB such that

$$\frac{AZ}{BZ} = \frac{AZ'}{BZ'} = \frac{r_1}{r_2};$$

then the segments AB and ZZ' divide each other harmonically, and the circle C, r_3 on ZZ' as diameter is termed their *Circle of Similitude*. The points Z and Z' are the *Internal* and *External Centres of Similitude*.

95. The circle of similitude has the following fundamental properties :—

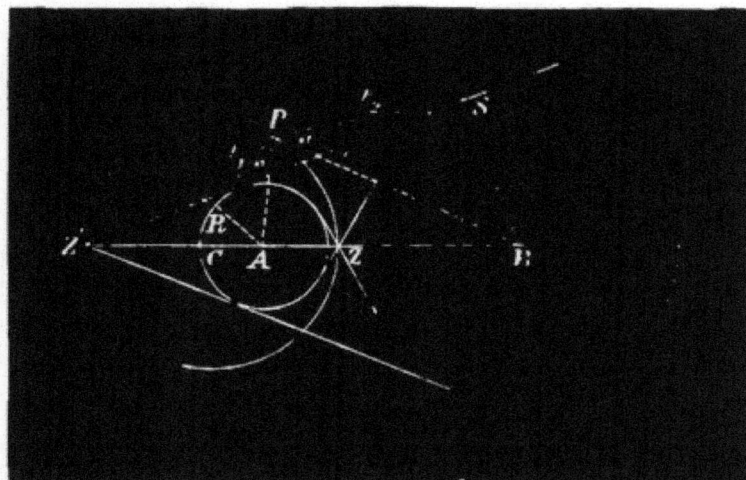

1°. Its centre C and radius r_3 are connected by the relation $CA \cdot CB = r_3^2$ (Art. 70), or *the centres of the given circles are inverse points with respect to their circle of similitude.*

2°. The points Z and Z' are the intersections of the transverse and direct common tangents.

3°. It is coaxal with the given circles.

[For Z and Z' are on the same circle coaxal with A and B, since the ratios of the tangents from them are each equal to the ratio of the radii, and only one circle coaxal with A, r_1 and B, r_2 can contain these points, viz. that on the line ZZ' as diameter.]

4°. From Cor. 3 it is the locus of a point such that the tangents drawn from it to the circle have the constant ratio of the radii.

[Cf. Art. 88, Cor. 2.]

This follows independently, since PZ and PZ' are the bisectors of the angle APB, hence

$$PA/PB = AZ/BZ = AR/BS ;$$

therefore, etc., by Euc. VI. 7.

5°. The circles subtend equal angles at any point on it. (By 4°.)

6°. In the particular case when the circle B, r_2 becomes a right line the centre B is at infinity, its inverse A (Cor. 1) coincides with C, therefore *the centres of similitude of a line and circle are the extremities of the diameter of the circle perpendicular to the line.*

EXAMPLES.

1. The circles of similitude of any three circles taken in pairs are coaxal.

[Their centres are collinear, Art. 72, Ex. 21 ; therefore, etc., Art. 88, Ex. 13, 2°.]

2. A circle cuts two at angles α and β ; find the angle it makes with their circle of similitude.

3. The tangents from any point P on the circle of similitude to the circles A, r_1 and B, r_2 meet them at R and S; prove ($a°$) the chords which the circles intercept on the line RS are equal to one another; ($\beta°$) The tangents from R and S to the circles B and A are equal.

[Compare Art. 89, Cor. 4.]

4. The circle on the third diagonal of a complete cyclic quadrilateral is the circle of similitude of those described on the remaining two.

[Let $ABCD$ be the quadrilateral, LMN its diagonal line, PP' the third diagonal, $BD = 2r_1$, $CA = 2r_2$, $PP' = 2r_3$. Join PM, PN.

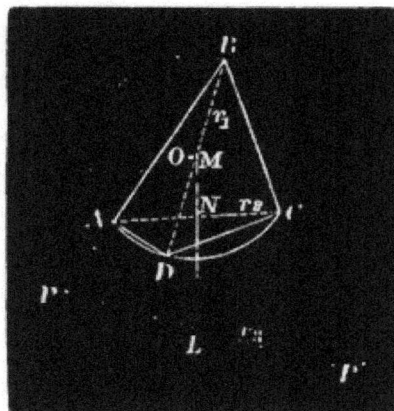

The triangles PAC and PBD are similar, Euc. III. 21; hence, since PN and PM are homologous lines, PBM and PCN are similar; therefore $PM/PN = r_1/r_2$. Similarly, $P'M/P'N = r_1/r_2$; therefore P and P' lie on a circle to which M and N are inverse points. Also the circles on the three diagonals are coaxal; therefore, etc. It follows also by 1° that $LM . LN = r_3^2$.]

5. Having given the three diagonals of a cyclic quadrilateral; to construct it.

[Let O be the centre of the circle and r_1, r_2, r_3 the diagonals. By Ex. 4 $LM . LN = r_3^2$, and is therefore known. Also $LM/LN = r_1^2/r_2^2$; hence the lines LM and LN are determined. $LM = r_1 r_3/r_2$, $LN = r_1 r_2/r_3$, and $MN = \dfrac{r_3}{r_1 r_2}\left(\dfrac{r_1^2 - r_2^2}{r_1 r_2}\right)$. But OM and ON are known (Euc. I. 47), consequently the triangle OMN is completely determined.]

6. Six circles pass through two points P and Q on the circumcircle of a triangle ABC and touch the sides; prove that the points of contact X, X'; Y, Y'; Z, Z' lie in threes on four lines.

[Let the line joining the points P and Q cut the sides of the triangle in L, M, and N respectively, and we have obviously $LX = LX'$ and $LB \cdot LC = LX^2 = LX'^2$, with similar relations on the remaining sides of the triangle; therefore, etc.]

7. From any point on a given line tangents are drawn to a circle; a circle is described touching the fixed circle and variable pair of tangents to it; prove that the envelope of the polar of its centre is a circle.

8. The circle of similitude of the circum- and nine-points-circle of a triangle is that described on the interval between the centroid and orthocentre as diameter.

[Let O be the circum-centre, H orthocentre, N the nine-points centre, and E the centroid. By a well-known property of these four collinear points $OE/NE = OH/NH = 2 =$ ratio of radii of circum- and nine-points-circles; therefore, etc.

[It is called the *Orthocentroidal Circle* of the triangle.]

MISCELLANEOUS EXAMPLES.

1. Prove that the equation of the two circles touching three given ones with contacts of similar species are

$$\overline{23}\sqrt{S_1} + \overline{31}\sqrt{S_2} + \overline{12}\sqrt{S_3} = 0,$$

where S_1, S_2, S_3 denote the powers of any point on either of the tangential circles with respect to the given ones.

2. If a variable chord AB of a circle is such that the sum of the tangents from A and B to another given circle is proportional to the length of AB, it envelopes a circle coaxal with the two.

3. If a variable circle touches two fixed circles and cuts a circle concentric with either in the points A and B: required to find the envelope of AB. (Dublin Univ. Exam. Papers, 1891.)

[Applying Casey's relation between the common tangents to four circles to the points A and B and the two given circles, it follows by Ex. 2 that the envelope of AB is a coaxal circle.]

4. Prove that the circles cutting three given ones orthogonally passing through their circles, and bisecting the circumferences are coaxal.

5. Reciprocate the following theorem from a limiting point :—The square of the distance of any point on a circle from a limiting point varies as its distance from the radical axis.

[The rectangle under the distances of the foci from any tangent to a conic is constant.]

6. Prove that the limiting points of any two circles lie on a pair of opposite connections of their common escribed quadrilateral.

7. If δ denote the distance between the limiting points and γ the length of their imaginary common chord, prove that $\delta = i\gamma$.

8. If two circles whose radii are r_1 and r_2 are so related that a hexagon can be inscribed to one and circumscribed to the other, then

$$\frac{1}{(r_1^2 - \delta^n)^2 + 4r_1r_2^2\delta} + \frac{1}{(r_1^2 - \delta^2)^2 - 4r_1r_2^2\delta} = \frac{1}{2r_2^2(r_1^2 + \delta^2) - (r_1^2 - \delta^2)^2}$$

9. If an octagon can be inscribed to one and circumscribed to the other,

$$\left\{ \frac{1}{(r_1^2 - \delta^2)^2 + 4r_2^2 r\delta} \right\}^2 + \left\{ \frac{1}{(r_1^2 - \delta^2)^2 - 4r_2^2 r\delta} \right\}^2 = \left\{ \frac{1}{2r_2^2(r_1^2 + \delta^2) - (r_1^2 - \delta^2)^2} \right\}^2.$$

10. The mean centre of the vertices of a cyclic quadrilateral lies on the circumference of the nine-points-circle of the harmonic triangle of the quadrilateral. (Russell.)

11. If a variable polygon is inscribed to one circle and escribed to another, the locus of the mean centre of any number (r) of consecutive points of contact is a circle. (Weill). Cf. Art. 53, Ex. 12.

12. Prove the following extension of Weill's theorems :—If a variable polygon of any order be inscribed in a circle of a coaxal

system having all its sides touching respectively fixed circles of the system; there exists a set of multiples for which the mean centre of the points of contact of the sides with the circles is a fixed point.

[Let any circle of the system be denoted by (O, r, δ) where δ is the distance of its centre from the circumcentre of the polygon, and let a, β, γ, and c be the displacements of the points of contact of the sides AB, BC, CD, etc. for consecutive positions. Then, by Art. 53, Ex. 12, we have

$$\frac{\frac{\sqrt{\delta_1}}{r_1} a}{AB} = \frac{\frac{\sqrt{\delta_2}}{r_2} \beta}{BC} = \frac{\frac{\sqrt{\delta_3}}{r_3} \gamma}{CD} = \text{etc.}$$

hence the mean centre of the points of contact remains fixed for the system of multiples $\sqrt{\delta_1}/r_1$, $\sqrt{\delta_2}/r_2$, $\sqrt{\delta_3}/r_3$, etc.]

12a. The locus of the mean centre of r consecutive points of contact for their respective multiples is a circle.

[For, join the extremities of the r sides thus forming a polygon of $r+1$ sides, and let the last side touch a fixed circle $(O_{r+1}, r_{r+1}, \delta_{r+1})$ of the system. (Art. 89, Ex. 4.) By Ex. 12, the mean centre of the $r+1$ points of contact for the corresponding multiples is a fixed point (X). Let Y be the mean centre for the r points and Z the point of contact of the last side. Then Y divides the line XZ in a constant ratio, and since Z describes a circle, therefore, etc.] *

---

* The following is an independent proof of the generalization of Weill's theorem.

Let $ABCD$.. and $A'B'C'D'$... be *any* two positions of the variable polygon; T_1, T_2, T_3, T_1', T_2', T_3' ... points of contact of the sides AB, BC, ... ; $A'B'$, $B'C'$, ... with the corresponding circles O_1, r_1, δ_1; O_2, r_2, δ_2, ... of the system; R the point of intersection of AB and $A'B'$ and θ the angle between them; S the intersection of AA' and BB', and ϕ the angle between them. Then AA', BB', CC' ... touch a circle (Ω, ρ, λ) coaxal with the given system. Let L, M, N ... be its points of contact with AA', BB', CC', etc. ... and we have

$$\frac{T_1 T_1'}{LM} = \frac{r_1 \sin \frac{1}{2}\theta}{\rho \sin \frac{1}{2}\phi} = \frac{r_1}{\rho} \cdot \frac{BM}{BT_1} = \frac{r_2}{\rho} \cdot \frac{\sqrt{\lambda}}{\sqrt{\delta_1}},$$

therefore $\quad \frac{\sqrt{\delta_1}}{r_1} \cdot T_1 T_1' \ / \ LM = \frac{\sqrt{\delta_2}}{r_2} \cdot T_2 T_2' \ / \ MN = \text{etc.}$

13. If the diagonals of a cyclic quadrilateral are conjugate lines and a homothetic quadrilateral be described with their intersection as homothetic centre ; prove that the consecutive pairs of sides of the one quadrilateral intersect the corresponding pairs of the other in eight points which lie on a circle coaxal with the circum-circles of the quadrilaterals.   See Art. 96.

[Use the theorem of Art. 92, Ex. 2.]

---

i.e.,   multiples $\sqrt{\delta_1}/r_1$, $\sqrt{\delta_2}/r_2$, $\sqrt{\delta_3}/r_3$ of the displacements $T_1 T_1'$, $T_2 T_2'$ ... are proportional to the sides of the polygon ; therefore, etc. Bowesman.]

CHAPTER IX.

SECTION I.

TWO SIMILAR FIGURES.

96. Two figures similar and similarly placed are said to be *Homothetic,* and their homologous parts are called *Corresponding Points, Lines,* etc. It is plain, if a line of either figure is displaced through an angle θ, that every line of it is displaced through the same angle. For let AB be displaced to $A'B'$. It follows (Euc. III. 21, 22), since $B = B'$, that the angle between BC and $B'C'$ is equal to θ.

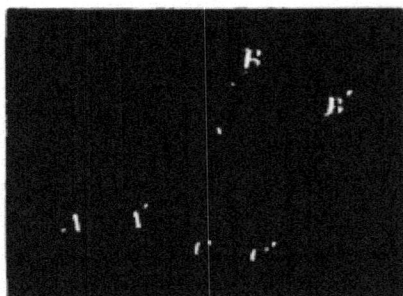

Also, since corresponding lines meet at equal angles, a variable pair of corresponding lines passing through a pair of corresponding points A and A' intersect on the circumference of a circle described on AA' containing an angle θ; and conversely.

Corresponding lines are made up of corresponding points; and the point of intersection of any two lines of either figure is the correspondent of the points of intersection of the corresponding lines of the other.

97. We have seen how to find a point S which, with the extremities of two linear segments AB and $A'B'$, forms similar triangles (Art. 25), and that it possesses the properties.

$a°$. A variable line XX' dividing the segments similarly $AX : BX = A'X' : B'X'$ subtends a constant angle at it; and

$\beta°$. Its distances from the lines are proportional to their lengths (Euc. VI. 19).

Now, if similar polygons be similarly described on AB and $A'B'$, it follows, as in Euc. VI. 20, that—

1°. The distances of S from each pair of corresponding lines are proportional to these lines.

2°. All pairs of corresponding points P and P' of the polygons subtend the same angle at it, and with it form a triangle of constant species.

3°. The polygons can be made homothetic by the revolution of either around it (2°).

For this reason it is called the *Homothetic Centre* of the Polygons, or their *Centre of Similitude.*

The ratio of SP to SP' is the *Ratio of Similitude* of the figures.

98. Since to each point P of one figure corresponds a point P' of the other such that PSP' is a triangle of constant species, if P coincides with S, P' also coincides with

it; and therefore S taken as a point of either figure is its own correspondent in the other.

Hence it is a *Double Point* of the polygons.

99. From these considerations we make the following inferences :—

I. If upon the lines joining a fixed point S to the vertices of any polygon F_1 similar and similarly situated triangles are constructed, their vertices form a polygon F_2 similar to the given one, and S is their double point.

II. If the lines joining corresponding points of two directly similar figures are divided in the same ratio, the points of section form a polygon similar to the given ones (H. Van Aubel).

III. If the vertices of a polygon, constant in species, move on curves of any nature, to each position of it there is a corresponding centre of similitude.

This is called the *Instantaneous Centre* for the position, and is such that the lines drawn from it to all points A, B, $C \ldots X$ of the figure make equal angles with the tangents at these points to their respective loci.

[This is seen by taking two indefinitely near positions of the polygon.]

IV. Reciprocally :—If the lines L, M, N of the figure, moving as in the previous case. envelope curves, the lines joining the contacts of any position to S make equal angles with L, M, N.

[For the points of contact are the intersections of two consecutive positions of the moveable figure and are therefore corresponding points.]

THREE SIMILAR FIGURES.

100. Let F_1, F_2, F_3 be any three directly similar figures; S_1 the double point of F_2 and F_3; S_2 and S_3 the double points of the remaining pairs F_3, F_1 and F_1, F_2; a_1, a_2, a_3 the lengths of corresponding lines d_1, d_2, d_3; a_1, a_2, a_3 the angles of the triangle $D_1 D_2 D_3$, whose sides are d_1, d_2, d_3.

Then, by Art. 96,

1°. The variable triangle $D_1 D_2 D_3$, formed by any three corresponding lines, is constant in species.

2°. The distances of S_1 from d_2 and d_3 are proportional to a_2 and a_3, and similarly for S_2 and S_3 (Art. 97 ($\beta°$)); therefore, the lines joining S_1, S_2, S_3 to the *corresponding* vertices of $D_1 D_2 D_3$ divide the angles D_1, D_2, D_3 each into constant parts, and are concurrent (Art. 65).

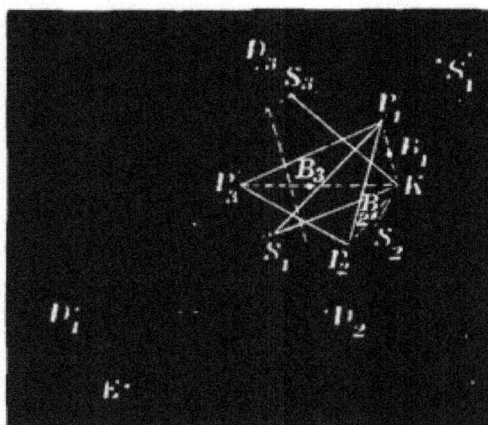

Hence *the triangle $S_1 S_2 S_3$, whose vertices are the centres of similitude of F_1, F_2, F_3 taken in pairs* ⟨Triangle of

Similitude), is in perspective with all homologous triangles $D_1D_2D_3$, *etc.* ; and the centre of perspective K is a point such that its distances from any triad of homologous lines are in the ratios $a_1 : a_2 : a_3$.

3°. Since the base angles of each of the triangles D_2D_3K, D_3D_1K, D_1D_2K are constant (Art. 100, 2°) as D_1, D_2, D_3 vary, the angles subtended by the sides of $S_1S_2S_3$ at K are each constant, and the locus of K is therefore the circum-circle ; hence,

Any triangle formed by three homologous lines is in perspective with $S_1S_2S_3$ *at a point on the circum-circle of the latter ; or the locus of the centre of perspective of* $S_1S_2S_3$ *and any triangle formed by three homologous lines is the circum-circle of the former.* This is called the *Circle of Similitude* of F_1, F_2, F_3.

4°. The chords KP_1, KP_2, KP_3 drawn parallel to d_1, d_2, d_3 are homologous lines, for they intersect at angles α, β, γ, and their distances from d_1, d_2, d_3 are in the ratios $a_1 : a_2 : a_3$.[*] Moreover, they meet the circle in fixed points, since the angle S_2KP_1 is constant and S_2 a fixed point ; therefore P_1 is fixed, and similarly P_2 and P_3 are fixed points.

They are termed the *Invariable Points*, and $P_1P_2P_3$ the *Invariable Triangle*, of F_1, F_2, F_3.

4°. May be enunciated as follows :—

All concurrent triads of homologous lines pass through the invariable points and intersect on the circle of similitude; and reciprocally :—*the lines joining* P_1, P_2, P_3 *to any three homologous points* B_1, B_2, B_3 *meet in a point on the*

---

[*] These lines are therefore the sides of an evanescent triangle $D_1D_2D_3$ of constant species.

circle of similitude; *and all triangles whose vertices are three homologous points are in perspective with* $P_1P_2P_3$ *and the locus of their centre of perspective is the circle of similitude.*

101. **Theorem.**—*The triangle of similitude and the invariable triangle are in perspective;* and the distances of the centre of perspective E from the sides of the latter are inversely, as the ratios $a_1 : a_2 : a_3$.

Since S_1 is its own correspondent with respect to F_2 and F_3, P_2S_1 and P_3S_1 are homologous lines and lengths of these figures, therefore

$$S_1P_2 : S_1P_3 = a_2 : a_3 \dots\dots\dots\dots(1)$$

but (Euc. III. 22) $S_1P_2 : S_1P_3$ as the distances of S_1 from P_1P_2 and $P_1P_3 = a_2 : a_3$ by (1), with similar relations for the points S_2 and S_3; therefore, etc., Art. 65.

102. **Theorem.**—*The invariable triangle is inversely similar to* $D_1D_2D_3$.

Follows by Euc. III. 22.

103. **Adjoint Points.***—Let S_1' be the point of F_1 which corresponds to S_1 of the figures F_2 and F_3.

Then $S_1'S_1S_1$ is a particular case of a triangle formed by three homologous points, and is therefore (Art. 100, 4°) in perspective with $P_1P_2P_3$ at a point on the circle of similitude; hence the lines P_1S_1', P_2S_1, P_3S_1 are concurrent. Their common point is therefore S_1; that is to say, P_1S_1' passes through E and S_1 (Art. 101); hence,

The lines $S_1S_1', S_2S_2', S_3S_3'$ *meet each other in* E *and the circle of similitude at the invariable points.*

---

* The theorems contained in Arts. 100-103 are due to Tarry. *Mathesis*, 1882, p. 72.

o

Defs. The point E is called the *Director Point*, and S_1', S_2', S_3' the *Adjoint Points* of F_1, F_2, F_3.

104. **Theorems.***—*In any three similar figures there exists an infinite number of triads of homologous points C_1, C_2, C_3 which are collinear.   2°. The loci of these points are circles passing through E.   3°. The variable line $C_1C_2C_3$ turns around E.*   Neuberg.

The triangles $S_1C_2C_3$, $S_2C_3C_1$, $S_3C_1C_2$ are constant in species (Art. 97, 2°); hence the angles $S_2C_1S_3, S_3C_2S_1, S_1C_3S_2$ are given, and therefore the loci of the points are circles passing through each pair of double points.

Again, since $S_2C_1C_2$ is a constant angle, the variable line C_1C_2 meets the locus of C_1 in a fixed point, and similarly it meets the loci of C_2 and C_3 in fixed points. Therefore the fixed points are coincident; that is to say, *the circular loci have a point in common.*

In the particular case of the collinear triads $S_1'S_1S_1$, $S_2S_2'S_2$, $S_3S_3S_3'$ it has been proved (Art. 103) that their lines of collinearity pass through E; therefore, etc.   The points S_1', S_2', S_3' are on the corresponding circles.

105. **Particular Cases.**—Let the three similar figures F_1, F_2, F_3 be described on the sides of a triangle ABC.   It has been shown that the middle points of the symmedian chords of the circum-circle † are the common vertices of directly similar triangles described on the sides, taken in pairs (Art. 25, Ex. 2), and they are therefore the three double points.   Hence,

1°. *Brocard's second triangle is the triangle of simili-*

---

tude, and the Brocard circle the circle of similitude, of three directly similar figures described on the sides of a triangle.

2°. Brocard's first triangle is their invariable triangle, Art. 29, Ex. 3.

3°. Brocard's second triangle and the given one are in perspective at a point on the circum-circle of the former whose distances from the sides of ABC are in the ratios of their lengths (Art. 100, 2°). See also Art. 16, Ex. 2.

4°. The centre of perspective is the symmedian point of ABC.

5°. The locus of the intersection of concurrent triads of homologous lines is the Brocard Circle, Art. 100, 4°.

6°. Brocard's first and second triangles are in perspective (Art. 101), and their centre of perspective E, or director points, is the centroid of ABC. (Art. 53, Ex. 6.)

7°. All collinear triads of homologous points lie on a variable line passing through E, and each point describes a circle passing through two vertices of Brocard's second triangle and the centroid of ABC.

M'CAY'S CIRCLES.

106. The loci in 7° of the previous Article are fully described by M'Cay in his memoir "On Three Circles related to a Triangle."* Amongst many other properties they possess those given in this and the following Article.

The notation employed is as follows:—ABC is the given triangle; $A_1B_1C_1$, $A_2B_2C_2$ Brocard's first and second triangles; E centroid; A', B', C' three homologous collinear points; M middle point of AB; H circum-centre;

A_3, B_3, C_3 the homologues of A_2, B_2, C_2 respectively as double points of F_1, F_2, F_3. P_{ac} the c correspondent of P regarded as an a point, and L_{ac} and L_{ab} the c and b

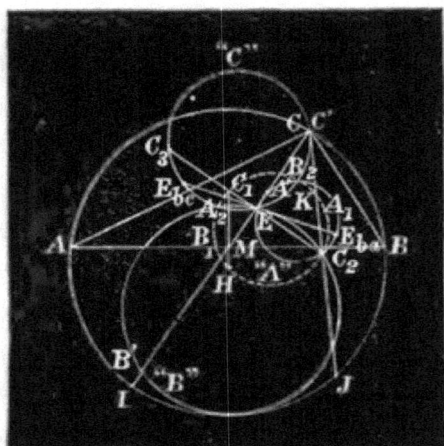

correspondents of any line L regarded as an a line; the circular loci the "A," "B," and "C" circles of the triangle.

1°. The mean centre of any three collinear homologous points is at E (Art. 53, Ex. 6).

2°. If one of them C' coincides with E, $A'B'$ is a tangent to the "C" circle and $EA' = EB'$ or $EF_{ca} = EE_{cb}$; similarly we have $EE_{ab} = EE_{ac}$ and $EE_{bc} = EE_{ba}$.

3°. If one of them coincides with a double point A_2, the line of collinearity is $A_2EA_1A_3$ (Art. 103) and $EA_3 = 2EA_2$.

Similarly the lines $B_1B_2B_3$ and $C_1C_2C_3$ each pass through E, which is the common point of trisection of the segments A_2A_3, B_2B_3, C_2C_3.

4°. The circles cut each other at angles A, B, and C.

5°. Their centres are on the perpendicular bisectors of the sides.

This is proved for the " C " circle as follows :—

On the sides of ABC construct three directly similar triangles BCA', CAB', ABC', each inversely similar to

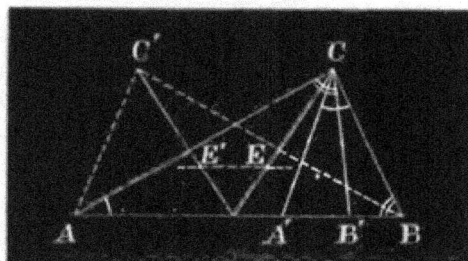

ABC. Their centres of gravity are therefore corresponding points. But they lie on a parallel through E to AB; hence E', the centroid of ABC', is on the " C " circle and E and E' are reflexions with respect to the perpendicular bisector of AB.

107. Problem.—*To find the Centres and Radii of M‘Cay's Circles.*

This is done by finding where the circles again cut the corresponding medians. We take, for example, the " C " circle and require to find C'. Let L denote the median CM, and take it an a line. Since it makes an angle BCM with the side a, we draw the corresponding b and c lines by making angles CAB' and ABC' equal to BCM.

From similar triangles MBC' and MCB we have $MC.MC' = MB^2 = MC.MI$; hence $MC' = MI$. This also follows, since the triangles ABI and BAC' are similar.

Again, the triangle CBC_2 is inversely similar to ABC', but it is (hyp.) directly similar to BAC_3. Hence BAC_3 and ABC' are inversely similar; therefore C' is the reflexion of C_3 with respect to the perpendicular bisector of the base.

The connection between three collinear points A', B', C' on the median to the side c of the given triangle and C_2, C_2, C_3 has thus been established.

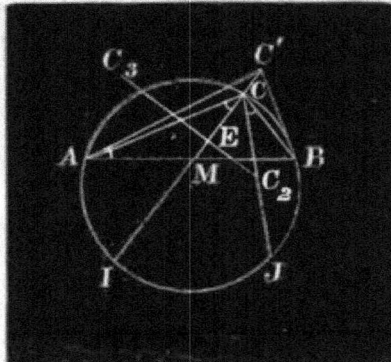

The triangles BCA', CAB', ABC' are similar to one another, and to CBC_2, ACC_2, and BAC_3; and therefore A', C_2; B', C_2; C', C_3 are reflexions of one another with respect to the corresponding perpendicular bisectors of the sides of ABC.

It follows that if the median and symmedian cut the circum-circle in I and J, and these points be joined to M, the lines MI and MJ produced through M pass through C' and C_3 respectively; $MJ = MC_3$ and $MI = MC'$, or C' and C_3 *are the reflexions of I and J with respect to the base AB.*

Let d be the distance of the centre of the " C " circle from AB, m the median, and θ the angle it makes with the base, t the tangent from M to the circle. Then

$$t^2 = ME \cdot MC' = ME \cdot MI = \frac{c^2}{12} \quad\ldots\ldots\ldots\ldots(1)$$

Again, $2d \sin \theta = ME + MC' = \dfrac{m}{3} + \dfrac{3t^2}{m} = \dfrac{a^2 + b^2 + c^2}{6m}$ ....(2)

by (1); whence $d = \frac{1}{6}c \cot \omega$, and the radius of the " C" circle is given by the equation

$$\rho = \sqrt{d^2 - t^2} = \tfrac{1}{6}c\sqrt{\cot^2\omega - 3} \text{ (cf. Art. 28, Ex. 19).}$$

Also, since the highest and lowest points of the circle are distant from the base $\rho + d$ and $\rho - \delta$, these quantities are the roots of the quadratic equation

$$12h^2 - 4c \cot \omega \cdot h + c^2 = 0 ; \dots\dots\dots\dots(3)$$

or, putting $h = \frac{1}{2}c \tan \phi$,

$$3 \tan^2\phi - 2 \cot \omega \cdot \tan \phi + 1 = 0, \dots\dots\dots\dots(4)$$

an equation which reduces by an easy transformation to

$$\sin (\omega + 2\phi) = 2 \sin \omega \dots\dots\dots(5)$$

The forms (4) and (5) are remarkable inasmuch as they express ϕ as a symmetric function of the angles; hence,

Three similar isosceles triangles may be constructed on the sides of ABC, whose vertices are a triad of collinear homologous points.

Let P, Q, R be the vertices of these triangles. Since

$$HR = HM - MR = R \cos A - \tfrac{1}{2}a \tan \phi = \dfrac{R \cos(A + \phi)}{\cos \phi},$$

with similar values for HP and HQ; also, from the collinearity of P, Q, R we have $\Sigma \dfrac{\sin A}{HP} = 0$.

By substitution, we obtain

$$\dfrac{\sin A}{\cos(A + \phi)} + \dfrac{\sin B}{\cos(B + \phi)} + \dfrac{\sin C}{\cos(C + \phi)} = 0, \dots\dots(6)$$

an equation which is therefore identical with the forms (4) and (5).

Let h_1 and h_2 be the roots of (3), then

$$\dfrac{1}{h_1} + \dfrac{1}{h_2} = \dfrac{4 \cot \omega}{c} = \dfrac{2}{\frac{1}{2}c \tan \omega} = \dfrac{2}{MC''}$$

where C' is the vertex of Brocard's first triangle; therefore

The vertices of Brocard's first triangle and the corresponding sides of ABC are pole and polar with respect to the "A," "B," and "C" circles.

Many other beautiful properties of these circles are given in the memoir from which the preceding are extracts.

108. If A', B', C' be the feet of the perpendiculars of ABC, the triangles $AB'C'$, $A'BC'$, and $A'B'C$ are similar, and may therefore be taken as portions of three directly similar figures F_1, F_2, F_3 whose double points are A', B', C', homologous lines in the ratios $\cos A : \cos B : \cos C$, the middle points of the segments of the perpendiculars towards the angles A'', B'', C'', the invariable points A''', B''', C''', points of concurrence of homologous lines middle points of sides, and the nine-points-circle the circle of similitude (Neuberg).

EXAMPLES.

1. If similar figures F_1, F_2, F_3 be described on the perpendiculars AA', BB', CC' of a triangle, their circle of similitude is the ortho-centroidal circle.

[For the orthocentre being the point of concurrence of three corresponding lines is on the circle of similitude (Art. 100, 4°). Also the parallels through the centroid E to the sides of the triangle trisect the perpendiculars at right angles, and are therefore also corresponding lines; therefore, etc.

We note that the parallels meet the corresponding perpendiculars in P, Q, R, the invariable points of F_1, F_2, F_3.]

2. The lines joining the in- and circum-centres of the *copedal* triangles $B'C'A$, $C'A'B$, $A'B'C$ meet at the point of contact of the nine-points and in-circle of ABC.

[By Art. 108, the three triangles being parts of similar figures have the nine-points-circle of ABC for circle of similitude, and the middle points of the segments of the perpendiculars for invariable points; hence (Art. 100, 4°), if I_1, I_2, I_3 and O_1, O_2, O_3 denote the in- and circum-centres of the triangles, the lines I_1O_1, I_2O_2, I_3O_3 correspond, and are concurrent on the circle of similitude.

Dr. Casey * proves the remainder of the property, which includes *Feuerbach's Theorem*, as follows :—

Let N be the nine-points-centre; then $NO_3 = \frac{1}{2}R$. Draw IP parallel to NO_3. Now, if PI is proved to be equal to the radius of the in-circle, the line I_3O_3 is the join of parallel radii, and therefore passes through a centre of similitude of the circles; similarly for I_1O_1 and I_2O_2.

Since COI and CO_3I_3 are corresponding parts of similar figures, they are similar; therefore the angle $DIO = II_3P$, and $ODI = OCI = CIP$, since NO_3 is parallel to OC. Hence the triangles ODI and PII_3 are similar, and

$$\frac{IP}{R} = \frac{II_3}{ID} = \frac{II_3 . IC}{2Rr} = \frac{2r^2}{2Rr}\left(= \frac{r}{R} \right),$$

since $CI/CI_3 = 1/\cos C$, the ratio of similitude of ABC and $A'B'C$ (Art. 108).]

3. If A and A', corresponding points of two similar figures, are conjugate points with respect to a fixed circle, required to find their loci.

---

* Casey's *Sequel to Euclid*, fifth edition, p. 202.

[Take S the double point, M the middle point of AA'. Then SAA' is a triangle of constant species; therefore SM/MA is a constant ratio. But $MA = t$, the tangent from M to the given circle (Art. 73, 2°). Hence SM/t is constant and M describes a circle (Art. 72, Ex. 3); therefore also A and A' describe circles.]

4. If $X_1X_2X_3$ be a triangle formed by joining a triad of corresponding points of three similar figures such that $X_1X_2 : X_1X_3$ = const., the locus of each vertex is a circle.

[The triangle $S_3X_1X_2$ is constant in species, Art. 97; similarly for $S_2X_1X_3$; hence S_3X_1/X_1X_2 and S_2X_1/X_1X_3 are constant ratios. Dividing one by the other, we have the base S_2S_3 and ratio of sides of the triangle $S_2S_3X_1$; therefore, etc.

It is to be noted that as the ratio S_2X_1/S_3X_1 varies in magnitude the vertex X_1 describes a coaxal system of which S_2 and S_3 are the limiting points.]

5. If the area of $X_1X_2X_3$ is given, each vertex describes a circle.

[For $X_1X_2 . X_1X_3 \sin X_1$ varies as $S_2X_1 . S_3X_1 \sin(X_1 - \theta)$; therefore, etc. (Art. 23, Ex. 3). X_2 and X_3 similarly describe circles.]

6. If a side or an angle of $X_1X_2X_3$ is given, its vertices describe circles.

7. If the area of a variable triangle formed by three corresponding lines be given, its sides envelope circles whose centres are the invariable points of F_1, F_2, F_3.

These and many other excellent illustrations of the theory of three directly similar figures are to be found in Casey's *Sequel to Euclid*, to which the student is referred. See fifth edition, Miscellaneous Examples, pp. 231-248.

CHAPTER X.

SECTION I.

CENTRES OF SIMILITUDE.

109. If A, r_1; B, r_2 be any two non-intersecting circles, P and Q the points of intersection of the direct and transverse common tangents, it is easily proved that A, B, P, Q are collinear, and that $AP/BP = AQ/BQ = r_1/r_2$; hence the *centres of similitude of two circles are the points of intersection of the direct and transverse common tangents.**

In the case of intersecting circles, if C be a point of intersection, we infer from these equations that the bisectors of the angle between the circles meet the line of centres in P and Q (Euc. VI. 3).

For the in- and ex-circles of a triangle taken in pairs the twelve centres of similitude are the vertices and the points where the bisectors of the angles meet the opposite sides.

The centres of similitude of a line L and circle A are the extremities of the diameter perpendicular to L.

For the common tangents to the circle and line are

---

* Therefore the common tangents, real or imaginary, to any two circles always intersect in real points.

parallel to the latter, and the line of centres is the diameter at right angles to L; therefore, etc.

110. It has been seen as a particular case of a general property of coaxal circles (Art. 93) that any line $A_1A_2B_1B_2$ through C, $a°$, cuts the circles at equal angles and, $\beta°$, that the intercepted chords A_1A_2 and B_1B_2 are in the ratio of the radii. These are obvious by the following method:—

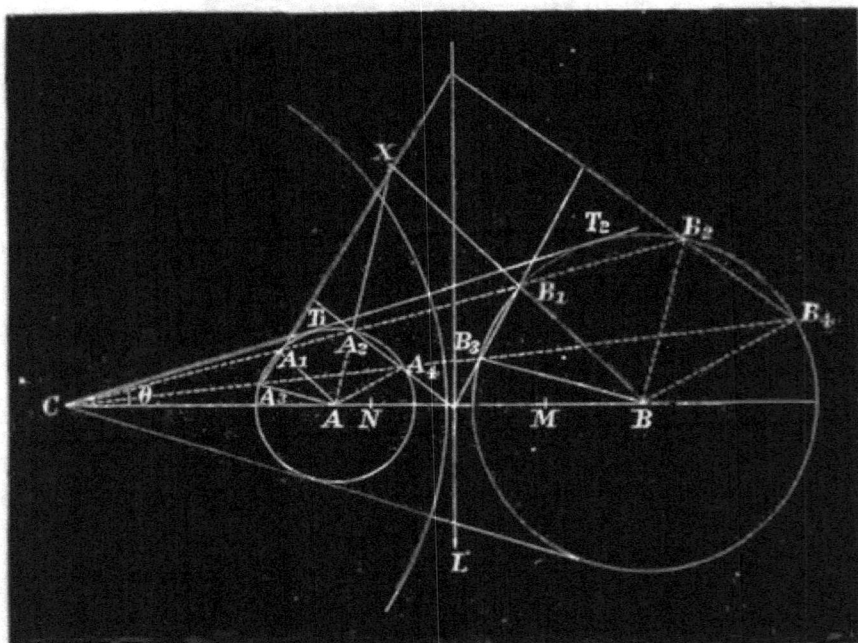

Join AA_1 and BB_1. Since $CA/CB = r_1/r_2 = AA_1/BB_1$ the triangles CAA_1 and CBB_1 are similar (Euc. VI. 7); therefore AA_1 is parallel to BB_1, and similarly AA_2 to BB_2. Hence the isosceles triangles AA_1A_2 and BB_1B_2 are similar, whence, $a°$, the angles A_1AA_2 and B_1BB_2 are equal, and, $\beta°$, $A_1A_2/B_1B_2 = r_1/r_2$.

Definitions. A_1 and B_1 are termed *Homologous Points;* and since the radii AA_1 and BB_1 through them are parallel, *the tangents at homologous points on the circles are* *parallel.* Thus the tangents at A_2 and B_2 are parallel. More generally any two points A_n and B_n which connect through C such that $CA_n/CB_n = r_1/r_2$ are homologous. A_1 and B_2 are termed *Antihomologous Points*, and since the radii AA_1 and BB_2 through them make equal angles with their line of connexion, *the tangents at antihomologous points meet on the radical axis.*

Let a second transversal through C meet the circles in $A_3A_4B_3B_4$. The chords A_1A_3 and B_1B_3 joining pairs of homologous points are termed *Homologous Lines*, and those joining pairs of antihomologous points *Antihomologous Lines*. Thus A_2A_4, B_1B_3, and A_1A_3, B_2B_4 are pairs of antihomologous lines.

111. Theorem.—*Homologous chords $(A_1A_3,\ B_1B_3)$ of any two circles are parallel.*

For it has been shown that AA_1 and BB_1, AA_2 and BB_2 are pairs of parallel lines; hence the two isosceles triangles AA_1A_3 and BB_1B_3 have equal vertical angles, and are therefore similar (Euc. VI. 6).

NOTE.—Since any line through C meets homologous lines A_1A_3 and B_1B_3 in homologous points A_n and B_n, therefore A_n, B_n are in general the corresponding intersections of pairs of homologous lines. The two points A_1A_3, A_2A_4 and B_1B_3, B_2B_4 are homologous.

112. Theorem.—*Antihomologous chords $(A_2A_4,\ B_1B_3)$ of any two circles meet on their radical axis.*

By Art. 111, we have $CA_1/CA_3 = CB_1/CB_3$, but (Euc. III. 36) $CA_1/CA_3 = CA_4/CA_2$; hence $CB_1/CB_3 = CA_4/CA_2$ or $CA_2 \cdot CB_1 = CA_4 \cdot CB_3$; thus :—*any two points are concyclic with the corresponding pair of antihomologous points;* therefore, etc. (Art. 88, Ex. 6).

PRODUCTS OF ANTISIMILITUDE.

113. By the previous Article, we have from the cyclic quadrilateral $A_2A_4B_1B_3$

$$CA_2 \cdot CB_1 = CA_4 \cdot CB_3.$$

We may therefore infer that *the rectangle under the distances of either centre of similitude from a pair of antihomologous points is constant.*

If the circles A, r_1; B, r_2 be regarded as portions of two geometrical figures, any point A_n of one is antihomologous to B_n of the other when the line A_nB_n passes through a centre of similitude C, and $CA_n \cdot CB_n$ is equal to the above constant, which is termed the *Product of Antisimilitude (External or Internal).*

To find the values of the products, we take the extreme positions of the variable line CA_2B_1 which for real intersections are the common tangents.

We have therefore

$$CA_2 \cdot CB_1 = CT_1 \cdot CT_2 \dots\dots\dots\dots\dots(1)$$

Again, since T_1T_2 subtends a right angle at each of the limiting points M and N (Art. 88, Cor. 3),

$$CT_1 \cdot CT_2 = CM \cdot CN \dots\dots\dots\dots\dots(2)$$

These constant values which may be expressed in terms of the distance (δ) between the centres of the given circles and their radii (r_1 and r_2) are of importance in the theory of coaxal circles, and will frequently be made use of in the next chapter.

Join AT_1 and BT_2. Let $ACT_1 = \theta$.

Then
$$CT_1 \cdot CT_2 = r_1r_2 \cot^2\theta = r_1r_2 \cdot \left(\frac{T_1T_2}{r_2-r_1}\right)^2$$

$$= \frac{r_1r_2}{(r_1-r_2)^2}[\delta^2 - (r_1-r_2)^2] \dots\dots\dots\dots\dots(3)$$

Similarly the internal product of antisimilitude is found
to be equal to

$$\frac{r_1 r_2}{(r_1+r_2)^2}[(r_1+r_2)^2 - \delta^2]\dots\dots\dots\dots(4)$$

NOTE.—It should be noticed when the two circles lie wholly out-
side each other $\delta > r_1+r_2$, if they intersect $\delta < r_1+r_2$ and $> r_1 \sim r_2$
(Euc. I. 20), and when one lies completely within the other
$\delta < r_1 \sim r_2$ (Euc. III. 12); hence it follows from (3) that the external
product of antisimilitude is *negative* only when one circle lies wholly
within the other. Also from (4) the internal product is *negative*
when the circles are external to one another and *positive* in every
other case. In the case where both products are positive $\delta > r_1 \sim r_2$
and $< r_1+r_2$; therefore δ, r_1, r_2 form a triangle (Euc. I. 20), or the
circles intersect in a pair of real points.

EXAMPLES.

1. If a variable circle touch two circles with contacts of similar
species, its points of contact are antihomologous points.

[By Art. 112, if AA_2 and BB_1 be produced to meet in X,
$XB_1 = XA_2$. In the case of internal contact the points of contact
are A_1, B_2.]

2. Describe a circle passing through a given point (P) and
touching two fixed circles (A, r_1) (B, r_2).

[By Art. 110, the required circle passes through an antihomologous point P', and the problem thus reduces to "*describe a circle
passing through two fixed points and touching a given circle.*"]

3. The polars of the external centre of similitude with respect to
two circles are equidistant from the radical axis, and therefore also
from the limiting points.

4. The line at infinity is an axis of perspective of two circles.

[Regard the circles as similar polygons of an infinite number of
sides, and join their corresponding vertices (*i.e.* the homologous
points). Thus the ex-centre of similitude is a *Centre of Perspective*
of the circles. Again, the corresponding sides (*i.e.* homologous lines)
intersect on the axis of perspective. In this case they are parallel.

Hence *the line at infinity is the axis of perspective of every two circles.* (Cf. Art. 87).]

5. The radical axis is also an axis of perspective of two circles.

[For since antihomologous points B_1, A_2 connect through a centre of similitude C, the circles may be regarded as polygons of an infinite number of sides whose corresponding vertices are antihomologous points and whose corresponding sides are therefore antihomologous lines ; but these latter intersect on the radical axis (Art. 112), which is therefore the axis of perspective.*

6. The poles A_n, B_n of the chords A_1A_2 and B_1B_2 are homologous points.

[For they are the intersections of pairs of homologous lines, viz. the tangents at A_1, A_2 and B_1, B_2 respectively.]

7. In Ex. 6 the lines A_1B_1 and A_nB_n are conjugate with respect to both circles.

8. If C, C' denote the centres of similitude of two circles which cut orthogonally at X ; the inverse (C'') of the point C' with respect to the circle A is the inverse of C with respect to the circle B.

[Since C' and C'' are inverse points, $AC''X=AXC'=45°$; hence $AC''X=BXC$, therefore $CB/BX=BX/BC''$, therefore etc.

9. A variable circle touches two equal circles with contacts of opposite species : show that the product of the intercepts on their transverse common tangents made by the perpendiculars from the centre and measured from their point of intersection is constant.

10. The centres of similitude, the centre of the circle of similitude, and the centre of either circle B are pairs of inverse points with respect to a circle concentric with A.

---

* Two circles are thus shown to be doubly in perspective to each centre of similitude ; the two axes of perspective forming the coaxal circle whose radius is infinitely great, viz., the radical axis and the line at infinity. It follows that " for every two circles in the same plane, however circumstanced as to magnitude and position, the radical axis and the line at infinity, being both axes of perspective, are both chords of intersection ; the corresponding points of intersection, real or imaginary, according to circumstances in the case of the former, being of course from the nature of the figures always imaginary in the case of the latter." (Townsend.)

11. The poles A_1, B_1 of the radical axis of two circles $(A, r_1; B, r_2)$ are inverse points with respect to their circle of similitude.

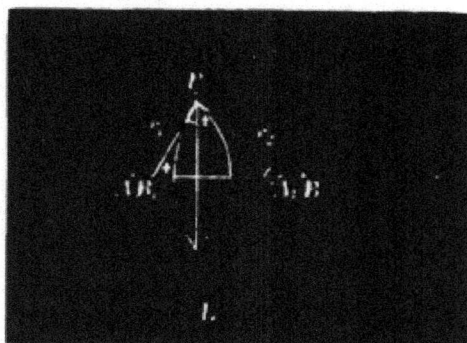

[For since $AA_1 . AL = r_1^2$, angle $APL = AA_1P$;
also since $BB_1 . BL = r_2^2$, angle $BPL = BB_1P$.
By addition $APB = PA_1B_1 + PB_1A_1 = \pi - A_1PB_1$.
Thus A_1, B_1 and A, B, since they subtend similar angles at P, are pairs of inverse points with respect to the circle of similitude (Art. 72, Cor. 8).]

12. If a variable circle V cut two circles A and B at constant angles, show that the centre of similitude of any two positions V_1 and V_2 is on L the radical axis of A and B.

[For V_1 and V_2 meet the line L at equal angles (Art. 88, Ex. 7); therefore it passes through their ex-centre of similitude.]

12a. Hence show that if the circles A and B each cut three fixed circles V_1, V_2, V_3 at the same angles α, β, γ, an axis of similitude of the three is the radical axis of the two.

13. Construct a quadrilateral, having given the four sides, and that two adjacent angles are equal. (*Mathesis*, 1881.)

14. **Feuerbach's Theorem.** To prove by an elementary method that the nine-points-circle touches the in-circle.

Draw $C'X$ the fourth common tangent to the in- and ex-circles to the side c of the triangle ABC. We shall prove that the line joining M, the middle point of the base, to the point of contact X passes through the point of contact Y of the in- and nine-points-circles.

Let T be the point of contact of the in-circle, P the foot of the perpendicular, and C' the foot of the internal bisector of C.

By Art. 71, Ex. 3, $MP \cdot MC' = \frac{1}{4}(a \sim b)^2 = MT^2 = MX \cdot MY$. Hence $XYPC'$ is a cyclic quadrilateral and angle $MC'X = MYP$; but

$MC'X = MC''C - XC'C = A \sim B$; hence $MYP = A \sim B$, and therefore *Y is on the nine-points-circle*, since the latter cuts the base AB at this angle. Therefore the circles cut or touch at Y. But the tangents at M and X to the circles are parallel, since they both meet the base at the same angle $A \sim B$. M and X are thus homologous points.

15. The straight lines joining the points of contact of the fourth common tangents to the in- and three ex-circles to the middle points of the corresponding sides are concurrent.   (*Dublin Univ. Exam. Papers.*)

[By Ex. 14, the point of concurrence is where the nine-points-touches the in-circle.]

16. A right line $ABCD$ is drawn across two circles cutting them at angles a and β respectively; show that if a variable circle cuts the given ones at the same angles in the points A', B', C', D', AA', BB', CC', DD' are concurrent; and find the locus of their point of concurrence.

[The given circles meet the line $ABCD$ and circle $A'B'C'D'$ at *equal* angles; hence AA' etc. are antihomologous points with respect to the external centre of similitude of the latter. Therefore AA' etc. meet on the circle $A'B'C'D'$ at a point (P) the tangent at which is parallel to $ABCD$. The locus of P is the radical axis of the fixed circles by Ex. 12.]

SECTION II.

CIRCLES OF ANTISIMILITUDE.

Definitions. The circle described with either centre of similitude of two given circles as centre, the square of whose radius is equal to the corresponding product (Art. 113) of antisimilitude, is known as a *Circle of Antisimilitude.*

Thus there are two circles of antisimilitude, *External* and *Internal*, according as the centre coincides with the external or internal centre of similitude of the given circles.

From the definition it is evident that all pairs of antihomologous points are *inverse points with respect to the circle of antisimilitude*, or, more generally, that *each of the two given circles is the inverse of the other with respect to either circle of antisimilitude.*

In the next chapter this latter circle, from this fundamental property, will be otherwise known as the *Circle of Inversion* of the two given ones.

114. The following theorems are of importance in the geometry of these circles.

1°. Any two circles A and B and their circles of antisimilitude are coaxal.

For the constant product $CA_2 \cdot CB_1$ (Art. 113) has been proved equal to $CM \cdot CN$; hence M and N are a common pair of inverse points to the four circles.

2°. The squares of the tangents t_1 and t_2 from any point of either circle of antisimilitude to A and B are in the ratio of the radii; or $t_1{}^2 : t_2{}^2 = r_1 : r_2$.

Since the circles are coaxal,

$$t_1{}^2 : t_2{}^2 = CA : CB = r_1 : r_2.\quad \text{(Art. 88, Cor. 2.)}$$

3°. The external circle of antisimilitude cuts ortho-gonally all circles cutting A and B at equal angles.

Since AA_2 and BB_1 are equally inclined to the line A_2B_1, if they are produced to meet in X, then XB_1A_2 is an isosceles triangle, and X is therefore the centre of a circle cutting A and B at equal angles.

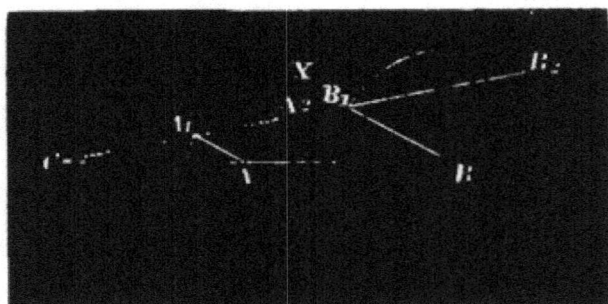

Thus any circle cutting A and B at equal angles passes through a pair of inverse points A_2 and B_1 with respect to the ex-circle of antisimilitude; therefore, etc.

See also the method of Art. 93, Cors. 3, 4. .

4°. Any circle intersecting A and B at supplemental angles is orthogonal to the internal circle of antisimili-tude.

[Proof similar to 3°.]

5°. Any circle intersecting A and B orthogonally is orthogonal to both their circles of antisimilitude.

For in this particular case A and B are cut at angles which are at once both equal and supplemental; there fore, etc. by 3° and 4° combined.

EXAMPLES.

1. A variable circle passing through a fixed point and cutting two given ones at equal angles passes through a second fixed point.

[In every position it passes through the inverse of the fixed point with respect to the ex-circle of antisimilitude.]

2. A variable circle passing through a fixed point and cutting two fixed circles at supplemental angles passes through a second fixed point.

[The inverse of the given one with respect to the in-circle of antisimilitude.]

3. Two circles X, Y intersecting two others A and B at equal angles have for radical axis a line passing through the centre C of the ex-circle of antisimilitude of A and B.

[For if X and Y intersect in a point P, each must pass through the inverse of P with respect to C.]

3a. If the angles are supplemental, the radical axis of X and Y passes through the in-centre of antisimilitude.

4. If three circles X, Y, Z meet two others A and B at equal or supplemental angles, the radical centre of the three coincides with the external or internal centre of similitude C or C' of the two.

[For by Ex. 3 the radical axes of Y, Z; Z, X; X, Y each pass through C or C' according as the angles of section are equal or supplemental; therefore, etc.]

Note.—In this example it may be noticed that in the first case the circles A and B each cut X, Y, and Z at equal angles; therefore they cut the ex-circles of antisimilitude of Y, Z; Z, X; X, Y at right angles (Art. 114). But the ex-circles of antisimilitude are coaxal; hence *a variable circle A cutting three others X, Y, Z at equal angles describes a coaxal system, the conjugate of that formed by the circles of antisimilitude of X, Y, Z taken two and two.* More generally, a variable circle cutting three others X, Y, Z at *similar* angles describes four coaxal systems whose radical axes are the four axes of similitude of X, Y, Z. Also, since the common orthogonal circle of the three cuts them at once at equal and supplemental angles, it belongs to each of the four coaxal systems.

5. If two circles A and B touch with similar contacts three others X, Y, Z, the radical axis of A and B is the line joining the ex-centres of similitude of X, Y, Z taken in pairs.

[A particular case of the foregoing.]

6. The eight circles that can be described to touch three given ones arrange themselves in pairs coaxal with the four axes of similitude of the given ones.

7. In Ex. 5 the chords of the three circles joining the points of contact with the two meet at the in-centre of similitude of A and B, and therefore at the radical centre of X, Y, Z.

8. The chords of contact pass through the poles of the radical axis of A and B with respect to each of the circles X, Y, Z.

[For the tangents at the extremities of the chord of contact of X being equal intersect on the radical axis of A and B.]

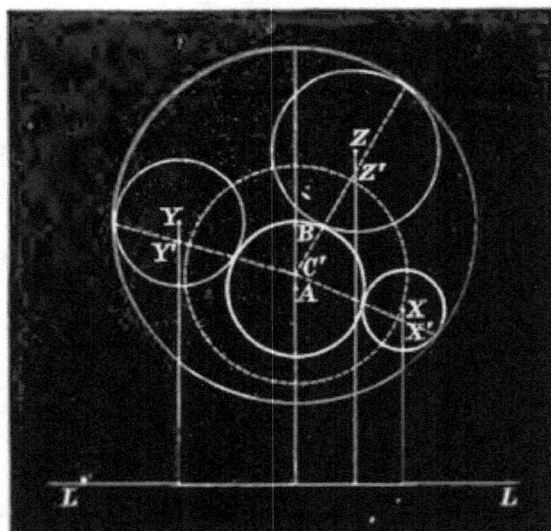

NOTE.—Gergonne deduces by means of the foregoing properties a simple geometrical construction for the eight circles of contact of three given ones X, Y, Z. The circles having similar contacts are found as follows :—Find the ex-centres of similitude of X, Y, Z taken in pairs ; the line L joining them is the radical axis of the required circles A and B. Next find C'' the centre of the common orthogonal circle of the given ones. C' is the in-centre of similitude of A and B. Now obtain the inverses X'', Y'', Z'' of L with respect to X, Y, Z respectively. Join $C'X'$, $C'Y'$, and $C'Z'$; these lines meet the given circles at the required points of contact ; therefore, etc. The remaining circles may be similarly found.

Otherwise, thus :—By Casey's relation in Art. 7, if we number the given circles 1, 2, 3 and let 4 be the required point of contact with 1, we have the ratio of the tangents from 4 to 2 and 3, a given quantity k. Similarly for the second circle which has the similar contacts with the three given ones, the ratio of the tangents from its point of contact (5) to 2 and 3 = the same ratio k ; therefore, etc. (Art. 88, Cor. 1).

9. Let A_1A_2, B_1B_2 be the extremities of the common diameter of two circles ; M, N their limiting points ; prove that the circles on A_1B_1, A_2B_2, MN as diameters are coaxal.

[For their centres are collinear, and they each cut the internal circle of antisimilitude orthogonally (Art. 114, 4°); therefore, etc.]

10. A variable circle cutting three given ones at equal angles passes through two fixed points, real or imaginary.

[For it cuts the external circles of antisimilitude of the given ones taken two and two orthogonally, and these (Art. 88, Ex. 13. 2°) are coaxal ; therefore the variable circle passes through their limiting points, real or imaginary.]

11. Two variable circles X and Y touch externally two fixed circles A, r_1 and B, r_2 at four points B_1, A_2 and A_2, B_1 in a right line ; prove that

$a°$. The line joining their centres passes through a fixed point.

$\beta°$. The sum of their radii is constant.

$\gamma°$. The foot of their radical axis describes a circle.

[$a°$. Since the diagonals of a parallelogram bisect each other, XF bisects and is bisected at the middle point Z of AB.

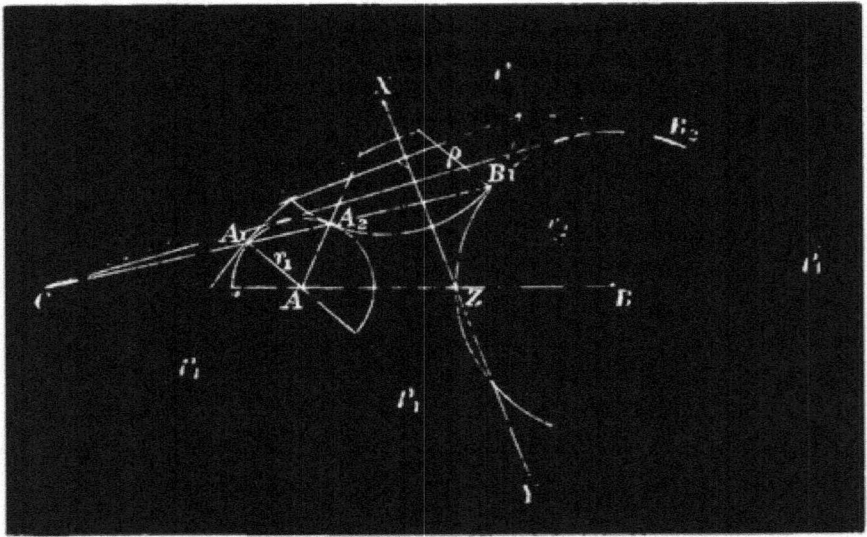

β . Let L be the radical axis of A, r_1, and B, r_2; then $XL/\rho = YL\,\rho_1$ = const. (Art. 88, Ex. 7), and therefore $\dfrac{XL + YL}{\rho + \rho_1}$ = const., but the numerator is constant by $a°$ ($=2ZL$); therefore, etc.

$\gamma°$. The circle on CZ is evidently the locus.]

12. Circles are described touching two fixed circles (as in Fig. of Ex. 8) ; find the locus of the limiting points of these circles taken in pairs.

[The internal circle of antisimilitude of the two given circles (Art. 114, $3°$).]

12a. Circles are described touching one another, and each touching two given circles ; find the locus of their points of contact.

[The points of contact are the coincident limiting points of the touching circles ; hence the required locus is the internal circle of antisimilitude of the two given ones.]

13. If n points be taken on a circle, prove that (1) the mean centres of the n systems of $n-1$ points formed by omitting each

point in succession, lie on a circle S_n ; (2) if another point be taken on the original circle, the centres of the $n+1$ circles (S_n) obtained by omitting each point in succession lie on an equal circle ; and so on *ad infinitum.*   (*St. Clair.*)*

[Let G be the mean centre of the system of n points. Produce AG to a, making $AG : Ga = n - 1 : 1$; then a is the mean centre of the $n-1$ points formed by excluding A. In the same manner we get $BG : Gb = n - 1 : 1$, etc. ; hence the points $a, b, \ldots$ lie on a circle ; and G is a centre of similitude of the locus circle and the given one.

---

* *Educational Times,* February, 1891.

CHAPTER XI.

INVERSION.

SECTION I.

INTRODUCTORY.

115. It has been seen (Art. 74) that the inverse of every point on a line with respect to a circle lies on a circle described on the line joining the centre of the given circle with the pole of the line.

This circle is said to be the inverse of the line with respect to the given circle; and it may be generally inferred that *the inverse of a line is a circle passing through the centre of the given circle;* and conversely. This latter is named the *Circle of Inversion*, and its centre the *Origin* or *Centre of Inversion*.

We shall now proceed to discuss the inversion of a system of points which are not collinear. Take the simplest case—the vertices of a triangle ABC. Let their inverses with respect to a circle of inversion O, r be respectively A', B', C'.

It is obvious that the three quadrilaterals $BCB'C'$, $CAC'A'$, $ABA'B'$ are cyclic; hence we have the angular relations :—

$$A'C'O = OAC, \ B'C'O = OBC, \text{ etc.} \quad \text{(Euc. III. 22)},$$

234

and thence by addition,

$$AOB = C + C' \dots\dots\dots\dots\dots\dots(1)$$

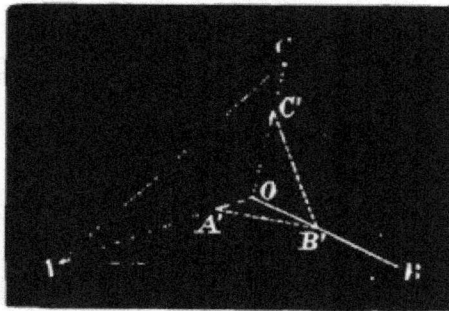

Similarly, $\qquad BOC = A + A' \dots\dots\dots\dots\dots\dots(2)$

and $\qquad\qquad COA = B + B' \dots\dots\dots\dots\dots\dots(3)$

If the base AB and origin O are fixed, and C given in magnitude, C' is also given in magnitude by (1); hence:
—*If a variable point (C) describes a circle (circum-circle of ABC), the locus of its inverse (C') is a circle (A'B'C').* [*]

Two circles or, more generally, any two curves so related that every point of one has a *Corresponding Point* on the other inverse to it with respect to a given circle, are *Inverse Figures* with respect to the circle of inversion.

It has thus been proved that in general a line or circle

---

[*] This statement is equivalent to the following :—

If a variable line OPP' is drawn from a fixed point O to a given circle and divided at X such that OP . OX = const. ; the locus of X is a circle, which may be thus proved independently. Since $OP . OP'$ and $OP . OX$ are both constant, $OX : OP' = $ const. Through X draw XC' parallel to CP'. From similar triangles $OX : OP' = OC' : OC = C'X : CP' = $ const. Hence C' is a fixed point, and $C'X$ is of constant length. The locus of X is therefore a known circle ; and the circle of inversion is obviously a circle of antisimilitude of the given one and its inverse.

inverts into a circle; and in the particular case when the origin is on the circle, its inverse is a line.

116. **Species of** $A'B'C'$. Let the points A, B, C be fixed. Since $AOB = C + C'$, C' may have any value depending on the position of the point O. The following particular cases are worthy of notice, and may be readily inferred :—

1°. If O is the circum-centre of ABC,
$$A = A', B = B', C = C'.$$

2°. If O is the right (or positive) Brocard point of ABC, $\qquad AOB = C + C' = \pi - B$;
hence $\qquad\qquad C' = A.$
Similarly $\qquad\qquad A' = B$ and $B' = C.$

3°. If O is the left (or negative) Brocard point, ABC and $A'B'C'$ are again similar.

4°. If O is one of the vertices (C_2) of Brocard's second triangle, $AOB = 2C = C + C'$, therefore $C = C'$; and also $B = A'$ and $A = B'$.

Hence the triangles are similar when the centre of inversion coincides with any of the six points O, Ω, Ω', A_2, B_2, C_2, or their inverses. (Art. 72, Ex. 22.)

5°. If O is on the circum-circle, $C' = 0°$, and the points A', B', C' are collinear.

6°. Let BOC, COA and AOB be equal respectively to $60° + A$, $60° + B$, $60° + C$. Then $A' = B' = C' = 60°$; therefore *the vertices of any triangle may be inverted into those of an equilateral; or one of any given species.*

117. In the preceding figure the point O has been taken inside the triangle. It is easy to verify the analogous angular relations when the centre of inversion is outside ABC.

It will be observed from the relations of Art. 116 that, if a variable triangle of the species of $A'B'C'$ be inscribed in the given one, the fixed point in connexion with the figure determined by the method of Art. 19 coincides with the centre of inversion.

118. Relations between the sides of ABC and $A'B'C'$. From similar triangles AOB and $A'OB'$,

$$AB^2/A'B'^2 = OA . OB/OA' . OB';$$

but

$$OA' = r^2/OA \text{ and } OB' = r^2/OB,$$

therefore by substitution

$$AB/A'B' = OA . OB/r^2,$$

or

$$c/c' = OA . OB/r^2.$$

By dividing the similar relations $a/a' = OB . OC/r^2$ and $b/b' = OC . OA/r^2$, we have

$$\frac{a}{b} \bigg/ \frac{a'}{b'} = \frac{OB}{OA} = \text{const.}$$

Hence :—*If the base and ratio of sides of a triangle are given, the base and ratio of sides after inversion are also known.* In each case the locus of the vertex is a circle having the extremities of the base for a pair of inverse points (Art. 70); and since the loci are inverse figures, we have the following important theorem :—

Every circle and a pair of inverse points invert into a circle and a pair of inverse points; and more generally, *A circle and a pair of figures each the inverse of the other with respect to it, retain this relation after inversion from any origin.*

119. Theorem.—*Any circle X, its inverse X' and the circle of inversion O are coaxal, i.e. have a pair of common points, real or imaginary.*

Let P and Q be the common pair of inverse points of the circles O and X. It is manifest that they are inverse points to X'. For X, P, Q invert respectively into X', Q, P, which by the last Article are a circle and pair of inverse points; therefore, etc,

The theorem requires no proof when the intersections of the circles are real, as the coaxal system is of the common point species.

COR. 1. The circle of antisimilitude is the circle of inversion of either of two given ones with respect to the other; hence, *Two circles and their circles of antisimili-tude are coaxal.*

COR. 2. The inverses of the vertices of any triangle with respect to the polar circle, real or imaginary, are the vertices of the pedal triangle; hence, *The circum- and nine-points-circles are inverse figures with respect to the polar circle of the triangle; and the three circles are coaxal.*

120. **Inversion of a System of Four Points.** Let A, B, C, D and A', B', C', D' be any four points and their inverses with respect to a given circle of inversion O, r.

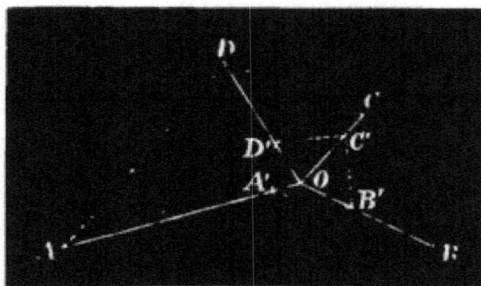

The quadrilaterals $BCB'C'$, $CDC'D'$,... are cyclic. Hence the angular relations :—

$$OA'D' = ODA, \quad OC'D' = ODC,$$

from which we obtain

$$AOC + D + D' = 2\pi \quad \dots\dots\dots\dots\dots\dots(1)$$

Also $AQC = B + B'$; therefore by substituting in (1),

$$B + B' + D + D' = 2\pi; \quad \dots\dots\dots\dots\dots\dots(2)$$

similarly, $\qquad A + A' + C + C' = 2\pi,$

or the sums of corresponding pairs of opposite angles of the two quadrilaterals are together equal to four right angles.

The following particular cases are noticed :—

1°. If $B + D = \pi$, then also $B' + D' = \pi$; *i.e., a cyclic system of points inverts into a cyclic system.* Cf. Art.115.

2°. If $B' = D'$ and $A' = C'$ simultaneously, $A'B'C'D'$ is a parallelogram, and its angles are given by the equations

$$B + D = 2(\pi - B') = 2(\pi - D')$$

and $\qquad A + C = 2(\pi - A') = 2(\pi - C').$

NOTE.—The centres of inversion in this case are easily found; for $AOC = B + B' = B + \pi - \frac{1}{2}(B + D)$, and BOD similarly equals $A + \pi - \frac{1}{2}(B + D)$; hence *there are two centres of inversion from which the vertices of any quadrilateral invert into the vertices of a parallelogram* in an assigned order, viz., the intersections of the known circles COA and BOD. Four other points might be similarly found from the intersections of pairs. of circles BOC, AOD, and AOB, COD.

3°. A cyclic. system of four points may be inverted into the vertices of a rectangle.

121. Relations between the sides of $ABCD$ and $A'B'C'D'$.—By Art. 118, $BC/B'C' = OB \cdot OC/r^2$ and $AD/A'D' = OA \cdot OD/r^2$. Multiplying these relations we have

$$\frac{BC \cdot AD}{B'C' \cdot A'D'} = \frac{OA \cdot OB \cdot OC \cdot OD}{r^4}; \quad \dots\dots\dots\dots(1)$$

similarly, $\dfrac{CA \cdot BD}{C'A' \cdot B'D'} = \dfrac{OA \cdot OB \cdot OC \cdot OD}{r^4},$ ....(2)

etc. etc. ; hence

$$BC \cdot AD : CA \cdot BD : AB \cdot CD$$
$$= B'C' \cdot A'D' : C'A' \cdot B'D' : A'B' \cdot C'D' ....(3)$$

Cor. 1. If A, B, C, D be a harmonic system of points on a circle; A', B', C', D' are also a harmonic cyclic system.

For if the ratios on the left side of (3) are equal, those on the right are also equal.

Cor. 2. Combining 3° of the last Article with the previous corollary, it follows that *a harmonic system of cyclic points may be inverted into the vertices of a square.*

Examples.

1. Any two triangles may be placed such that the vertices of the one may be inverses of those of the other taken in any assigned order.

2. Any four points may be inverted into an orthocentric system.

[For the latter quadrilateral has the following angles :— A', $90-A'$, $180+A'$, $90-A'$; hence since $BOD = A+A'$, $COA = B+90°-A'$, and $A+C+A'+\pi+A' = 180°$; the centres of inversion are the intersections of two known circles BOD and COA.]

3. Each side of a triangle divided by the perpendicular on it from any origin remains unchanged by inversion.

3a. If the origin is the symmedian point of the one triangle, it is also the symmedian point of the other.

4. If a, β, γ denote the perpendiculars from any point on a circle, on the sides of an inscribed triangle, then

$$\beta\gamma \sin A + \gamma a \sin B + a\beta \sin C = 0.$$

[For let $A'B'C'$ be any three points on a line L, and O the origin ;

since　　　　　　$\dfrac{B'C' + C'A' + A'B'}{OL} = 0,$

after inversion O is on the inverse circle L' and

$$\frac{BC}{a}+\frac{CA}{\beta}+\frac{AB}{\gamma}=0, \quad \text{or} \quad \frac{a}{a}+\frac{b}{\beta}+\frac{c}{\gamma}=0 \, ;$$

therefore, etc.]

5. Prove generally for any cyclic polygon that

$$\Sigma(a/a)=0. \qquad\qquad \text{(Casey.)}$$

6. The inverse of a figure with respect to a line is its reflexion with respect to the line, and is equal in every respect to the given one.

7. The inverses A', B', C'... of the points of intersection A, B, C, D of any two figures are the corresponding points of intersection of the inverse figures; and the lines AA', BB', CC'... are concurrent at the centre of inversion.

7a. If two curves touch at A, their inverses touch at A' the inverse of A.

8. A circle coincides with its inverse when the circle of inversion is orthogonal to it.

9. A variable chord AB of a circle, the inverse C' of a fixed point C on it and the centre O are concyclic.

[Since the points A, B, C, ∞ are collinear; their inverses with respect to the given circle are concyclic; *i.e.*, $ABC'O$ is a cyclic quadrilateral.]

10. From any point P on the circum-circle a line is drawn through the symmedian point K, cutting the sides of the triangle ABC in A', B', C', prove the relation $\Sigma 1/PA' = 3/PK$.

[Employ the properties of Ex. 4 and Art. 15, Ex. 1 (3).]

122. Theorem. *The inverse of the circum-circle of a triangle ABC with respect to the in-circle is the nine-points-circle of the triangle PQR formed by joining the points of contact.*

Let X, Y, Z be the middle points of the sides of PQR. From similar triangles we get

$$OA \cdot OX = OB \cdot OY = OC \cdot OZ = r^2 \, ;$$

therefore, etc.

Mr. Piers C. Ward has applied this property in the following elegant proof of *Mannheim's Theorem* :—

Inverting with respect to the in-circle, the circum-circle inverts into XYZ, that is, a circle passing through

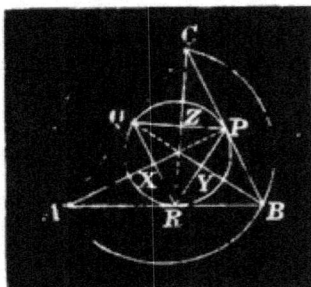

a fixed point Z and of constant radius $(=\frac{1}{2}r)$. It therefore envelopes a circle concentric with Z whose radius is equal to the diameter of XYZ; therefore, etc., by Art. 121, Ex. 7*a*.

EXAMPLES.

1. A variable triangle ABC is inscribed to one and escribed to another circle ; prove that the mean centre of the points of contact P, Q, R is a fixed point.

[This particular case of *Weill's Theorem* (Art. 53, Ex. 12) is easily seen. For the mean centre of P, Q, R is the point of trisection of the line joining its circum- and nine-points-centres, both of which are fixed ; therefore, etc.]

2. If a quadrilateral $ABCD$ be inscribed to one circle and circumscribed to another ; prove that the mean centre of its points of contact P, Q, R, S with the inner circle is a fixed point.

[Let W, X, Y, Z be the middle points of the sides of the cyclic quadrilateral P, Q, R, S. Then W, X, Y, Z is a cyclic parallelogram, and is therefore a rectangle. The mean centre of P, Q, R, S is evidently that of the system W, X, Y, Z, or the centre of the circle inverse to $ABCD$ with respect to the other given circle.]

3. The four nine-points-circles of the four triangles formed by taking the vertices of a cyclic quadrilateral in threes pass through a point.

[For the nine-points-circles invert into the circum-circles of the triangles formed by drawing tangents to the circle at the vertices of the quadrilateral; therefore, etc. The more general property for any quadrilateral has been independently demonstrated. Art. 79, Ex. 15.]

SECTION II.

ANGLES OF INTERSECTION OF FIGURES AND OF THEIR INVERSES.

123. *The general relations* existing between the centres and radii of a circle, its inverse, and the circle of inversion are as follows:—

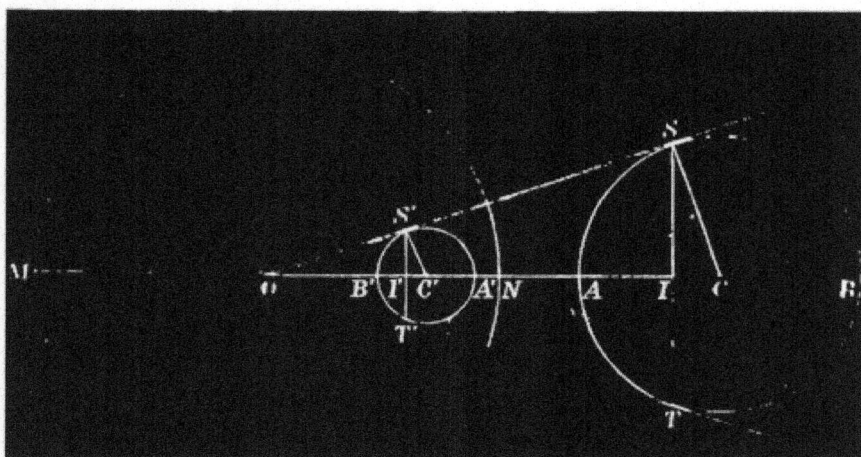

Let C, C', O be the centres of the three circles; AB, $A'B'$, MN the extremities of their common diameter; SS' and TT' the direct common tangents intersecting in O. Join ST and $S'T'$.

Since AB and $A'B'$ are inverse segments with respect to the circle of inversion, the three circles are coaxal. (Art. 114, Ex. 9.)

Let I and I' denote the points of intersection of ST and $S'T'$ with the line of centres; by comparing equal triangles OIS and OIT, etc., it follows that ST and $S'T'$ are both perpendicular to AB. The quadrilateral $CSS'I'$ is therefore cyclic; hence the inverse of C is I'; and similarly the inverse of C' is I with respect to the circle of inversion, and therefore :—

The centre C of any circle inverts into the inverse I' of the centre of inversion O with respect to the inverse circle C'; and

The inverse I of the centre of inversion O with respect to any circle C inverts into the centre C' of the circle inverse to the circle C.[*]

In the particular case when the inverse circle is a line, the inverse of the centre of a given circle *is the reflexion of the origin with respect to the line.*

The inverse of ST is the circle on OC' as diameter.

Again, by similar triangles $OC/OC' = OS/OS' = CS/C'S'$, or, say
$$d/d' = t/t' = r/r' \quad\quad\quad\quad\quad\quad (1)$$

To find d', t', and r', we have
$$d'/d = tt'/t^2 = R^2/(d^2 \sim r^2),$$
where R is the radius of inversion.

Hence
$$d' = \frac{R^2 d}{d^2 \sim r^2} \quad\quad\quad\quad\quad\quad (2)$$

a relation which gives the position of the centre C' of the inverse circle.

---

[*] Townsend, *Modern Geometry of the Point, Line, and Circle,* 1863, p. 373.

From (1) we have therefore generally

$$\frac{d'}{d} = \frac{r'}{r} = \frac{R^2}{d^2 \sim r^2} \dots \dots \dots \dots \dots \dots (3)$$

from which the position of the centre and magnitude of the radius of the inverse circle may be determined.

Cor. If the centre of inversion is on the circle; $d = r$ and $r' = \infty$, thus verifying that the inverse of a circle from any origin on its circumference is a right line.

124. Problem.—*To invert two circles such that the ratio of the radii of their inverses may be a given quantity κ.*

Let r_1, r_2 be the radii of the given circles; d_1, d_2 the distances of their centres from the origin O; R the radius of inversion; t_1, t_2 the tangents, real or imaginary, from O to the given circles. Then if ρ_1, ρ_2 denote the radii of the inverse circles, we have, by Art. 123,

$$\rho_1 = R^2 r_1 / t_1^2 \text{ and } \rho_2 = R^2 r_2 / t_2^2.$$

Dividing these equations,

$$\frac{\rho_1}{\rho_2} = \frac{r_1}{r_2} \cdot \frac{t_2^2}{t_1^2} = \kappa.$$

The centre of inversion is therefore on a locus such that tangents drawn from any point on it to the given circles have a constant ratio; *i.e. a circle coaxal with them.*

Cor. Any two circles may be inverted into equal circles; and the locus of the centre of inversion is either circle of antisimilitude.

For when $\rho_1 = \rho_2$; $t_1^2/t_2^2 = r_1/r_2$; therefore, etc. (Art. 114, 2°.)

Otherwise thus:—Since a circle and two inverse figures invert into a circle and two inverse figures; if the origin be taken on either circle of antisimilitude this circle inverts into a line. Therefore *any two figures the inverse of each other with respect to a circle invert into reflexions of each other with respect to a line.* (Art. 121, Ex. 6.)

1. Show how to invert any three circles into equal circles.

[The centres of inversion are the points of section of the circles of antisimilitude of the given ones taken in pairs.]

2. How many centres of inversion are there in the solution of Ex. 1?

[The three external circles of antisimilitude are coaxal (Art. 88, Ex. 13), and therefore meet in two real or imaginary points. Also since every two internal and one external circles of anti-similitude are coaxal, there are in all *eight* centres of inversion real or imaginary.]

3. Any three circles are unaltered by inversion with respect to their common orthogonal circle. For this reason the latter has been named the *Circle of Self-Inversion* of the given ones.

4. To invert the sides of a triangle into

     $a°$. Three equal circles.

     $\beta°$. Three circles whose radii have any given ratios $p : q : r$.

[$a°$. The centres of the in- and ex-circles are the four origins. $\beta°$. The distances of the origin from the sides are in the inverse ratios $p : q : r$.]

125. Theorem.—*The tangents at corresponding points A and A′ of two inverse figures make equal angles with their line of connexion AA′.*

For take the corresponding points B and B' on the curves which are consecutive to A and A'. Join AA' and BB'; they each pass through O.

The lines AB and $A'B'$ joining consecutive points may be regarded as tangents to the respective curves; also since $ABA'B'$ is a cyclic quadrilateral and the angle at O indefinitely small, we have (Euc. III. 22)

$$BAO = OB'A' = AA'B';$$

therefore TAA' is an isosceles triangle.

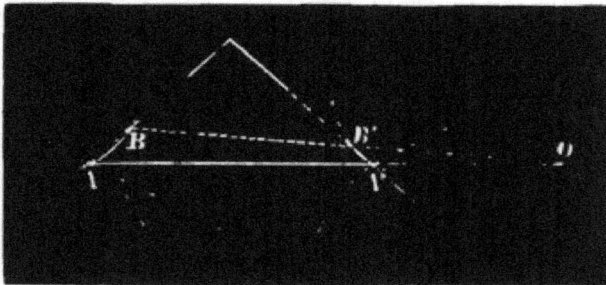

126. **Theorem.**—*The angle of intersection of two curves is similar* to that of their inverses at the corresponding point.*

For the angle between any two curves is the angle between the tangents at their points of intersection.

But the tangents determine two isosceles triangles (Art. 125) on the line AA'; therefore, etc.

If the centre of inversion is external or internal to both circles the angle remains unaltered; if on the other hand it is external to either and internal to the other, the angles of intersection before and after inversion are supplemental.

---

* "The angle of intersection of two circles undergoes as a *figure* no change of form under the process of inversion, but often does as a *magnitude*, change into its supplement, under that process.

"In the application of the theory of inversion to the geometry of the circle, this circumstance must always be attended to.

"The two cases of contact, external and internal, come of course under it as particular cases; and in but one case alone, that of *orthogonal* intersection, which presents no ambiguity, can the precaution ever be entirely dispensed with." Townsend's *Modern Geometry of the Point, Line, and Circle*, Art. 407.

127. Amongst the various results which follow from the preceding Articles, we note

1°. Any two circles meeting at an angle a invert from either point of intersection into two lines inclined at the same angle, *e.g.* two orthogonal circles into two lines at right angles.

2°. Three mutually orthogonal circles, *e.g.* the three *real* polar circles of the triangles formed from an orthocentric system of points, invert from any of their points of intersection into a circle and two perpendicular diameters.

3°. Any three circles invert from any centre on their common orthogonal circle into three others whose centres are collinear; the line of collinearity being the inverse of the common orthogonal circle.

4°. A system of circles having more than one orthogonal circle inverts into a system having more than one orthogonal line.

5°. In 4° the intersections of the common orthogonal circles are evidently the limiting points of the given system which is coaxal. (Art. 86.)

Hence for any centre of inversion :—

a°. *A coaxal system inverts into a coaxal system;* or

b°. *A circle and a pair of inverse points invert into a circle and a pair of inverse points;*

and for a centre of inversion at either of the limiting points :—

c°. *A coaxal system inverts into a concentric system,* the common centre being the inverse of the second limiting point with respect to the circle of inversion.

6°. *A system of concurrent lines inverts into a coaxal system of the common point species,* the common points being the centre of inversion and the inverse of the point of concurrence.

7°. An angle and its bisectors invert into two circles and their circles of antisimilitude. (Art. 109.)

8°. If two circles, concentric with the extremities of the third diagonal of a cyclic quadrilateral, are described cutting the given one orthogonally; they are mutually orthogonal, and their points of intersection O_1 and O_2 are therefore inverse points with respect to the given circle. Hence if we take O_1 and O_2 as centres of inversion we arrive at the following results :—The three circles invert into a circle and two rectangular diameters; the vertices of the quadrilateral, which are inverse points with respect to the circles, invert into inverse points in the same order with respect to the lines, *i.e.* form the vertices of a rectangle. Thus *the vertices of any cyclic quadrilateral may be inverted into those of a rectangle, and the centres of inversion are inverse points with respect to the circle.*

9°. A circle may invert into a circle having its centre at a given point A.

For let A' the inverse of A be the centre, and AA' the radius of inversion. Then the given circle and pair of points A and A' inverse to it, invert into a circle and a pair of inverse points; but the inverse of the centre of inversion A' is at infinity; therefore A is the centre of the inverse circle.

10°. Two parallel lines invert into two circles touching externally if the origin is between the lines; and internally if the lines are on the same side of the origin.

11°. If a quadrilateral $ABCD$ inverts into a parallelogram from an origin O; the pairs of circles BOC, AOD and COA, BOD touch at O.*

SECTION III.

ANHARMONIC RATIOS UNALTERED BY INVERSION.

128. **Theorem.**—*If A, B, C, D be any four concyclic points and A', B', C', D' their inverses with respect to any circle of inversion, then*
$$BC.AD:CA.BD:AB.CD = B'C'.A'D':C'A'.B'D':A'B'.C'D'.$$

This property has been shown to hold for *any* four points and their inverses, and is therefore true in the particular case when they lie on a circle; hence the anharmonic ratios of four concyclic points are equal to the anharmonic ratios of their inverses with respect to any circle of inversion. Particular cases have been noticed in Art. 121, Cors. 1, 2.

129. **Problem.**—*To invert a regular cyclic polygon $ABC\ldots$ from any origin P.*

The circumcircle $ABC\ldots$ inverts into a circle $\alpha\beta\gamma\ldots$; the diameters AA', BB', $CC'\ldots$ into circles passing through the origin P and cutting $\alpha\beta\gamma\ldots$ orthogonally in aa', $\beta\beta'$, $\gamma\gamma'\ldots.$

---

* Hence a construction for the required centres of inversion.

They therefore pass through Q the inverse of P with respect to the inverse circle and thus form a coaxal system of the common point species. (Art. 127, 6°.) Also the chords aa', $\beta\beta'$, $\gamma\gamma'$... meet in a point K on PQ (Art. 72, Ex. 6).

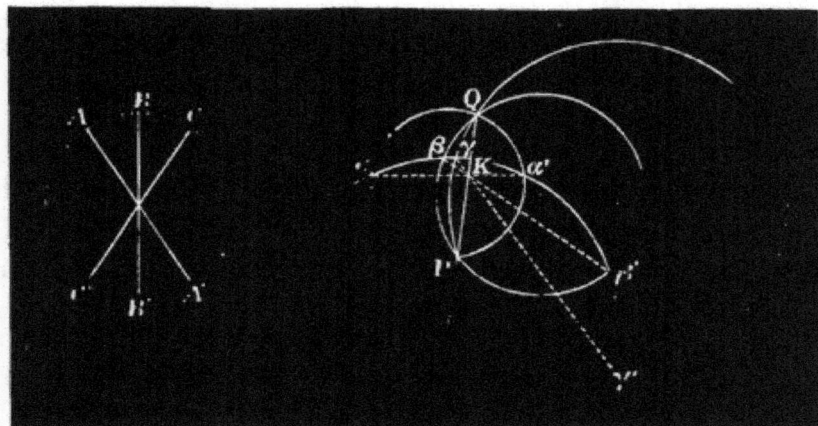

On the primitive figure any side BC of the polygon and any diameter AA' meet the circle in a harmonic row of points; therefore (Art. 128) on the inverse figure $\beta\gamma aa'$ is an harmonic row; hence $\beta a/\gamma a = \beta a'/\gamma a'$, or, by Euc. III. 22, the diagonal aa' of the quadrilateral is the locus of a point such that its distances from either pairs of sides which meet at its extremities are proportional to the lengths of the sides; similarly for the quadrilaterals $\gamma\delta\beta\beta'$, etc. Therefore *the distances of the point K from the sides of the polygon $a\beta\gamma$... are proportional to the sides.*

For an *Harmonic Quadrilateral, K* is evidently at the intersection of the diagonals; and the inverse of the regular polygon possessing, as has been shown, a corresponding and more general property has been termed by Casey an *Harmonic Polygon.*

Definitions.—The point K is called the *Symmedian Point of the Polygon;* and if the ratio of any perpendicular from K to half the side on which it falls is tan ω, then ω is the *Brocard Angle of the Polygon.*

For the properties of harmonic polygons the reader is referred to Casey's *Sequel to Euclid,* Supplementary Chapter, Section VI.

130. Cosymmedian Triangles.—Let ABC be a triangle K, its symmedian point, and let the lines $AK, BK,$ CK meet the circum-circle again in A', B', C'. If the circle of inversion be K, ρ where

$$KA \cdot KA' = KB \cdot KB' = KC \cdot KC' = -\rho^2,$$

the vertices of ABC invert into A', B', C'.

Also since $BCAA'$ is a harmonic quadrilateral, therefore $B'C'A'A$ is harmonic, or $A'A$ is a symmedian of the triangle $A'B'C'$; similarly the other symmedians are $B'B$ and $C'C$.

It appears thus that the two triangles have the same symmedian lines, symmedian point, Brocard Circle, Brocard Angle, Brocard Points, etc. On account of these relations they have been termed *Cosymmedian Triangles.**

---

* Their properties were first stated by Casey before the Royal Irish Academy in December, 1885. A further account of them will be found in Milne's *Companion.*

Example.

1. If ABC be a triangle and G its centroid; AA', BB', CC' chords of the circum-circle passing through G; the symmedian point of $A'B'C'$ is on the diameter which contains Tarry's point. (Vigarié.)

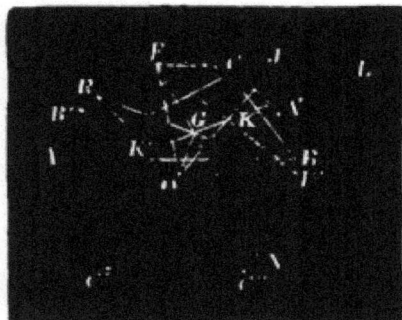

[Let the circle be self-inverted from G as origin and the points A, B, C invert into A', B', C'' respectively. Let AA'', BB'', CC'' be the symmedian chords meeting in K.

If a circle CGC'' meet GK in the point L then

$$KG \cdot KL = KC \cdot KC'';$$

and similar relations hold for the circles AGA'' and BGB''; therefore these three circles meet in a second common point L, which is the inverse of K', the symmedian point of $A'B'C'$.

Let J be the inverse of K with respect to the circum-circle ABC, and it follows that $KO \cdot KJ = KG \cdot KL =$ the power of K with respect to the circum-circle. Hence $OGJL$ is a cyclic figure, and the angle $GOK = L$.

It has been shown (Art. 67, Ex. 18) that Tarry's point on the circum-circle corresponds to O the circum-centre on the Brocard Circle with respect to ABC and Brocard's first triangle, and that G is their common centroid; hence angle $GNO = GOK$ and $GRO = GKO = GF'O$. Therefore $OGKF'$ is a cyclic quadrilateral, and (Euc. III. 21) *the points F, K, F' are collinear*. Therefore $KO \cdot KJ = KG \cdot KL = KF \cdot KF'$ or F, J, L are collinear, the line being the inverse of the circle $OGKF'$ with respect to K as origin.

Now the circum-circles of ABC and GFL cut each other orthogonally since the angle $OFG = L$; hence the inverse of the latter from G is the diameter NR, and therefore L inverts into a point K' on it; therefore, etc.

This solution is due to M'Cay.*]

MISCELLANEOUS EXAMPLES.

1. The six circles that can be described to touch three given ones A, B, C, two externally and one internally and two internally and one externally, are in pairs the inverses of one another with respect to the common orthogonal circle of A, B, C.

[Invert with respect to the common orthogonal circle of A, B, C, and since A, B, C remain unaltered after inversion, three of the circles of contact invert into the remaining three; therefore, etc.]

2. The eight circles of contact with A, B, C have a common circle of antisimilitude.

[As in Ex. 1 they are in pairs the inverses of each other with respect to the common orthogonal circle of A, B, and C.]

3. Three circles are described touching the ex-circles of a triangle, two externally and one internally; prove that they each pass through the centre of Taylor's Circle.

[Invert with respect to Taylor's Circle and the circles in question invert into the remaining circles of contact, which in this case are the sides of the triangle; and since the circles invert into lines they each pass through the centre of inversion.]

4. If ABC be a triangle; C, ρ a circle of inversion, A' and B' the inverses of A and B; to prove that

$$2s = \rho^2 \sin C / r'$$

where r' is the radius of the in-circle of $A'B'C$.

[We have $AC = \rho^2/A'C$, $BC = \rho^2/B'C$ and $AB \cdot A'B' = \rho^2/A'C \cdot B'C$, hence by addition

$$2s = \rho^2 \cdot \frac{A'C + B'C + A'B'}{A'C \cdot B'C} = \rho^2 \sin C / r'.]$$

5. **Mannheim's Theorem.*** Having given the vertical angle C and radius r' of the in-circle of a triangle $A'B'C$; the envelope of the circum-circle is a fixed circle.

[From Ex. 4 by inverting from the vertex with respect to a circle of inversion C, ρ, the inverse of the circum-circle is the base AB of a triangle of known perimeter; and since the inverse envelopes a circle, viz., the ex-circle of the triangle ABC; therefore, etc.]

6. A variable circle touches the base of an isosceles triangle at its middle point; prove that the chords of intersection with the sides that meet within the circle envelope a fixed circle. (M'Vicker.)
[See the property of Art. 61, Ex. 1.]

6a. By inverting from the vertex derive *Mannheim's Theorem.*

7. Two circles meet at an angle ω, and are such that $2 \cos \omega = \sqrt{r/R}$; prove that a triangle may be inscribed to one and circumscribed to the other. Hence find the locus of a point from which two circles may be inverted into two others, so that a triangle may be inscribed to one and circumscribed to the other.

8. A variable chord XX' of a circle O, r passes through a fixed point Q; to prove that the circum-circles of the triangles QOX and QOX' envelope coaxal systems.

[Let P be the inverse of Q with respect to the given circle. The circles in question invert into the right lines PX and PX', which by Art. 72. Cor. 5, touch each of two concentric systems, viz., the in- and ex-circles of the triangle PXX'.]

9. Prove that the vertices of a triangle and the reflexions O_1, O_2, O_3 of any point O with respect to the sides may be inverted into the vertices of a triangle and three collinear points on the sides. (Russell.)

[The circle BCO_1, CAO_2, ABO_3 meet in a point P (Art. 79, Ex. 15), which is seen from Euc. III. 22 to be on the circum-circle of $O_1O_2O_3$. Inverting from P; therefore, etc.]

---

* This well-known property is thus seen to be the inverse of :—*Having given the vertical angle C and either of the quantities s or $s-c$; the envelope of the base is a circle.*

10. Any triangle ABC and a Simson line XYZ may be inverted from the *pole of the line* into a triangle $X'Y'Z'$ and Simson line $A'B'C'$.

11. If four circles be mutually orthogonal, and if any figure be inverted with respect to each in succession; the fourth inversion will coincide with the original figure.

[The following proof has been given by M'Cay :—Invert the four orthogonal circles from a point of intersection of any two of them. The latter invert into rectangular lines ; a third circle becomes one ρ, cutting these lines at right angles ; and the fourth after inversion (ρ'), since it cuts the third at right angles and is concentric with it, satisfies the relation $\rho^2 + \rho'^2 = 0$ or $\rho^2 = -\rho'^2$.

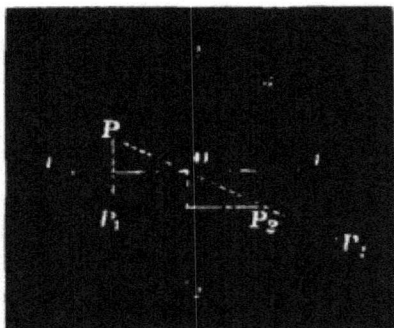

Let P_1, P_2, P_3 denote the successive inversions of the point P on the inverse figure ; since $OP_2 \cdot OP_3 = \rho^2$ and $OP_2 = -OP$, therefore $OP \cdot OP_3 = -\rho^2$, or the inverse of P_3 with respect to the imaginary circle of radius $i\rho$, whose centre is at O, coincides with P ; therefore, etc.]

12. "The centres of the four circles circumscribed about the four triangles formed by four right lines are concyclic." Prove this theorem by inversion from the point P common to the four circumcircles, and show that the circle passes through P.

[It is evident that, 1°, the four lines invert into four circles passing through P ; 2°, the four circles into lines joining the remaining pairs of intersections of the circles in 1° ; 3°, the centres of the four circles into the reflexions of P with respect to the four lines on the inverse figure by Art. 123 ; but these are collinear ; therefore, etc.]

13. Let T be a common tangent to two circles, t and t' the tangents to them from any point O ; if the circles are inverted from O as origin prove that T^2/tt' is unaltered.

14. The vertex C of a given angle ACB is fixed ; required to find the envelope of the circle ACB where A and B are points on a given line.

15. A chord AB of a circle passes through a fixed point P ; find the locus of the point of intersection of the circles passing through P and touching the given one at A and B.

16. If two circles be inverted into any two others ; for each pair the square of the common tangent divided by the product of the diameters are equal.

[Compare Art. 126 and Art. 4, footnote.]

17. Prove Casey's relation among the common tangents to four circles all of which are touched by a fifth (Art. 7) by the inversion of a system of four circles touching a line.

18. Draw two parallel lines and describe a number of circles touching the lines and each other in succession. Invert this system from a point on a diameter of any circle perpendicular to the lines and deduce the following theorem :—

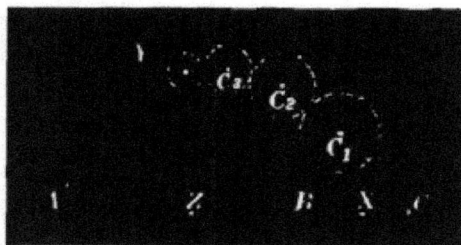

A, B, C are three collinear points, and circles X, Y, Z are described on the segments BC, CA, AB respectively. A system of circles is drawn as in figure to touch each other and the given ones, if C_n, ρ denote the nth circle to prove that the distance of its centre from $AB = 2n\rho$. (Pappus.)

R

19. If three circles Ar_1, Br_2, Cr_3 touch one another in pairs; prove by inversion that the radii of the circles which touch them with contacts of similar species are

$$\frac{r_1 r_2 r_3}{\Sigma r_1 r_2 \pm 2\Delta}$$

where 2Δ is the area of the triangle ABC.

[Invert from the point of contact of Br_2, Cr_3 with a radius equal to the tangent to Ar_1; etc.]

20. The rectangle under the distances of the ex-centre of similitude of two circles from their radical axis and in-centre of similitude is equal to the constant product of antisimilitude.

[The circle of similitude inverts from either centre of similitude into the radical axis of the given circles.

20a. Prove that the poles of the radical axis of two circles with respect to the circles are harmonic conjugates with respect to the centres of similitude.

[This is the inverse of the theorem:—*The polars of either centre of similitude with respect to two circles are equidistant from their radical axis;* the circle of antisimilitude being taken as circle of inversion.]

21. A variable circle $ABCD$ touching two fixed circles externally meets their radical axis in L and O and the pair of transverse common tangents in A, C and B, D respectively; prove the following properties of the figure:—

1°. The limiting points M and N of the circles are the middle points of the parallel sides of the quadrilateral $PQRS$.

2°. The lines AB and CD move parallel to the direct common tangents PQ and RS respectively.

3°. The vertices of $ABCD$ lie on the lines joining O and L to the limiting points.

4°. BC and AD envelope circles concentric with M and N respectively.

To prove 1°. Since the four common tangents to the two given circles form a common escribed quadrilateral, the diagonals of which are concurrent with the diagonals of the corresponding inscribed quadrilaterals; therefore, etc. See Art. 67, Cor. 6.

2°. Let the points A and B and the given circles be numbered 1, 2, 3, 4. Apply Casey's relation connecting the common tangents to four circles all touched by a fifth and reduce, it follows that $AZ + BZ \varpropto AB$. Hence AB is constant in direction and PQ is a particular position of it, therefore AB and PQ are parallel; similarly CD and RS are parallel.

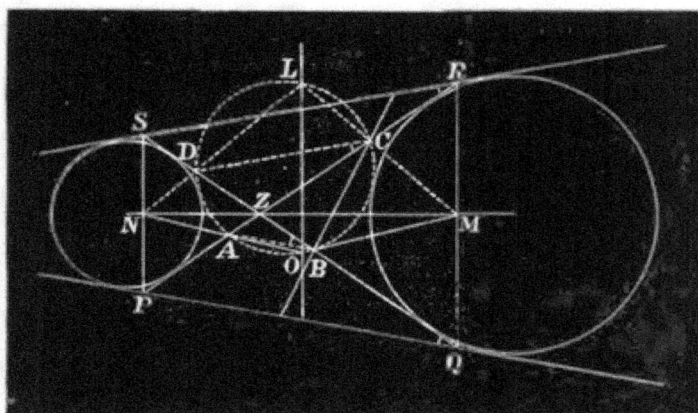

3°. To prove that the points D, L, N are collinear. Invert the figure from D as origin. The circles, their radical axis and pair of inverse points invert into three coaxal circles, one of which passes through the origin, and their limiting points; also the circle $ABCD$ inverts into the direct common tangent of the latter system. It follows easily (Art. 92, Ex. 5) that the inverses of N and L pass through D : therefore, etc.

4°. BM bisects externally the base angle B of the triangle ZBC, since LO bisects internally the vertical angle of the isosceles triangle LMN; similarly CM bisects externally the other base angle, therefore M is the ex-centre of BCZ.

Note.—This property, communicated by Mr. Charles M'Vicker, is a manifest extension of Mannheim's Theorem. For if either of the circles is reduced to a point Z, we have of the triangle BCZ the vertical angle Z fixed in magnitude and position and the ex-circle ; since the variable circum-circle BCZ (*i.e.* $ABCD$) envelopes a circle to which the vertex and centre of the ex-circle are a pair of inverse points ; therefore, etc.

22. Prove the converse of *Casey's Theorem* (Art. 7), showing the relation which holds between the common tangents to four circles, all of which are touched by a fifth.

[Invert the circles 1, 2, 3 into equal circles (Art. 124) $A, r; B, r; C, r;$ and find the inverse D, r_1 of 4 with respect to the same circle of inversion. The relation $\Sigma \overline{23} . \overline{14} = 0$ holds for the four circles after inversion (Art. 126) ; also the tangents $\overline{23}, \overline{31}, \overline{12}$ are equal to the sides of the triangle ABC formed by joining the centres of the equal circles. Now describe a circle concentric with D and a radius equal to $r \sim r_1$, and the tangents from A, B, C to it are respectively equal to $\overline{14}, \overline{24}, \overline{34}$. Hence the general relation has been reduced to the corresponding one for three points and a circle. It is easy to see that the circum-circle of ABC touches $D, r \sim r_1$; for by the converse of Ptolemy's Theorem the limiting points of the two circles are on ABC ; therefore, etc. Fry.]

NOTE.—The method of inversion so useful in Modern Geometry was discovered by the Rev. Dr. Stubbs of Trinity College, Dublin, in the year 1843. His valuable memoir on the subject is to be found in the *Philosophical Magazine*, Nov., 1843, p. 338. About the same time, Dr. Ingram published his researches in the *Transactions of the Dublin Philosophical Society*. See vol. i., p. 145.

CHAPTER XII.

GENERAL THEORY OF ANHARMONIC SECTION.

SECTION I.

ANHARMONIC SECTION.

131. Definitions.—Let a line AB be divided by two variable points C and D such that $AC/BC \div AD/BD$ is a constant ratio ($=\kappa$). The value of κ is thus

$$- CA . BD/BC . AD,$$

and is termed the *Anharmonic Ratio* in which the segment AB is divided by the points C and D. Similarly the anharmonic ratio of CD divided at A and B is

$$CA/DA \div CB/DB \text{ or } - CA . BD/BC . AD.$$

The points C and D are *Conjugate* or *Corresponding Points* in the *Row* A, B, C, D, and AB and CD are *Conjugate Segments*. It is obvious that conjugate segments divide each other *Equianharmonically*, *i.e.* the anharmonic ratio of AB divided at C and D is equal to that of CD divided at A and B.

132. Let the four points A, B, C, D be divided into three pairs of opposite segments BC, AD; CA, BD; AB, CD; then the anharmonic ratios of

261

$$BC \text{ divided in } A \text{ and } D = BA/CA \div BD/CD = \lambda, \quad (1)$$
$$CA \text{ divided in } B \text{ and } D = CB/AB \div CD/AD = \mu, \quad (2)$$
$$\text{and } AB \text{ divided in } C \text{ and } D = AC/BC \div AD/BD = \nu, \quad (3)$$

or their reciprocals; since a segment divided in A and D is divided in the reciprocal anharmonic ratio by D and A.

These three fractions λ, μ, ν and their reciprocals are the six anharmonic ratios of the four points A, B, C, D.

NOTE.—Let a line AB be divided internally in a variable point X and externally in X' such that $AX/BX = k \cdot AX'/BX'$. As X approaches B, AX/BX increases; therefore the conjugate point X' approaches B simultaneously. For let $AX' = a$ and $BX' = b$ and we have

$$\frac{a-x}{b-x} > \text{ or } < \frac{a}{b} \text{ according as } a > \text{ or } < b.$$

but $a > b$, thus it follows that as X' moves towards B the ratio AX'/BX' continually increases, and becomes infinitely great when the variable point coincides with B. Here also it coincides with its conjugate X, and the point B is thus a *Double Point* of the systems described by the variables X and X'. Similarly A is a double point.

Again, as X' recedes from B on the line produced, X approaches M the middle point of AB In the limit when X' is at infinity and AX'/BX' therefore equal to unity, its conjugate $X(=P)$ divides the line in the simple ratio $AP/PB = k$. Similarly when X moves to infinity, its conjugate $X'(=Q)$ gives the relation $AQ/BQ = 1/k$; and the two points whose conjugates are at infinity are isotomic conjugates with respect to AB.

We may note here, and we shall see presently, that when the corresponding points of the two systems move in the same direction the *double points are imaginary.*

133. Problem.—*To express all the Anharmonic Ratios of ABCD in terms of any one of them* (λ).

Since $BC \cdot AD + CA \cdot BD + AB \cdot CD = 0$;
dividing by $AB \cdot CD$, we have

$$\frac{BC \cdot AD}{AB \cdot CD} + \frac{CA \cdot BD}{AB \cdot CD} + 1 = 0,$$

whence on substituting from Art. 132

$$-\mu - 1/\lambda + 1 = 0.$$

Thus generally it follows, by dividing the above equation by each of its terms, that

$$\mu + 1/\lambda = 1; \quad \nu + 1/\mu \doteq 1; \quad \lambda + 1/\nu = 1.$$

The six ratios are therefore

$$\lambda, \quad 1/\lambda, \quad (\lambda-1)/\lambda, \quad \lambda/(\lambda-1), \quad 1-\lambda, \quad 1/(1-\lambda).$$

These may be expressed as trigonometrical functions of an angle. For let $\lambda = \sec^2\theta$. Then the ratios taken in the above order reduce to the following :—

$$.\sec^2\theta, \cos^2\theta, \sin^2\theta, \csc^2\theta, -\tan^2\theta, -\cot^2\theta.$$

If two of the ratios are equal, e.g. $\lambda = (\lambda-1)/\lambda$, then $\lambda^2 - \lambda + 1 = 0$ and $\lambda = \omega$ or ω^2, the imaginary cube roots of unity. In this case the three pairs of ratios have the values ω and ω^2.

If $\lambda = -1$ the points form an harmonic row, and the remaining ratios are $-1, -2, -1/2, 2, 1/2$.

In speaking of *the* anharmonic ratio of four points on a line the order in which the points are taken is to be understood. Dr. Salmon introduced the convenient notation $[ABCD]$ to denote the ratio into which AB is divided by C and D. $[ABCD]$ is equivalent to $AC/BC \div AD/BD$, and $[ABCD] \cdot [ABDC] = 1$.

EXAMPLES.

1. To prove that $[ABCD] = [BADC] = [DCBA] = [CDAB]$; and hence when any two constituents of four points are interchanged, the anharmonic ratio of the system remains unaltered, provided the remaining pair be likewise interchanged.

2. If $[ABCD] = [ABDC] = \kappa$; find the value of κ.

[It is plain that κ is equal to its reciprocal, and is therefore unity. The four points form in this case an harmonic system.]

3. To prove for any collinear system of points A, B, C, D, E ...
that $[ABCE]/[ABCD]=[ABDE]$.

[Expanding the ratios on the left side and reducing ; therefore, etc.]

4. For any two collinear systems of points A, B, C, D, E ...
A', B', C', D', E' ... having given $[ABCD]=[A'B'C'D']$ and
$[ABCE]=[A'B'C'E']$, to prove that
$[BCDE]=[B'C'D'E']$, $[CADE]=[C'A'D'E']$, $[ABDE]=[A'B'D'E']$.
[By Ex. 3.]

5. If $[ABCD]=[ABC'D']$, prove that $[ABCC']=[ABDD']$.

[Expanding the ratios the required result follows by alternation.]

6. If in Ex. 4 $[ABCD]=[A'B'C'D']$, $[ABCE]=[A'B'C'E']$,
$[ABCF]=[A'B'C'F']$, etc., etc. ; prove that
$$[ADEF]=[A'D'E'F'], [BDEF]=[B'D'E'F'], \text{etc.} \ldots\ldots\ldots(1)$$
and $[DEFG ...]=[D'E'F'G' ...]$.

7. If a segment MN is divided equianharmonically by pairs of
points A, A', B, B', C, C', etc. ; to prove that
   1°. $[MABC ...]=[MA'B'C' ...]$ and $[NABC ...]=[NA'B'C']$.
   2°. $[ABCD ...]=[A'B'C'D' ...]$.
[Since $[MNAA']=[MNBB']=[MNCC']= ...$ etc., by Ex. 5.
$[MNAB]=[MNA'B']$; $[MNAC]=[MNA'C']$, etc. Hence by division
we have $[MABC]=[MA'B'C']$, etc. ...
   To prove 2°. We have by 1° $[MABC]=[MA'B'C']$ and
$[MABD]=[MA'B'D']$, therefore by division $[ABCD]=[A'B'C'D']$.

8. If a segment MN is divided harmonically by points A and A',
B and B', C and C' ; to prove that the anharmonic ratio of four of
the six points taken in any order is equal to that of their four
conjugates, $[ABCC']=[A'B'C'C]$.

[By Ex. 7. $[MABC]=[MA'B'C']$; but (hyp.) C and C' are interchangeable, therefore $[MABC']=[MA'B'C]$; dividing these equations, therefore, etc., as in Ex. 4.]

9. To prove the converse of Ex. 8, i.e., for any six collinear points
A, B, C, A', B', C', if the anharmonic ratio of any four is equal to
that of their four conjugates $[CABA']=[C'A'B'A]$ then
   1°. The anharmonic ratio of every four is equal to that of their
four conjugates.

2°. The segments AA', BB', CC' have a common segment of harmonic section.

[To prove 1°. By hyp. since $[CABA']=[\dot{C}'A'B'A]$; on rearranging, by Ex. 1, we get $[AA'BC]=[A'AB'C']=[AA'C'B']$. Therefore by alternation (Ex. 5) $[AA'BC']=[AA'CB']=[A'AC'B]$; similarly for all other combinations. To prove 2°. Let MN divide the segments AA' and BB' harmonically, it divides CC' also harmonically. For $[MABA']=[MA'B'A]$ (by Ex. 7) and $[NABA']=[NA'B'A]$; also by 1° $[CABA']=[C'A'B'A]$ and $[C'ABA']=[C'A'B'A]$, hence (Ex. 6) $[MNCC']=[MNC'C]$; therefore, etc. (Ex. 2)].

10. Show generally for two equianharmonic systems if any two conjugates A and A' are interchangeable, e.g., if $[ABCD]=[A'B'C'D']$ and $[A'BCD]=[AB'C'D']$ that

1°. Every four are equianharmonic with their four opposites;

2°. The segments AA', BB', CC', DD' have a common segment of harmonic section.

[By the method of Ex. 9.]

SECTION II.

ANHARMONIC SECTION OF AN ANGLE.

134. It has been explained in Art. 3 that the anharmonic ratio of four points A, B, C, D is equal to that of the pencil $O \cdot ABCD$ formed by joining them to any point O. It follows then that all the properties of four collinear points stated in the previous section involve correlative properties of a pencil of rays, and that the latter are immediately derived from the former by aid of the equation

$BC \cdot AD : CA \cdot BD : AB \cdot CD$

$= \sin \widehat{BC} \cdot \sin \widehat{AD} : \sin \widehat{CA} \cdot \sin \widehat{BD} : \sin \widehat{AB} \cdot \sin \widehat{CD}$.

Also by describing a circle through the vertex O of the

pencil $O.ABCD$, and denoting by A, B, C, D the points where it meets the legs of the pencil again; since the sines of the angles at O are in the ratios of the chords opposite to them we may further obtain from the anharmonic properties of collinear points corresponding relations amongst points which lie on a circle.

135. The following properties will appear evident:—

1°. All transversals to a pencil of rays are cut equianharmonically.

2°. A transversal to a pencil drawn parallel to one of its rays D is divided by the remaining three in the simple ratio AC/BC; which is the anharmonic ratio of the pencil.

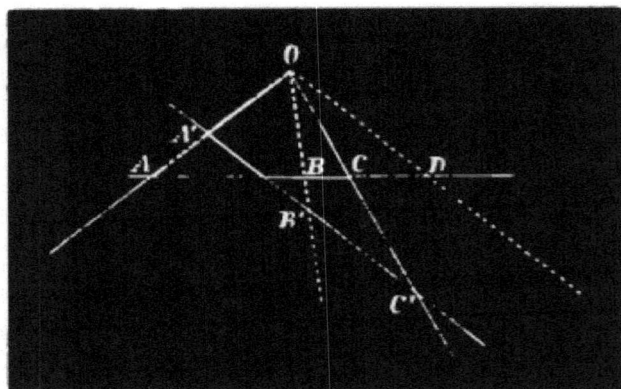

3°. In 2°, if the pencil is harmonic, any transversal $A'B'C'$ parallel to D is such that $A'B' = B'C'$.

4°. For any two equianharmonic rows of points $A, B, C, D, \dots$ and $A', B', C', D', \dots$, if the lines AA', BB', and CC' are concurrent at O; DD' and all other lines joining corresponding points of the given systems pass through O.

[This important property is the converse of 1° and follows easily by an indirect proof.]

136. Theorem.—*If two lines be divided equianhar-monically such that a pair of corresponding points coincide at their intersection* $[OABC...]=[OA'B'C'...]$ *the systems are in perspective;* and reciprocally *if two equianharmonic pencils are such that a pair of corresponding rays coincide on the lines joining their vertices they are in perspective.*

Let AA' and BB' meet in P. Join PC, and if possible let PC cut the other axis in C''. Then

$$[OABC]=[OA'B'C''],$$

since the rows are in perspective. But

$$[OABC]=[OA'B'C'] \text{ (hyp.)};$$

therefore $[OA'B'C']=[OA'B'C'']$, *i.e.* C' and C'' coincide. Reciprocally for any two pencils $P \cdot ABC$, ... and $P' \cdot A'B'C'$, ... if the rays A, A' and B, B' intersect respec-

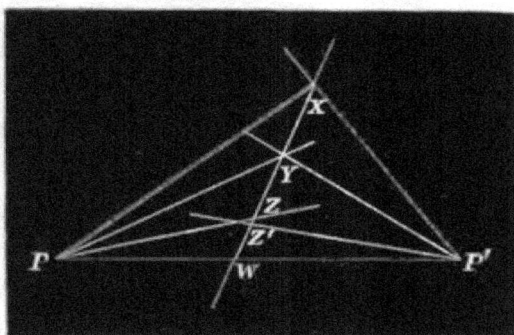

tively in X and Y, it follows that C and C' meet on the line XY.

Otherwise thus :—The rows $[XYZW]$ and $[XYZ'W]$ are equi-anharmonic ; therefore Z and Z' coincide.

Cor. 1. If two pencils are equianharmonic, any two rows passing through the intersection of a pair of corresponding rays are in perspective.

COR. 2. Through a given point P a line may be drawn across a triangle ABC, cutting its sides in the points Q, R, S, such that $[PQRS]=$ a given anharmonic ratio.

[For the pencil $(A . PQRS)$ formed with the row at any vertex A of the triangle is given, and since three of its rays are given the fourth is known.]

Def. Lines divided equianharmonically are also said to be divided *Homographically.* The term homographic is applied in general to the equianharmonic division of figures of the same kind, *e.g.* lines, circles, etc., etc.

EXAMPLES.

1. Every tangent to a circle is cut harmonically by the sides of the escribed square.

[In the limiting position when the variable tangent coincides with a side of the square the row of points determined on it are harmonic; therefore, etc., Art. 81, Ex. 3.]

2. To express the anharmonic ratios in which a variable tangent is divided by four fixed tangents, in terms of the chords of contact of the tangents.

[Let P, Q, R, S denote the points of contact of the sides of the escribed quadrilateral, which meet the variable tangent at O in A, B, C, D; O' the centre of the circle. Then $ABCD=O'.ABCD$ $=O.PQRS$, since $O'A$, OP; $O'B$, OQ ... are four pairs of perpendicular lines; therefore the required expressions are

$$QR . PS : RP . QS : PQ . RS.]$$

3. For any quadrilateral escribed to a circle at the points P, Q, R, S, each pair of diagonals and a corresponding pair of opposite connectors of the inscribed quadrilateral $PQRS$ are concurrent. (See Art. 67, Cor. 8.)

[To prove that the sets of lines

QR, PS, YY', ZZ'

RP, QS, ZZ', XX'

PQ, RS, XX', YY'

are each concurrent.

Consider each of the four tangents at the points P, Q, R, S a transversal to the quadrilateral $XX'YY'ZZ'$. Since consecutive tangents meet on the circle, the tangents at P and Q are cut *in the same order* at the points P, Z, Y, X' and Z, Q, X, Y''; therefore $[PZYX']=[ZQXY'']=[QZY'X]$. Hence PQ, YY', XX' are concurrent. Similarly RS, YY' and XX' are concurrent; therefore, etc.]

Note.—As the above properties are more generally true for the Conic, we consider an interesting case which arises in the parabola when the fourth tangent is at infinity (Art. 81). Let tangents AC and BC be drawn to a parabola at the points A and B, and a third tangent XY meeting BC and CA in X and Y respectively. Then the equianharmonic relations easily reduce to $BX/CX=CY/AY$; or a variable tangent divides two fixed tangents in the same ratio. It also subtends a constant angle at the focus. Therefore *the foci of the three parabolas described to touch each pair of sides (b, c, etc.) of a triangle ABC at the extremities of the third side (BC) are the vertices of Brocard's second triangle.*

4. If a circle touch four others the anharmonic ratios of the points of contact are equal to

$$\overline{23}.\overline{14}:\overline{31}.\overline{24}:\overline{12}.\overline{34}.$$

[By Art. 7.]

5. The anharmonic ratios of the points of contact of the nine-points-circle with the in- and three ex-circles of the triangle ABC are

$$\frac{a^2-b^2}{a^2-c^2}\quad \frac{b^2-c^2}{b^2-a^2}\quad \frac{c^2-a^2}{c^2-b^2}$$

[As in Ex. 4.]

6. If the anharmonic ratios of four points A, B, C, D on a circle (or conic) be denoted by λ, μ, ν, etc., to prove that the anharmonic ratios of the pencil $P.ABCD$ are λ^2, μ^2, ν^2, etc., where P is the pole of the line AB.

[Let PC, PD meet the conic again in C', D', and AB in E, G; then CD', DC', and AB are concurrent at F; and since

$C'. ABCD=D'. ABCD$, $[ABCD]=[ABEF]=[ABFG]=\lambda$ (say);

therefore

$$\frac{AE}{BE}\bigg/\frac{AF}{BF}=\frac{AF}{BF}\bigg/\frac{AG}{BG}=\lambda,$$

whence $\qquad \dfrac{AE}{BE}\Big/\dfrac{AG}{BG}$ or $[ABEG]=\lambda^2$.

But $[ABEG] = P\,.\,ABEG = P\,.\,ABCD$; therefore, etc.]

137. Directive Axis.—For any two homographic rows of points $ABC\,...,\ A'B'C'\,...$ on different axes L and L', if any pair of corresponding points A and A' be each joined to all the points on the other axis, the two pencils $A\,.\,A'B'C'\,...,\ A'\,.\,ABC\,...$ are in perspective (Art. 136), *i.e.* the intersections of the pairs of lines $AB',A'B\,(C'')$; AC', $A'C\,(B'')$; $AD',A'D$, etc., are collinear. We are thus enabled *to find a point P' on the line L' corresponding to a given point P on L.*

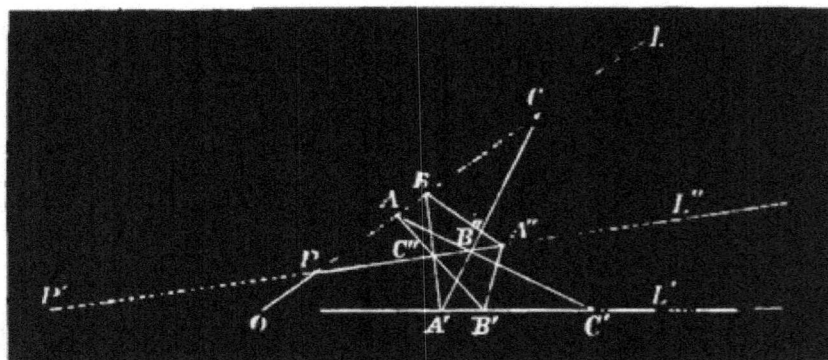

For having obtained the line $B''C''$, join $A'P$ and let it meet $B''C''$ in P''; then AP'' meets the axis L' in the required point.

An important point arises out of the consideration of the correspondents to the intersections O, P, and P' of the axes L, L', L'' taken in pairs. By means of the general method given above we find that P on the axis L corresponds to O on the axis L', and that P' on the axis L' corresponds to O on the axis L. This shows that the axis L'' of perspective of the pencils

$$A\,.\,A'B'C'\,...,\quad A'\,.\,ABC\,...,$$

whose vertices A and A' were arbitrarily chosen as any pair of correspondents of the given homographic systems, *is a fixed line, since it meets each axis in a point corresponding to their intersection O regarded as a point on the other.* Hence: *all pairs of corresponding connectors* $(XY', X'Y)$ *of pairs of non-corresponding points lie on a line.* This line is called the *Directive Axis* of the given homographic systems.

Otherwise thus : Take the two homographic pencils at A'' and L and L' as transversals to them respectively, then

$$[BCPO]=[C'B'P'O] ;$$

similarly for the vertex B'' it follows that $[CAPO]=[A'C''P'O]$, therefore by division (Art. 133, Ex. 3) $[ABPO]=[B'A'P'O]$, *i.e.* the lines AB', $A'B$, PP' are concurrent.

The same proof applies to the more general case of two systems of points on a conic.

138. **Directive Centre.**—The following property of two homographic pencils is derived from Art. 137 by

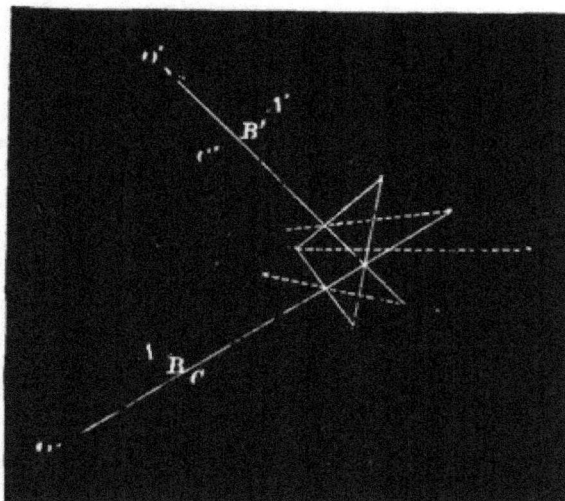

reciprocation :—*For any two homographic pencils of rays* $O.ABC...$ *and* $O'.A'B'C'...$ *the lines joining pairs of cor-*

responding intersections $(AB', A'B)$ *of non-corresponding rays* $(A, B'$ *and* $A', B)$ *are concurrent.*

The point of concurrence is termed the *Directive Centre* of the systems, and its property just stated may be proved by methods analogous to either of those given in Art. 137 for the directive axis. These are left as useful exercises for the student.

139. **Problem.**—*To find a point* X *on either axis* L *whose correspondent on the other is at infinity* (∞').

Since the lines joining A, ∞' and A', X meet on the directive axis, we have the following construction:— through A draw a parallel to L', join A' to its point of intersection with the directive axis; this line meets L in the required point.

EXAMPLES.

1. Having given two homographic pencils of rays at different vertices; to find a ray of either corresponding to a given one of the other.

[By means of their directive centre.]

2. If two homographic rows of points are such that the points ∞, ∞' at infinity on the axis correspond, the lines are divided similarly.

[For $[ABC\infty]=[A'B'C'\infty']$, hence $AB:BC=A'B':B'C'$; therefore, etc.]

3. Having given the vertical angle in magnitude and position of a triangle of constant species, the extremities of the base divide the sides homographically.

4. If the lines AA', BB', CC' connecting the corresponding vertices of two triangles ABC and $A'B'C'$ are concurrent at a point O, the intersections X, Y, Z of the pairs of sides BC, $B'C'$, etc., are collinear (cf. Art. 66).

[Join XY and let it meet the lines AA', BB', CC' in X', Y', Z' respectively. Then

$$X \cdot OBY'B' = X \cdot OCZ'C' = Y \cdot OCZ'C' = Y \cdot OAX'A';$$

therefore $[OBY'B]=[OAX'A']$, and since the point O is common to both rows the pairs of connectors AB, $X'Y'$, $A'B'$ are concurrent.

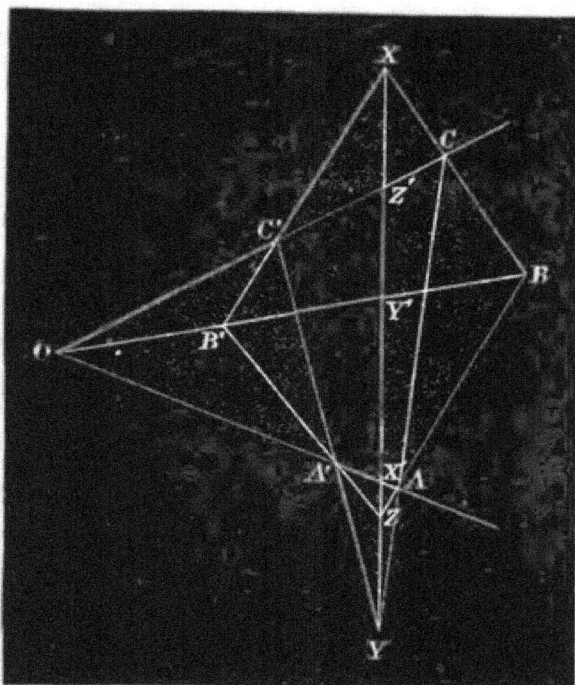

Therefore also the centre O and axis of perspective L of the two triangles divide the corresponding segments AA', BB', CC' equian-harmonically.]

5. A variable triangle moves with its vertices on three concurrent lines such that two of its sides pass through fixed points X and Y; then the third side passes through a fixed point on the line XY.

[By Ex. 4.]

6. The lines joining pairs of corresponding points of any two figures in perspective are cut homographically by the centre and axis of perspective.

7. Any line passing through either centre of perspective of two circles is cut in a constant anharmonic ratio by their radical axis.

8. Every four of the six points X, Y, Z, X', Y', Z' in Ex. 4 are equianharmonic with their four opposites.

9. In the figure of Art. 137 prove the relations

1°. $[BCPO]=[B'C'OP']=[B''C''P'P]$,
$[CAPO]=[C'A'OP']=[C''A''P'P]$.

2°. $[ABCP]=[A'B'C'O]=[A''B''C''P]$,
$[ABCO]=[A'B'C'P']=[A''B''C''P']$.

NOTE.—It will be seen that the triangle $AB'C''$ is inscribed to $A'BC$ and escribed to $B'C'A''$, and more generally that of this system of three triangles *each is inscribed to one and escribed to the other of the remaining two.*

The vertex A and opposite side $B'C''$ of the triangle $A'B'C''$ form with the extremities B and C of the corresponding side of $A'BC$ to which it is inscribed a row of points B, C, A, P. Similarly the vertex A' and opposite side BC of $A'BC$ form with the corresponding side $B'C''$ of the triangle $A''B'C'$ to which it is inscribed a row B', C'', A', O. But these rows are equianharmonic (Ex. 8, 2°); hence for such a system of triangles *the vertex and the opposite side of each divide homographically the corresponding side of the triangle to which it is inscribed.*

Again, $B'C''PP'$ is the row of points formed by the extremities of the base $B'C''$ and its intersections with the corresponding sides BC and $B'C'$ of the remaining triangles. But

$$B'C''PP'=BCPO=B'C'OP' ;$$

hence *the sides of each are cut homographically by the corresponding sides of the other two.*

Let the point C' vary along the axis L'. Then the lines AC' and BC'' turn around the fixed points A and B; A'' and B'' move on the lines $A'C$ and BC', and the directive axis passes through the fixed point C''. In this case $A''B'C''$ is a variable triangle inscribed to $A'BC$ and escribed to ABC'', both of which are fixed. Hence *for a variable triangle $A''B'C'$ inscribed to a given one $A'B'C$, if two of its sides pass through the vertices A and B of a triangle escribed to the latter, its third side passes through the third vertex C''.*

Let us now consider two positions of the variable triangle $A''B'C'$. Since its sides pass through the fixed points A, B, C'' respectively, ABC'' is a common inscribed triangle. Hence *when two triangles are each inscribed to a third $A'B'C$, if the sides $A''B''$, etc., and opposite vertices C'', etc., divide the corresponding side $A'B'$ of $A'B'C$ in a constant* *anharmonic ratio* $[A'B'C'P]$, *the intersections of their corresponding sides determine a common inscribed triangle* ABC'' *which is escribed to* $A'B'C'$.

And the vertex C''' and opposite side AB cut the corresponding sides $B'C'''$, etc., in the above constant anharmonic ratio.

140. Theorem.—*For any two homographic rows of points* $ABC \ldots X$ *and* $A'B'C' \ldots X'$, *if* X *and* X' *be the points whose correspondents* ∞' *and* ∞ *are at infinity; to prove the relations*

$$AX \cdot A'X' = BX \cdot B'X' = CX \cdot C'X' = \text{etc.}$$

Since A, A'; B, B'; X, ∞'; ∞, X' are four pairs of corresponding points $[ABX\infty] = [A'B'\infty'X']$. Expanding and reducing, this relation becomes $AX/BX = 1 \div A'X'/B'X'$; therefore $AX \cdot A'X' = BX \cdot B'X'$, etc., etc.; or :—*If variable points* A *and* A' *be taken on fixed lines* L *and* L' *respectively such that the rectangle under the distances from two fixed points* X *and* X' *on the lines is constant, they describe homographic systems.*

Cor. 1. When the vertical angle of a triangle of constant area is given in magnitude and position, the extremities of the base divide the sides homographically.

In this case the points X and X', whose correspondents are ∞' and ∞, are supposed to coincide at the intersection of the axes.

By Art. 81, Ex. 3, we see that the envelope of the base is a conic; and by Ex. 29 of the same article the curve is a hyperbola whose asymptotes are the given axes.

Cor. 2. Any two homographic rows of points may be so placed that the corresponding segments AA', BB', etc., may have a common segment of harmonic section.

Place the systems so that the axes L and L' and the

points X and X' are coincident. The equations of the article are then written

$$XA \cdot XA' = XB \cdot XB' = XC \cdot XC' = \pm \rho^2.$$

Describe a circle with X as centre having A, A'; B, B'; etc., pairs of inverse points, and let it cut the axis in M and N. MN is the common segment of harmonic section by Art. 70, but it is imaginary when A and A' lie in opposite directions from X.

Def. Two homographic systems of points on any axis which have a common segment of harmonic section are said to be in *Involution*, and the corresponding points A, A'; B, B'; etc., are *Conjugate Points* of the Involution. We have seen in Cor. 2 that there always exists a pair of points, real or imaginary, each of which regarded as belonging to either system is coincident with its correspondent of the other. These are the *Double Points* (M, N) of the involution, and are connected with the systems by the equations

$$[MNBC] = [MNB'C'], \quad [MNCD] = [MNC'D'], \text{ etc., etc.,}$$
$$[MABC\ldots] = [MA'B'C'\ldots] \text{ and } [NABC\ldots] = [NA'B'C'\ldots].$$

See Art. 133, Ex. 7.

Cor. 3. In any two homographic rows of points on a common axis the double points M and N are found from the equations *

$$XA \cdot X'A' = XB \cdot X'B' \ldots = XM \cdot X'M = XN \cdot X'N;$$

they are therefore equidistant from X and X'.

---

* If the distances OA, OA' from any point O on the axis be x, x', it follows that $(x - OX)(x' - OX') = \text{const.}$, a result of the form

$$Axx' + Bx + Cx' + D = 0 \quad \text{(cf. Art. 143)}.$$

141. For any two homographic rows of points we have seen how to find the correspondent P' of any point P, $a°$, by means of the directive axis, Art. 137, and $\beta°$ by the formula $XP \cdot X'P' = $ const. It will now be proved that two given homographic rows can be generated by the revolution of either of two determinate angles around fixed vertices, the positions of the latter and the magnitude of the angles depending on the equal values $[ABCD...]$ and $[A'B'C'D'...]$ and the positions of the axes.

142. **Problem.**—*If $ABC...$ and $A'B'C'...$ be any two homographic rows of points; to find two points such that the angles subtended at them by the segments AA', BB', etc., joining pairs of corresponding points are equal.*

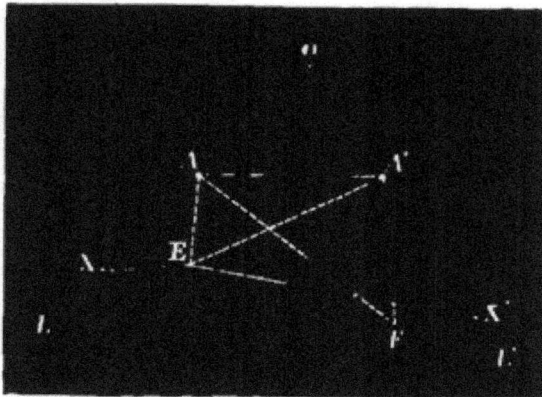

Let E and F be the required points; X, X' the correspondents of ∞' and ∞ (Art. 139). Since AEA' is a constant angle, if any point P on L coincides with X, EP' is parallel to the axis L'. Similarly if Q' and X' coincide, EQ is parallel to L. Hence the lines EX and EX' are equally inclined to L and L', or the angles AXE and $A'X'E$ are equal.

Again, the angles subtended at E by any two points A and X and their correspondents A' and ∞' are equal (hyp.); therefore in the two triangles AEX and $EA'X'$ we also have the angles AEX and $EA'X'$ equal, and the triangles are similar. Hence (Euc. VI. 4)

$$AX/XE = EX'/X'A'$$

and $EX \cdot EX' = AX \cdot A'X' = \text{const.}$ (Art. 140).

Now in the triangle XEX' we are given the base XX' fixed, the difference of base angles and rectangle under the sides; therefore the vertex E is one or other of two fixed points E or F, which are obviously the opposite vertices of a parallelogram with XX' as diagonal.

Cor. 1. The angles AEA', AXF, and $A'X'F$ are equal.

For if A' and X' coincide, EA is parallel to L; therefore AEA' is equal to the angle between EX' and L or between FX and L, since EX' and FX are parallel.

Cor. 2. The triangles AEA', AXF, and $EX'A'$ are similar.

[For by similar triangles AEX and $EA'X'$ we have $AX/AE = EX'/EA'$, but $EX' = FX$, hence

$$AX/AE = FX/EA',$$

or by alternation $AX/XF = AE/EA'$; therefore, etc. (Euc. VI. 6).]

Cor. 3. If O denote the point of intersection of the axes L and L', the points E and F are isogonal conjugates with respect to the variable triangle OAA'.

[By Cor. 2, $FAX = EAA'$ and $FA'X' = EA'A$; therefore, etc.]

Cor. 4.* The product of the perpendiculars p and p' from E and F on the variable line AA' is constant ($pp' = k^2$).   [By Cor. 3.]

Cor. 5.* The locus of the intersection of every two rectangular positions of AA' is a circle the square of whose radius (ρ) is given by the equation $\rho^2 = 2k^2 + \delta^2$, where $2\delta = EF$.

Cor. 6. A variable line cutting two fixed lines homographically cuts all positions of itself in a system of points $A''B''C''...$ such that

$$[ABCD...] = [A'B'C'D'...] = [A''B''C''D''...].$$

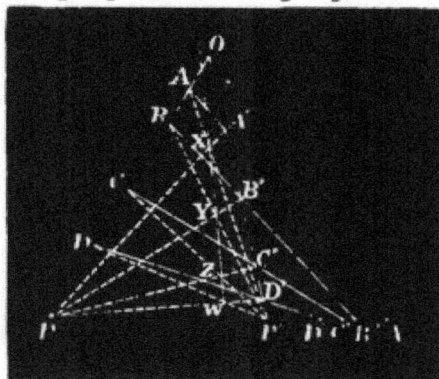

Draw the directive axis $XYZ...$ of the system as in figure. Then OX and OA'', divide the angle LOL' of the quadrilateral $PXP'O$ harmonically (Art. 68). Similarly for OY and $OB''...$. Hence we have $[O.XY...] = [O.A''B''...]$ Art. 133, Ex. 7. But

$$[O.XY...] = [P.XY...] = [P.A'B'...].$$

Therefore $[A'B'C'...] = [A''B''C''...]$.

---

  * These properties respectively may be otherwise stated :—A variable line AA' cutting two fixed axes homographically envelopes a conic of which E and F are the foci. The locus of intersection of rectangular tangents is a circle (the Director Circle).

Cor. 7. If a variable line meet two fixed circles in a harmonic row of points, it intersects all positions of itself homographically.

[For the rectangle under its distances from the centres of the circles is constant, Art. 78, Ex. 12; therefore, etc., Cor. 4.]

Cor. 8. A variable line meeting two fixed circles such that the chords intercepted by them are in a fixed ratio cuts all positions of itself homographically.

[By Art. 90, Ex. 8.]

143. If the distances of any point O from four points A, B, C, D on a line L passing through it be denoted by a, β, γ, x, and the distances of any point O' measured along another line L' to A', B', C', D' be similarly a', β', γ', x', the two systems of points are homographic if

$$\frac{(\beta-\gamma)(a-x)}{(\gamma-a)(\beta-x)} = \frac{(\beta'-\gamma')(a'-x')}{(\gamma'-a')(\beta'-x')},$$

which when multiplied out is of the form

$$Axx' + Bx + Cx' + D = 0, \dots\dots\dots\dots(1)$$

an equation which enables us to determine the position of any point of either system corresponding to a given one in the other. (See Art. 140, Cor. 3.)

We have seen that the lines joining corresponding points envelopes a conic touching L and L'. In the particular case when $x = \infty$ in (1) the simultaneous value of x' is also ∞, and the corresponding conic is therefore touched by the line at infinity. It follows obviously that when $A = 0$ in the above equation the conic is a parabola.

Thus *if a variable line be drawn cutting the sides a*

and b of a triangle ABC in X and Y such that

$$lAY + mBX = const.,$$

it envelopes a parabola to which the two sides of the triangle are tangents.

If the axes L and L' are coincident and $B = C$ in (1), x and x' are interchangeable in the equation and, as will be more fully explained in the next chapter, the two systems are in Involution.

The double points of two systems on a common axis are found from (1) by putting $x = x'$, in which case the equation reduces to the form $Ax^2 + (B + C)x + D = 0$.

Examples.

1. If the distances of two pairs of collinear points A, B and A', B' from an origin O on the line be denoted by the roots of the equations $ax^2 + 2bx + c = 0$ and $a'x^2 + 2b'x + c' = 0$, they form a harmonic row if $ac' + a'c - 2bb' = 0$.

2. Having given two of the anharmonic ratios of four collinear points equal, prove that

$$(\beta - \gamma)^2(\alpha - \delta)^2 + (\gamma - \alpha)^2(\beta - \delta)^2 + (\alpha - \beta)^2(\gamma - \delta)^2 = 0.$$

CHAPTER XIII.

INVOLUTION.

144. When of two systems of points A, B, C, ...; A', B', C', ... on any line or circle any three pairs A, A'; B, B'; C, C' which correspond are connected by a relation of the form $[BCAA'] = [B'C'A'A]$, it has been proved in Art. 133, Ex. 9, 1°. that every four and their four opposites are equianharmonic; 2°. that AA', BB', CC', ... have a common segment of harmonic section.

By Art. 140, Def., we may therefore regard either of these properties as a criterion of points in Involution.

Now since $[BCA'B'] = [B'C'AB]$, by expanding and reducing we get

$$1 - \frac{BA'}{CA'} \cdot \frac{CB'}{AB'} \cdot \frac{AC'}{BC'} \dots\dots\dots\dots(1)$$

a result previously arrived at in Art. 64, where it was shown by the application of Ceva's Theorem that a straight line drawn across a quadrilateral is cut in involution; the conjugate points A, A', etc., being the intersections of the line with the pairs of opposite connectors of the figure.

Again, if a pencil of six rays be taken and a circle described through the vertex cutting the rays in points

A, A'; B, B'; C, C', they form a system in involution if

$$\frac{\sin BOA'}{\sin COA'} \cdot \frac{\sin COB'}{\sin AOB'} \cdot \frac{\sin AOC'}{\sin BOC'} = 1 \dots\dots\dots(2)$$

The criteria (1) and (2) are called *Equations of Involution.*

145. It has been noticed in Art. 134, Ex. 10, that when any two conjugates A and A' of two homographic systems are interchangeable, every two are interchangeable, and AA', BB', CC' ... have a common segment or angle of harmonic section.

It follows that " when any one point on an axis, or ray through a vertex, has the same correspondent to whichever system it be regarded as belonging, then every point on the axis or ray through the vertex possesses the same property." *

In illustration of this theorem, let the correspondents be joined in pairs to any point (A'') on the directive axis of the systems (Art. 137).

Then the corresponding rays $A''B$, $A''B'$ are interchangeable, their productions through A'' being $A''C'$, $A''C$; therefore

The locus of a point at which two homographic rows subtend a pencil in involution is their directive axis; and similarly, or by reciprocation, *a variable line meeting two homographic pencils at a system of points in involution passes through their directive centre.*

146. A system of points in involution on a line is completely determined when two pairs of its conjugates A, A'; B, B' are given; and the conjugate C' of any point

---

* Townsend, *Modern Geometry*, vol. ii. p. 276.

C is its inverse with respect to the circle described with AB and $A'B'$ as a pair of inverse segments.

If the radius of the circle is indefinitely great, one of the double points (N) is at infinity, and therefore (Art. 72, Cor. 3) $MA = MA'$, $MB = MB'$, etc., etc.; that is, *if one of the double points of a system in involution is at infinity, the segments AA', BB', CC' ... have a common centre, viz., the other double point.*

Also a variable segment AA' of constant length moving along a given axis determines two systems of points in involution the double points of which are imaginary.

147. Theorem.—*If two chords AA', BB' of a circle meet in C, any line through C which meets the circle in O and O' determines a system of points A, A'; B, B'; O, O' in involution.*

Let AB and OO' meet in Z (Art. 64, IV. fig.). Then the pencil $B \cdot AB'OO'$ is equianharmonic with the row of points $ZCOO'$ it determines on the transversal to it through C. For a similar reason

$$[ZCOO'] = A \cdot BA'OO' = [A'BO'O],$$

from which relation it follows that every four of the six cyclic points and their four opposites are equianharmonic.

The concurrency of the chords AA', BB', OO', being involved in this relation, furnishes a geometrical explanation of the theorem of Art. 133, Ex. 9 (1).

The following generalized statement is a direct inference of the preceding :—

If through any point P, inside or outside a circle (or conic) a number of chords be drawn to cut the curve in A, A'; B, B'; C, C', ..., the two systems ABC ..., $A'B'C'$...

are in involution, and (Art. 64, III.) *the polar of P meets the circle in the double points, real or imaginary.**

EXAMPLES.

1. A variable line passing through either centre of similitude of two circles cuts them in four equianharmonic systems of points.

2. A variable circle cutting two given ones at equal or supplemental angles divides them equianharmonically.

3. If two circles V_1, V_2 cut two others at the same angles a and β in the points A, B, C, D and A', B', C', D', prove that
$$[ABCD]=[A'B'C'D'].†$$
[AA', BB', CC', DD' are concurrent at the external centre of similitude of V_1, V_2. Cf. Art. 113, Ex. 12.]

4. More generally for any number of circles V_1, V_2, ... V_n, prove that $[AA'A''...]=[BB'B''...]=[CC'C''...]=[DD'D''...]$.

5. In Ex. 3, if the angles a and β are right, the anharmonic ratio of the four points of intersection of the variable circle is equal to that of the four points on their common diameter.

6. If two triangles ABC, $A'B'C'$ inscribed in the same circle are in perspective at O, and from any point P on the circle lines PA', PB', PC' are drawn meeting the sides of ABC in X, Y, Z, the points X, Y, Z, O are collinear.
[The Pascal hexagons $PB'BACC'$, $PC'CBAA'$, $PA'ACBB'$ have YOZ, ZOX, XOY as Pascal lines; therefore, etc.]

7. If P' denote the point on the circle corresponding to P in the perspective, and the lines $P'A$, $P'B$, $P'C$ meet the sides of $A'B'C'$ in X', Y', Z', 1°. X', Y', Z' are collinear with X, Y, Z and the six points are in involution; 2°. $[XYZO]=[X'Y'Z'O]$.
(Townsend, vol. ii. p. 208.)

---

* When the point is outside its polar cuts the circle in real points M and N which divide AA', BB', CC'... harmonically, and are therefore the double points of the involution $ABC...$, $A'B'C'...$.

† It follows directly that the anharmonic ratio of four points on a circle is unaltered by inversion; the circle of inversion in this case being either circle of antisimilitude of V_1 and V_2.

8. A variable circle cutting three fixed circles at equal or similar angles determines six homographic systems of points on the circles.

[Take two positions of the variable circles cutting the given ones at equal angles a and β respectively ; then each of the given ones cuts a coaxal system (Art. 114, Ex. 10) at the same angles a and β ; therefore, etc. It is evident that the three pairs of double points of the homographic systems on each circle are the points of contact of the corresponding circles of contact.]

9. Describe a circle touching three given ones with contacts of assigned species.   [By Ex. 7.]

10. Describe a circle passing through a fixed point and cutting two given arcs on each of two circles equianharmonically.

11. Describe a circle cutting three pairs of arcs on three given circles equianharmonically.

12. The line joining the centres of perspective of any two chords of a circle is divided harmonically both by the circle and the chords.

13. Equal arcs of a circle are divided equianharmonically by the two circular points at infinity.

Desargues' Theorem.

148. *Any transversal to a cyclic quadrilateral ABCD meets the three pairs of opposite connectors BC and AD,*

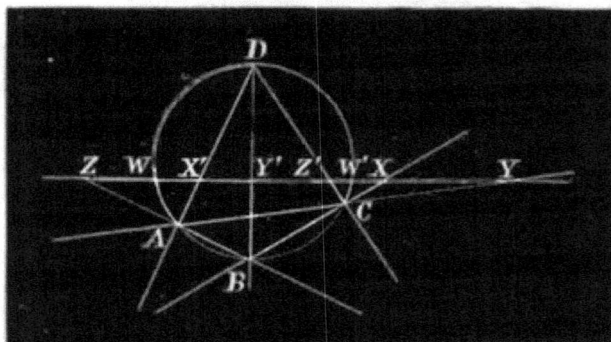

etc., etc., in X, X′; Y, Y″; Z, Z′ and the circle in W and W′ in eight points in involution.

For the pencils $B \cdot ADWW'$ and $C \cdot ADWW'$ are equal, and therefore $[ZY'WW']=[YZ'WW']=[Z'YW'W]$, or the two triads Y, Z, W; Y', Z', W' are in involution.

Again, because $C \cdot BDWW' = A \cdot BDWW'$ it follows similarly that Z, X, W and Z', X', W' are in involution; and since $A \cdot CDWW' = B \cdot CDWW'$, X, Y, W and X', Y', W' are in involution; therefore, etc., Art. 144.

COR. 1. By reciprocation with respect to the given circle we obtain the correlative theorem :—

For any escribed quadrilateral the lines joining any point P to the three pairs of opposite intersections X, X'; Y, Y'; Z, Z' and the pair of tangents PW, PW' are in involution.

COR. 2. By reciprocation from any origin it follows that the theorem and Cor. 1 are more generally true for a quadrilateral inscribed or escribed to conic.

COR. 3. In the particular case when a pair of opposite sides of a cyclic quadrilateral, or one inscribed in a conic, coincide, the remaining pair become tangents, and the

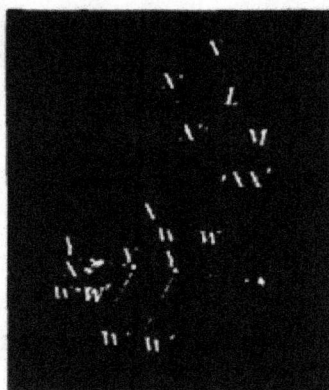

transversal (L) meets their chord of contact in a double point.

Also the line (M) passing through their point of intersection, which is therefore a double point, is divided
harmonically; *i.e. A variable chord of a conic passing
through a fixed point is divided harmonically by the
point and its polar.*

COR. 4. When the transversal (N) is a tangent to the
conic, the points of contact (WW') and (YY') are the
double points.

COR. 5. As a particular case of Cor. 4, let the transversal
be parallel to the chord of contact. Then one of the
double points (YY') is at infinity, and the other is there-

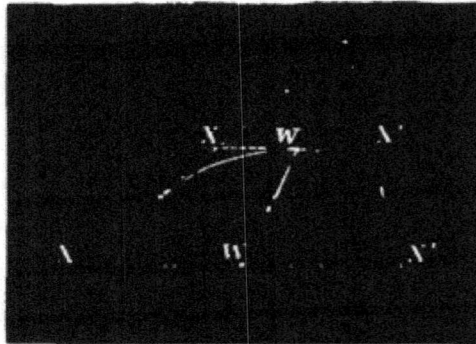

fore the middle point of XX', hence we have the following
property :—

The chord of contact of two parallel tangents (*i.e.* a
diameter) bisects every parallel chord of the conic, or *the
locus of the middle points of parallel chords of a conic is
a right line.*

COR. 6. Since a parabola touches the line at infinity
(Art. 81) and the chord of contact of any tangent and the
line at infinity is a diameter, any chord (WW') of a parabola meets a tangent at a point X, which is the centric,

and the diameter through its point of contact at a double
point (YY') of the involution.   Hence also
$$XW \cdot XW' = XY^2,$$

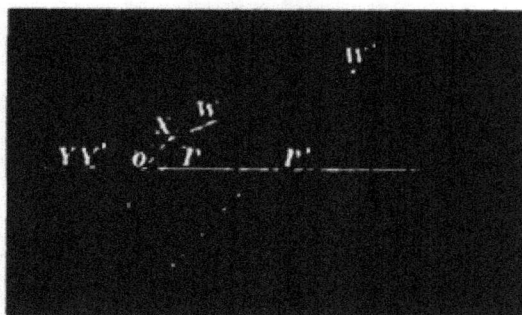

or by drawing the ordinates WP, $W'P'$,
$$OP \cdot OP' = OY^2.$$

Cor. 7. Since the asymptotes of a hyperbola and the
line at infinity are a particular case of a quadrilateral
inscribed in a conic, *any transversal WW' is divided*

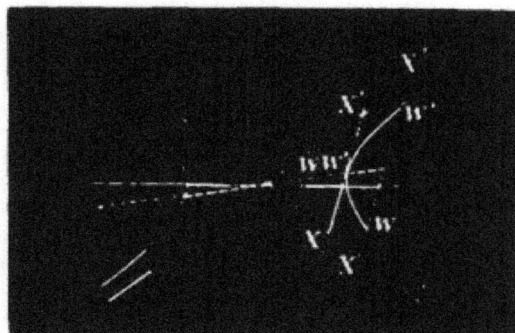

*similarly at X and X', because one of the double points
(YY') is at infinity.* The other double point is therefore
the middle point of WW', *and the intercepts WX and
$W'X'$ between the curve and the asymptotes are equal.*

Also, *the portion of any tangent to a hyperbola intercepted by the asymptotes is bisected at the point of contact.*

T

Cor. 8. If the point P in Cor. 1 is such that two pairs of opposite connectors PX, PX'; PY, PY' are at right angles, the tangents from P to the circle are likewise at right angles. But the circle reciprocates from P as origin into an equilateral hyperbola; therefore *if an equilateral hyperbola be circumscribed to a triangle, it passes through the orthocentre.*

More generally, *if an equilateral hyperbola be described about a quadrilateral, it passes through the orthocentre of the four triangles formed by taking the vertices in triads.*

The property of Art. 68, Ex. 8, will now appear obvious.

It follows also that *the locus of the centres of equilateral hyperbolas described about a triangle is its nine-points-circle.*

Cor. 9. If the sides of the quadrilateral be numbered 1, 2, 3, 4, and the perpendiculars from W and W'' on them be denoted by p_1, p_2, p_3, p_4; q_1, q_2, q_3, q_4, since

$$[WW'XX'] = [WW'YY'] = [WW'ZZ'],$$

and therefore $\dfrac{W'X}{WX} \cdot \dfrac{WZ'}{W'Z'} = \dfrac{WZ}{W'Z} \cdot \dfrac{WX'}{W'X''}$ etc., etc.,

we have $\dfrac{p_2 p_3}{q_2 q_3} = \dfrac{p_1 p_4}{q_1 q_4}$;

hence $p_2 p_3 / p_1 p_4$ is of **constant value for all points on the conic**, or *the locus of a point such that the products of the perpendiculars from it to the three pairs of opposite sides of a quadrilateral have constant ratios is a conic passing through its vertices;* and by reciprocation we derive the correlative theorem:—*If a quadrilateral is circumscribed to a conic, the rectangles under the distances of the pairs* *of opposite vertices from a variable tangent have are to each other in constant ratios.**

Cor. 10. If either asymptote of a hyperbola be taken as a transversal to an inscribed quadrilateral, the double points of the involution are both at infinity, and the segments XX', YY', ZZ' have a common middle point; therefore *the lines joining a variable point on a hyperbola to a pair of fixed points on it intercept segments of constant length on each of the asymptotes.*

This property is thus stated in Townsend's *Modern Geometry*, Art. 340 :—

"For every two homographic pencils of rays through different vertices there exist two lines, real or imaginary, on each of which the several pairs of corresponding rays intercept equal segments."

EXAMPLES.

1. A pencil whose rays are parallel to the three pairs of opposite connectors of a quadrilateral determines a system in involution.

[Since the line at infinity is a transversal cut in involution by the sides of the quadrilateral ; therefore, etc.]

2. The three pairs of parallels drawn through the vertices and the extremities of the third diagonal of a quadrilateral cut any transversal in a system of points in involution.

3. If the fourth vertex D of the quadrilateral $ABCD$ is the orthocentre of ABC, prove the following particular case of the general theorem of Art. 148 :—*For any pencil of rays in involution, if two pairs of conjugates are at right angles, then all pairs of conjugates are at right angles.*

4. Hence deduce "*The circles on the diagonals of a complete quadrilateral are coaxal.*"

---

* Chasles, *Sectiones coniques*, Art. 26.

5. Any line or circle intersects a coaxal system at points in involution.

6. The parallels through any point to the sides of a triangle and the lines connecting that point to the vertices form an involution.

7. Every two circles and their two centres of perspective subtend at any point a pencil in involution.

8. For every two self-reciprocal triangles with respect to the same circle any two vertices connect equianharmonically with the remaining four.

CHAPTER XIV.

DOUBLE POINTS.

149. The solutions of a large number of problems of every variety in Geometry are frequently made to depend on the finding of the double points of two homographic systems. On account of the great importance of these points various constructions have been given for them. Thus in the last corollary they are easily found when we have obtained the points whose conjugates are at infinity on the axis by the equations

$$XA \cdot X'A' = XM \cdot X'M = XN \cdot X'N.$$

We give in the following article two additional constructions for homographic rows on an axis and append a sufficient number of examples, some of which have apparently no connexion with our present subject, to enable the student to form an idea of their extensive applications.

150. For any two systems of points on a circle (Art. 67, Ex. 6) the pairs of lines BC', $B'C$; CA', $C'A$; AB', $A'B$ intersect respectively in points X, Y, Z, which are collinear; and the line of collinearity meets the circle in points M and N, real or imaginary, given by the equations

$$[ABCM] = [A'B'C'M] \text{ and } [ABCN] = [A'B'C'N].$$

But since the anharmonic ratios are unaltered by inversion, if the origin O be taken on the circle, the cyclic system inverts into points lying on a line and the double points of the former invert into the double points of the latter system.

Hence the following construction for the double points of two homographic systems $ABC...$ and $A'B'C'...$ on a line.

Take any arbitrary point O and describe the circles BOC', $B'OC$ meeting again in X; COA', $C'OA$ in Y; and AOB', $A'OB$ in Z. Then O, X, Y, Z lie on a circle which meets the axis in the required points M and N, real or imaginary. (Chasles.)

Otherwise thus:—Since $[BCAM] = [B'C'A'M]$, we have

$$\frac{BA}{CA} \Big/ \frac{BM}{CM} = \frac{B'A'}{C'A'} \Big/ \frac{B'M}{C'M},$$

which gives on reduction the ratios $MB \cdot MC'/MB' \cdot MC$, a known quantity.

But the numerator and denominator are respectively the squares of the tangents from M to the circles described on the segments BC' and $B'C$ as diameters; therefore, etc., by Art. 88, Cor. 2.

It should be noticed that two homographic systems *whose double points are imaginary* may be generated by the revolution of a constant angle about either of two fixed vertices which are reflexions of one another with respect to the axis. For if AA', BB', and CC' subtend equal angles at a point P (Art 72, Cor. 8), then

$$DPD' = APA' = \text{etc.},$$

since $[ABCD...] = [A'B'C'D'...]$.

1. Through a given point P draw a line meeting two given lines L and L' divided homographically in corresponding points X, X'.

[Join PA, PB, PC, and let these lines meet the axis L' in A'', B'', C'', then $ABC... = A''B''C''...$ since the systems are in perspective at P, therefore $A''B''C''... = A'B'C'...$, and if any point of either coincides with its correspondents of the other, what is required is done ; hence lines joining P to the double points of these systems give the two solutions of the problem.]

2. Draw a line through a point P cutting four lines L_1, L_2, L_3, L_4 in a row of points A, B, C, D having a given anharmonic ratio k.

[Take points A_1, A_2, A_3, ... on the axis L_1, and draw lines cutting the remaining axes in systems of points such that

$$A_1B_1C_1D_1 ... = A_2B_2C_2D_2 ... = A_3B_3C_3D_3 .. \ .$$

The angle L_1L_2 is thus divided homographically by the pairs of rays through C_1, D_1 ; C_2, D_2 ; C_3, D_3 ..., etc., and the systems $C_1C_2C_3 ...$, $D_1D_2D_3 ...$ are therefore equianharmonic.* Join PC_1, PC_2, PC_3, ..., and let the joining lines meet L_4 in D_1', D_2', D_3', ... It follows, as in Ex. 1, that $D_1D_2D_3 ... = D_1'D_2'D_3'...$, and the lines joining their double points to P are those required.]

3. Draw a line intersecting five lines such that the anharmonic ratio of any four of the points of intersection is equal to that of any other four.

4. Given two homographic pencils, find the pairs of corresponding rays which intersect on a given line L.

[Let the line meet the pencils in points ABC, $A'B'C'$; the required rays therefore pass through the double points of the homographic rows so determined.]

5. In Ex. 4 find the pair of corresponding rays which intersect at a given angle.

[Join the vertices O and O' of the pencils, and on OO' describe a segment of a circle containing the given angle ; let this circle cut the pencils in the points $ABC...$, $A'B'C''...$, and find the double points of these homographic systems ; therefore, etc.]

---

* This is otherwise evident as all the lines touch the same conic.

6. Find the direction of the parallel rays; and hence draw a transversal to two homographic pencils which shall be divided similarly by them.

7. Find two points on a given line which shall be isogonal conjugates with respect to a given triangle.

8. Construct a triangle with its sides passing through given points and its vertices on given lines, or on a circle.

9. Let the line L joining the vertices of two homographic pencils regarded as a ray of each system have for conjugates L_1 and L_2; prove that any transversal through the point L_1L_2 is cut in involution (cf. Art. 145).

10. Through a given point P draw a line intersecting five lines in the points A, A'; B, B'; P' in any assigned order forming with P an involution.

[Let the lines containing A, B meet in O; those containing A', B' in O'. Since (hyp.) $O . ABPP' = O' . A'B'P'P = O' . B'A'PP'$ and the pairs of rays which correspond $OB, O'B'$; $OB, O'A'$ are fixed; therefore the variable rays OP' and $O'P'$ divide the fifth line L homographically and the double points give the required solutions.]

11. Find a point on a given line such that if joined to five given points any two pairs of connectors shall be in involution with the line and fifth.

12. Describe a circle touching three circles with contacts of assigned species.

INDEX.

PRINTED BY ROBERT MACLEHOSE AT THE UNIVERSITY PRESS, GLASGOW.

EUCLID AND PURE GEOMETRY.

A TREATISE ON GEOMETRICAL CONICS. In accordance
with the Syllabus of the Association for the Improvement of Geometrical
Teaching. By A. Cockshott, M.A., Assistant Master at Eton, and Rev. F.
B. Walters, M.A., Principal of King William's College, Isle of Man. Cr. 8vo.
5s.

GEOMETRICAL EXERCISES FOR BEGINNERS. By Samuel
Constable. Cr. 8vo. 3s. 6d.

EUCLIDIAN GEOMETRY. By Francis Cuthbertson, M.A.,
LL.D. Ex. fcap. 8vo. 4s. 6d.

*PROPERTIES OF CONIC SECTIONS PROVED GEOMETRI-
CALLY.* By Rev. H. G. Day, M.A. Part I. The Ellipse,
with an ample collection of Problems. Cr. 8vo. 3s. 6d.

RIDER PAPERS ON EUCLID. BOOKS I. AND II. By Rupert
Deakin, M.A. 18mo. 1s.

GEOMETRICAL TREATISE ON CONIC SECTIONS. By
W. H. Drew, M.A. New Ed., enlarged. Cr. 8vo. 5s.

*ELEMENTARY SYNTHETIC GEOMETRY OF THE POINT,
LINE AND CIRCLE IN THE PLANE.* By N. F. Dupuis,
M.A., Professor of Pure Mathematics in the University of Queen's College,
Kingston, Canada. Gl. 8vo. 4s. 6d.

A TEXT-BOOK OF EUCLID'S ELEMENTS. Including Alter-
native Proofs, together with additional Theorems and Exercises, classified
and arranged. By H. S. Hall, M.A., and F. H. Stephens, M.A., Masters of
the Military and Engineering Side, Clifton College. Gl. 8vo. Book I., 1s.;
Books I. and II., 1s. 6d.; Books I.-IV., 3s.; Books III.-IV., 2s.; Books
III.-VI., 3s.; Books V.-VI. and XI., 2s. 6d.; Books I.-VI. and XI., 4s. 6d.;
Book XI., 1s.   [KEY. *In preparation.*

THE ELEMENTS OF GEOMETRY. By G. B. Halsted, Pro-
fessor of Pure and Applied Mathematics in the University of Texas. 8vo.
12s 6d.

THE ELEMENTS OF SOLID GEOMETRY. By R. B. Hayward, M.A., F.R.S. Gl. 8vo. 3s.

EUCLID FOR BEGINNERS. Being an Introduction to existing
Text-Books. By Rev. J. B. Lock, M.A.   [*In the Press.*

GEOMETRICAL CONICS. Part I. The Parabola. By Rev. J.
J. Milne, M.A., and R. F. Davis, M.A. Cr. 8vo. 2s.

THE PROGRESSIVE EUCLID. BOOKS I. and II. With Notes,
Exercises, and Deductions. Edited by A. T. Richardson, M.A., Senior
Mathematical Master at the Isle of Wight College. Gl 8vo. 2s. 6d.

SYLLABUS OF PLANE GEOMETRY (corresponding to Euclid,
Books I.-VI.)—Prepared by the Association for the Improvement of Geometrical Teaching. Cr. 8vo. Sewed, 1s.

SYLLABUS OF MODERN PLANE GEOMETRY. Prepared
by the Association for the Improvement of Geometrical Teaching. Cr. 8vo.
Sewed. 1s.

THE ELEMENTS OF EUCLID. By I. Todhunter, F.R.S.
18mo. 3s. 6d. Books I. and II. 1s. KEY. Cr. 8vo. 6s. 6d.

WORKS BY VEN. ARCHDEACON WILSON, M.A.
Formerly Headmaster of Clifton College.

ELEMENTARY GEOMETRY. Books I.-V. Containing the
Subjects of Euclid's first Six Books. Following the Syllabus of the Geometrical Association. Ex. fcap. 8vo. 4s. 6d.

SOLID GEOMETRY AND CONIC SECTIONS. With Appen-
dices on Transversals and Harmonic Division. Ex. fcap. 8vo. 3s. 6d.

MACMILLAN & CO., LONDON.

HIGHER PURE MATHEMATICS.

ELEMENTARY TREATISE ON PARTIAL DIFFERENTIAL EQUATIONS. With Diagrams. By Sir G. B. Airy, K.C.B., formerly Astronomer-Royal. 2d Ed. Cr. 8vo. 5s. 6d.

ON THE ALGEBRAICAL AND NUMERICAL THEORY OF ERRORS OF OBSERVATIONS AND THE COMBINATION OF OBSERVATIONS. By the same. 2d Ed., revised. Cr. 8vo. 6s. 6d.

THE CALCULUS OF FINITE DIFFERENCES. By G. Boole. 3d Ed., revised by J. F. Moulton, Q.C. Cr. 8vo. 10s. 6d.

THE DIFFERENTIAL CALCULUS. By Joseph Edwards, M.A. With Applications and numerous Examples. Cr. 8vo. 10s. 6d.

AN ELEMENTARY TREATISE ON SPHERICAL HARMONICS, AND SUBJECTS CONNECTED WITH THEM. By Rev. N. M. Ferrers, D.D., F.R.S., Master of Gonville and Caius College, Cambridge. Cr. 8vo. 7s. 6d.

A TREATISE ON DIFFERENTIAL EQUATIONS. By Andrew Russell Forsyth, F.R.S., Fellow and Assistant Tutor of Trinity College, Cambridge. 2d Ed. 8vo. 14s.

AN ELEMENTARY TREATISE ON CURVE TRACING. By Percival Frost, M.A., D.Sc. 8vo. 12s.

GEOMETRY OF POSITION. By R. H. Graham. Cr. 8vo. 7s. 6d.

DIFFERENTIAL AND INTEGRAL CALCULUS. By A. G. Greenhill, Professor of Mathematics to the Senior Class of Artillery Officers, Woolwich. New Ed. Cr. 8vo. 10s. 6d.

APPLICATIONS OF ELLIPTIC FUNCTIONS. By the same. [In the Press.

INTRODUCTION TO QUATERNIONS, with numerous examples. By P. Kelland and P. G. Tait, Professors in the Department of Mathematics in the University of Edinburgh. 2d Ed. Cr. 8vo. 7s. 6d.

HOW TO DRAW A STRAIGHT LINE: a Lecture on Linkages. By A. B. Kempe. Illustrated. Cr. 8vo. 1s. 6d.

DIFFERENTIAL CALCULUS FOR BEGINNERS. By Alexander Knox. Fcap. 8vo. 3s. 6d.

THE THEORY OF DETERMINANTS IN THE HISTORICAL ORDER OF ITS DEVELOPMENT. Part I. Determinants in General. Leibnitz (1693) to Cayley (1841). By Thomas Muir, Mathematical Master in the High School of Glasgow. 8vo. 10s. 6d.

AN ELEMENTARY TREATISE ON THE DIFFERENTIAL CALCULUS. Founded on the Method of Rates or Fluxions. By J. M. Rice, Professor of Mathematics in the United States Navy, and W. W. Johnson, Professor of Mathematics in the United States Naval Academy. 3d Ed., revised and corrected. 8vo. 18s. Abridged Ed. 9s.

WORKS BY ISAAC TODHUNTER, F.R.S.

AN ELEMENTARY TREATISE ON THE THEORY OF EQUATIONS. Cr. 8vo. 7s. 6d.

A TREATISE ON THE DIFFERENTIAL CALCULUS. Cr. 8vo. 10s. 6d. KEY. Cr. 8vo. 10s. 6d.

A TREATISE ON THE INTEGRAL CALCULUS AND ITS APPLICATIONS. Cr. 8vo. 7s. 6d. KEY. Cr. 8vo. 10s. 6d.

AN ELEMENTARY TREATISE ON LAPLACE'S, LAMÉ'S AND BESSEL'S FUNCTIONS. Cr. 8vo. 10s. 6d.

MACMILLAN & CO., LONDON.

TRIGONOMETRY.

AN ELEMENTARY TREATISE ON PLANE TRIGONO-METRY. With Examples. By R. D. Beasley, M.A. 9th Ed., revised and enlarged. Cr. 8vo. 3s. 6d.

FOUR-FIGURE MATHEMATICAL TABLES. Comprising Logarithmic and Trigonometrical Tables, and Tables of Squares, Square Roots, and Reciprocals. By J. T. Bottomley, M.A., Lecturer in Natural Philosophy in the University of Glasgow. 8vo. 2s. 6d.

THE ALGEBRA OF CO-PLANAR VECTORS AND TRIGON-OMETRY. By R. B. Hayward, M.A., F.R.S., Assistant Master at Harrow. [In preparation.

A TREATISE ON TRIGONOMETRY. By W. E. Johnson, M.A., late Scholar and Assistant Mathematical Lecturer at King's College, Cambridge. Cr. 8vo. 8s. 6d.

ELEMENTS OF TRIGONOMETRY. By Rawdon Levett and A. F. Davison, Assistant Masters at King Edward's School, Birmingham. Cr. 8vo. [In the Press.

A TREATISE ON SPHERICAL TRIGONOMETRY. With applications to Spherical Geometry and numerous Examples. By W. J. M'Clelland, M.A., Principal of the Incorporated Society's School, Santry, Dublin, and T. Preston, M.A. Cr. 8vo. 8s. 6d., or; Part I. To the End of Solution of Triangles, 4s. 6d. Part II., 5s.

MANUAL OF LOGARITHMS. By G. F. Matthews, B.A. 8vo. 5s. nett

TEXT-BOOK OF PRACTICAL LOGARITHMS AND TRIGON-OMETRY. By J. H. Palmer, Headmaster, R.N., H.M.S. Cambridge, Devonport. Gl. 8vo. 4s. 6d.

EXAMPLES FOR PRACTICE IN THE USE OF SEVEN-FIGURE LOGARITHMS. By Joseph Wolstenholme, D.Sc., late Professor of Mathematics in the Royal Indian Engineering College, Cooper's Hill. 8vo. 5s.

THE ELEMENTS OF PLANE AND SPHERICAL TRIGON-OMETRY. By J. C. Snowball. 14th Ed. Cr. 8vo. 7s. 6d.

WORKS BY REV. J. B. LOCK, M.A.,
Senior Fellow and Bursar of Gonville and Caius College, Cambridge.

THE TRIGONOMETRY OF ONE ANGLE. Gl. 8vo. 2s. 6d.

TRIGONOMETRY FOR BEGINNERS, as far as the Solution of Triangles. 3d Ed. Gl. 8vo. 2s. 6d. KEY. Cr. 8vo. 6s. 6d.

ELEMENTARY TRIGONOMETRY. 6th Ed. (in this edition the chapter on logarithms has been carefully revised). Gl. 8vo. 4s. 6d. KEY. Cr. 8vo. 8s. 6d.

HIGHER TRIGONOMETRY. 5th Ed. Gl. 8vo. 4s. 6d. Both Parts complete in One Volume. Gl. 8vo. 7s. 6d.

WORKS BY ISAAC TODHUNTER, F.R.S.

TRIGONOMETRY FOR BEGINNERS. 18mo. 2s. 6d. KEY. Cr. 8vo. 8s. 6d.

PLANE TRIGONOMETRY. Cr. 8vo. 5s. A New Edition revised by R. W. Hogg, M.A. Cr. 8vo. 5s. KEY. Cr. 8vo. 10s. 6d.

A TREATISE ON SPHERICAL TRIGONOMETRY. Cr. 8vo. 4s. 6d.

MACMILLAN & CO., LONDON.

Sept. 1891

A Catalogue

OF

Educational Books

PUBLISHED BY

Macmillan & Co.

BEDFORD STREET, STRAND, LONDON

CONTENTS

A

CLASSICS.

Elementary Classics; Classical Series; Classical Library, (1) Texts, (2) Translations; Grammar, Composition, and Philology; Antiquities, Ancient History, and Philosophy.

*ELEMENTARY CLASSICS.

18mo, Eighteenpence each.

The following contain Introductions, Notes, and Vocabularies, and in some cases Exercises.

ACCIDENCE, LATIN, AND EXERCISES ARRANGED FOR BEGINNERS.—By W. WELCH, M.A., and C. G. DUFFIELD, M.A.

AESCHYLUS.—PROMETHEUS VINCTUS. By Rev. H. M. STEPHENSON, M.A.

ARRIAN.—SELECTIONS. With Exercises. By Rev. JOHN BOND, M.A., and Rev. A. S. WALPOLE, M.A.

AULUS GELLIUS, STORIES FROM.—Adapted for Beginners. With Exercises. By Rev. G. H. NALL, M.A., Assistant Master at Westminster.

CÆSAR.—THE HELVETIAN WAR. Being Selections from Book I. of The Gallic War. Adapted for Beginners. With Exercises. By W. WELCH, M.A., and C. G. DUFFIELD, M.A.

THE INVASION OF BRITAIN. Being Selections from Books IV. and V. of The Gallic War. Adapted for Beginners. With Exercises. By W. WELCH, M.A., and C. G. DUFFIELD, M.A.

SCENES FROM BOOKS V. AND VI. By C. COLBECK, M.A.

THE GALLIC WAR. BOOK I. By Rev. A. S. WALPOLE, M.A.

BOOKS II. AND III. By the Rev. W. G. RUTHERFORD, M.A., LL.D.

BOOK IV. By CLEMENT BRYANS, M.A., Assistant Master at Dulwich College.

BOOK V. By C. COLBECK, M.A., Assistant Master at Harrow.

BOOK VI. By the same Editor.

BOOK VII. By Rev. J. BOND, M.A., and Rev. A. S. WALPOLE, M.A.

THE CIVIL WAR. BOOK I. By M. MONTGOMERY, M.A.   [In the Press.

CICERO.—DE SENECTUTE. By E. S. SHUCKBURGH, M.A.

DE AMICITIA. By the same Editor.

STORIES OF ROMAN HISTORY. Adapted for Beginners. With Exercises. By Rev. G. E. JEANS, M.A., and A. V. JONES, M.A.

EURIPIDES.—ALCESTIS. By Rev. M. A. BAYFIELD, M.A.

MEDEA. By A. W. VERRALL, Litt.D., and Rev. M. A. BAYFIELD, M.A.
          [In the Press.

HECUBA. By Rev. J. BOND, M.A., and Rev. A. S. WALPOLE, M.A.

EUTROPIUS.—Adapted for Beginners. With Exercises. By W. WELCH, M.A., and C. G. DUFFIELD, M.A.

HERODOTUS. TALES FROM HERODOTUS. Atticised by G. S. FARNELL, M.A.   [In the Press.

HOMER.—ILIAD. BOOK I. By Rev. J. BOND, M.A., and Rev. A. S. WALPOLE, M.A.

BOOK XVIII. By S. R. JAMES, M.A., Assistant Master at Eton.

ODYSSEY. BOOK I. By Rev. J. BOND, M.A., and Rev. A. S. WALPOLE, M.A.

HORACE.—ODES. BOOKS I.-IV. By T. E. PAGE, M.A., Assistant Master at the Charterhouse. Each 1s. 6d.

LIVY.—BOOK I. By H. M. STEPHENSON, M.A.

BOOK XXI. Adapted from Mr. Capes's Edition. By J. E. MELHUISH, M.A.

BOOK XXII. By the same.

THE HANNIBALIAN WAR. Being part of the **XXI.** and **XXII. BOOKS OF LIVY** adapted for Beginners. By G. C. MACAULAY, M.A.

THE SIEGE OF SYRACUSE. Being part of the **XXIV.** and **XXV. BOOKS OF LIVY**, adapted for Beginners. With Exercises. By G. RICHARDS, M.A., and Rev. A. S. WALPOLE, M.A.

LEGENDS OF ANCIENT ROME. Adapted for Beginners. With Exercises. By H. WILKINSON, M.A.

LUCIAN.—EXTRACTS FROM LUCIAN. With Exercises. By Rev. J. BOND, M.A., and Rev. A. S. WALPOLE, M.A.

NEPOS.—SELECTIONS ILLUSTRATIVE OF GREEK AND ROMAN HISTORY. With Exercises. By G. S. FARNELL, M.A.

OVID.—SELECTIONS. By E. S. SHUCKBURGH, M.A.

EASY SELECTIONS FROM OVID IN ELEGIAC VERSE. With Exercises. By H. WILKINSON, M.A.

STORIES FROM THE METAMORPHOSES. With Exercises. By Rev. J. BOND, M.A., and Rev. A. S. WALPOLE, M.A.

PHÆDRUS. — SELECT FABLES. Adapted for Beginners. With Exercises. By Rev. A. S. WALPOLE, M.A.

THUCYDIDES.—THE RISE OF THE ATHENIAN EMPIRE. BOOK I. CHS 89-117 and 228-238. With Exercises. By F. H. COLSON, M.A.

VIRGIL.—SELECTIONS. By E. S. SHUCKBURGH, M.A.

BUCOLICS. By T. E. PAGE, M.A.

GEORGICS. BOOK I. By the same Editor.

BOOK II. By Rev. J. H. SKRINE, M.A.

ÆNEID. BOOK I. By Rev. A. S. WALPOLE, M.A.

BOOK II. By T. E. PAGE, M.A.

BOOK III. By the same Editor.

BOOK IV. By Rev. H. M. STEPHENSON, M.A.

BOOK V. By Rev. A. CALVERT, M.A.

BOOK VI. By T. E. PAGE, M.A.

BOOK VII. By Rev. A. CALVERT, M.A.

BOOK VIII. By the same Editor.

BOOK IX. By Rev. H. M. STEPHENSON, M.A.

BOOK X. By S. G. OWEN, M.A.

XENOPHON.—ANABASIS. Selections, adapted for Beginners. With Exercises. By W. WELCH, M.A., and C. G. DUFFIELD, M.A.

BOOK I. With Exercises. By E. A. WELLS, M.A.

BOOK I. By Rev. A. S. WALPOLE, M.A.

BOOK II. By the same Editor.

BOOK III. By Rev. G. H. NALL, M.A.

BOOK IV. By Rev. E. D. STONE, M.A.

SELECTIONS FROM BOOK IV. With Exercises. By the same Editor.

SELECTIONS FROM THE CYROPÆDIA. With Exercises. By A. H. COOKE, M.A., Fellow and Lecturer of King's College, Cambridge.

The following contain Introductions and Notes, **but no Vocabulary :—**

CICERO.—SELECT LETTERS. By Rev. G. E. JEANS, M.A.

HERODOTUS.—SELECTIONS FROM BOOKS VII. AND VIII. THE EXPEDITION OF XERXES. By A. H. COOKE, M.A.

HORACE.—SELECTIONS FROM THE SATIRES AND EPISTLES. By Rev. W. J. V. BAKER, M.A.

SELECT EPODES AND ARS POETICA. By H. A. DALTON, M.A., Assistant Master at Winchester.

PLATO.—EUTHYPHRO AND MENEXENUS. By C. E. GRAVES, M.A.

TERENCE.—SCENES FROM THE ANDRIA. By F. W. CORNISH, M.A., Assistant Master at Eton.

THE GREEK ELEGIAC POETS.—FROM CALLINUS TO CALLIMACHUS. Selected by Rev. HERBERT KYNASTON, D.D.

THUCYDIDES.—BOOK IV. CHS. 1-41. THE CAPTURE OF SPHACTERIA. By C. E. GRAVES, M.A.

CLASSICAL SERIES
FOR COLLEGES AND SCHOOLS.

Fcap. 8vo.

ÆSCHINES.—IN CTESIPHONTA. By Rev. T. GWATKIN, M.A., and E. S. SHUCKBURGH, M.A. 5s.

ÆSCHYLUS.—PERSÆ. By A. O. PRICKARD, M.A., Fellow and Tutor of New College, Oxford. With Map. 2s. 6d.

SEVEN AGAINST THEBES. SCHOOL EDITION. By A. W. VERRALL, Litt.D., Fellow of Trinity College, Cambridge, and M. A. BAYFIELD, M.A., Headmaster of Christ's College, Brecon. 2s. 6d.

ANDOCIDES.—DE MYSTERIIS. By W. J. HICKIE, M.A. 2s. 6d.

ATTIC ORATORS.—Selections from ANTIPHON, ANDOCIDES, LYSIAS, ISOCRATES, and ISAEUS. By R. C. JEBB, Litt.D., Regius Professor of Greek in the University of Cambridge. 5s.

*CÆSAR.—THE GALLIC WAR. By Rev. JOHN BOND, M.A., and Rev. A. S. WALPOLE, M.A. With Maps. 4s. 6d.

CATULLUS.—SELECT POEMS. Edited by F. P. SIMPSON, B.A. 3s. 6d. The Text of this Edition is carefully expurgated for School use.

*CICERO.—THE CATILINE ORATIONS. By A. S. WILKINS, Litt.D., Professor of Latin in the Owens College, Victoria University, Manchester. 2s. 6d.

PRO LEGE MANILIA. By Prof. A. S. WILKINS, Litt.D. 2s. 6d.

THE SECOND PHILIPPIC ORATION. By JOHN E. B. MAYOR, M.A., Professor of Latin in the University of Cambridge. 3s. 6d.

PRO ROSCIO AMERINO. By E. H. DONKIN, M.A. 2s. 6d.

PRO P. SESTIO. By Rev. H. A. HOLDEN, Litt.D. 3s. 6d.

SELECT LETTERS. Edited by R. Y. TYRRELL, M.A. [In the Press.

DEMOSTHENES.—DE CORONA. By B. DRAKE, M.A. 7th Edition, revised by E. S. SHUCKBURGH, M.A. 3s. 6d.

ADVERSUS LEPTINEM. By Rev. J. R. KING, M.A., Fellow and Tutor of Oriel College, Oxford. 2s. 6d.

THE FIRST PHILIPPIC. By Rev. T. GWATKIN, M.A. 2s. 6d.

IN MIDIAM. By Prof. A. S. WILKINS, Litt.D., and HERMAN HAGER, Ph.D., of the Owens College, Victoria University, Manchester. [In preparation.

EURIPIDES.—HIPPOLYTUS. By Rev. J. P. MAHAFFY, D.D., Fellow of Trinity College, and Professor of Ancient History in the University of Dublin, and J. B. BURY, M.A., Fellow of Trinity College, Dublin. 2s. 6d.

MEDEA. By A. W. VERRALL, Litt.D., Fellow of Trinity College, Cambridge. 2s. 6d.

IPHIGENIA IN TAURIS. By E. B. ENGLAND, M.A. 3s.

ION. By M. A. BAYFIELD, M.A., Headmaster of Christ's College, Brecon. 2s. 6d.

BACCHAE. By R. Y. TYRRELL, M.A., Regius Professor of Greek in the University of Dublin. [In preparation.

HERODOTUS.—BOOK III. By G. C. MACAULAY, M.A. 2s. 6d.

BOOK V. By J. STRACHAN, M.A., Professor of Greek in the Owens College, Victoria University, Manchester. [In preparation.

BOOK VI. By the same. 3s. 6d.

BOOK VII. By Mrs. A. F. BUTLER. 3s. 6d.

HESIOD.—THE WORKS AND DAYS. By W. T. LENDRUM, M.A., Assistant Master at Dulwich College. [In preparation.

HOMER.—ILIAD. BOOKS I., IX., XI., XVI.-XXIV. THE STORY OF ACHILLES. By the late J. H. Pratt, M.A., and Walter Leaf, Litt.D., Fellows of Trinity College, Cambridge. 5s.

ODYSSEY. BOOK IX. By Prof. John E. B. Mayor. 2s. 6d.

ODYSSEY. BOOKS XXI.-XXIV. THE TRIUMPH OF ODYSSEUS. By S. G. Hamilton, B.A., Fellow of Hertford College, Oxford. 2s. 6d.

HORACE.—*THE ODES. By T. E. Page, M.A., Assistant Master at the Charterhouse. 5s. (BOOKS I., II., III., and IV. separately, 2s. each.)

THE SATIRES. By Arthur Palmer, M.A., Professor of Latin in the University of Dublin. 5s.

THE EPISTLES AND ARS POETICA. By A. S. Wilkins, Litt.D., Professor of Latin in the Owens College, Victoria University, Manchester. 5s.

ISAEOS.—THE ORATIONS. By William Ridgeway, M.A., Professor of Greek in Queen's College, Cork. [In preparation.

JUVENAL.—*THIRTEEN SATIRES. By E. G. Hardy, M.A. 5s. The Text is carefully expurgated for School use.

SELECT SATIRES. By Prof. John E. B. Mayor. X. and XI. 3s. 6d. XII.-XVI. 4s. 6d.

LIVY.—*BOOKS II. and III. By Rev. H. M. Stephenson, M.A. 3s. 6d.

*BOOKS XXI. and XXII. By Rev. W. W. Capes, M.A. With Maps. 4s. 6d.

*BOOKS XXIII. and XXIV. By G. C. Macaulay, M.A. With Maps. 3s. 6d.

*THE LAST TWO KINGS OF MACEDON. EXTRACTS FROM THE FOURTH AND FIFTH DECADES OF LIVY. By F. H. Rawlins, M.A., Assistant Master at Eton. With Maps. 2s. 6d.

THE SUBJUGATION OF ITALY. SELECTIONS FROM THE FIRST DECADE. By G. E. Marindin, M.A. [In preparation.

LUCRETIUS.—BOOKS I.-III. By J. H. Warburton Lee, M.A., Assistant Master at Rossall. 3s. 6d.

LYSIAS.—SELECT ORATIONS. By E. S. Shuckburgh, M.A. 5s.

MARTIAL.—SELECT EPIGRAMS. By Rev. H. M. Stephenson, M.A. 5s.

*OVID.—FASTI. By G. H. Hallam, M.A., Assistant Master at Harrow. With Maps. 3s. 6d.

*HEROIDUM EPISTULÆ XIII. By E. S. Shuckburgh, M.A. 3s. 6d.

METAMORPHOSES. BOOKS I.-III. By C. Simmons, M.A. [In preparation. BOOKS XIII. and XIV. By the same Editor. 3s. 6d.

PLATO.—LACHES. By M. T. Tatham, M.A. 2s. 6d.

THE REPUBLIC. BOOKS I.-V. By T. H. Warren, M.A., President of Magdalen College, Oxford. 5s.

PLAUTUS.—MILES GLORIOSUS. By R. Y. Tyrrell, M.A., Regius Professor of Greek in the University of Dublin. 2d Ed., revised. 3s. 6d.

AMPHITRUO. By Arthur Palmer, M.A., Professor of Latin in the University of Dublin. 3s. 6d.

CAPTIVI. By A. Rhys-Smith, M.A. [In the Press.

PLINY.—LETTERS. BOOKS I. and II. By J. Cowan, M.A., Assistant Master at the Manchester Grammar School. 3s.

LETTERS. BOOK III. By Prof. John E. B. Mayor. With Life of Pliny by G. H. Rendall, M.A. 3s. 6d.

PLUTARCH.—LIFE OF THEMISTOKLES. By Rev. H. A. Holden, Litt.D. 3s. 6d.

LIVES OF GALBA AND OTHO. By E. G. Hardy, M.A. 5s.

POLYBIUS.—THE HISTORY OF THE ACHÆAN LEAGUE AS CONTAINED IN THE REMAINS OF POLYBIUS. By W. W. Capes, M.A. 5s.

PROPERTIUS.—SELECT POEMS. By Prof. J. P. Postgate, Litt.D., Fellow of Trinity College, Cambridge. 2d Ed., revised. 5s.

SALLUST.—*CATILINA and JUGURTHA. By C. Merivale, D.D., Dean of Ely. 3s. 6d. Or separately, 2s. each.

*BELLUM CATULINÆ. By A. M. Cook, M.A., Assistant Master at St. Paul's School. 2s. 6d.

JUGURTHA. By the same Editor. [In preparation.

TACITUS.—THE ANNALS. BOOKS I. and II. By J. S. Reid, Litt.D. [*In prep.*
THE ANNALS. BOOK VI. By A. J. Church, M.A., and W. J. Brodribb, M.A. 2s.
THE HISTORIES. BOOKS I. and II. By A. D. Godley, M.A., Fellow of Magdalen College, Oxford. 3s. 6d. BOOKS III.-V. By the same. 3s. 6d.
AGRICOLA and GERMANIA. By A. J. Church, M.A., and W. J. Brodribb, M.A. 3s. 6d. Or separately, 2s. each.
TERENCE.—HAUTON TIMORUMENOS. By E. S. Shuckburgh, M.A. 2s. 6d. With Translation. 3s. 6d.
PHORMIO. By Rev. John Bond, M.A., and Rev. A. S. Walpole, M.A. 2s. 6d.
THUCYDIDES.—BOOK I. By C. Bryans, M.A. [*In preparation.*
BOOK II. By E. C. Marchant, M.A., Assistant Master at St. Paul's. [*In the Press.*
BOOK III. By C. Bryans, M.A. [*In preparation.*
BOOK IV. By C. E. Graves, M.A., Classical Lecturer at St. John's College, Cambridge. 3s. 6d.
BOOK V. By the same Editor. [*In the Press.*
BOOKS VI. and VII. THE SICILIAN EXPEDITION. By Rev. Percival Frost, M.A. With Map. 3s. 6d.
BOOK VIII. By Prof. T. G. Tucker, Litt.D. [*In the Press.*
TIBULLUS.—SELECT POEMS. By Prof. J. P. Postgate, Litt.D. [*In preparation.*
VIRGIL.—ÆNEID. BOOKS II. and III. THE NARRATIVE OF ÆNEAS. By E. W. Howson, M.A., Assistant Master at Harrow. 2s.
XENOPHON.—*THE ANABASIS. BOOKS I.-IV. By Profs. W. W. Goodwin and J. W. White. Adapted to Goodwin's Greek Grammar. With Map. 3s. 6d.
HELLENICA. BOOKS I. and II. By H. Hailstone, B.A. With Map. 2s. 6d.
CYROPÆDIA. BOOKS VII. and VIII. By A. Goodwin, M.A., Professor of Classics in University College, London. 2s. 6d.
MEMORABILIA SOCRATIS. By A. R. Cluer, B.A., Balliol College, Oxford. 5s.
HIERO. By Rev. H. A. Holden, Litt.D., LL.D. 2s. 6d.
OECONOMICUS. By the same. With Lexicon. 5s.

CLASSICAL LIBRARY.

Texts, Edited with **Introductions and Notes,** for the use of Advanced Students ; Commentaries and Translations.

ÆSCHYLUS.—THE SUPPLICES. A Revised Text, with Translation. By T. G. Tucker, Litt.D., Professor of Classical Philology in the University of Melbourne. 8vo. 10s. 6d.
THE SEVEN AGAINST THEBES. With Translation. By A. W. Verrall, Litt.D., Fellow of Trinity College, Cambridge. 8vo. 7s. 6d.
AGAMEMNON. With Translation. By A. W. Verrall, Litt.D. 8vo. 12s.
AGAMEMNON, CHOEPHOROE, AND EUMENIDES. By A. O. Prickard, M.A., Fellow and Tutor of New College, Oxford. 8vo. [*In preparation.*
THE EUMENIDES. With Verse Translation. By Bernard Drake, M.A. 8vo. 5s.
ANTONINUS, MARCUS AURELIUS.—BOOK IV. OF THE MEDITATIONS. With Translation. By Hastings Crossley, M.A. 8vo. 6s.
ARISTOTLE.—THE METAPHYSICS. BOOK I. Translated by a Cambridge Graduate. 8vo. 5s.
THE POLITICS. By R. D. Hicks, M.A., Fellow of Trinity College, Cambridge. 8vo. [*In the Press.*
THE POLITICS. Translated by Rev. J. E. C. Welldon, M.A., Headmaster of Harrow. Cr. 8vo. 10s. 6d.
THE RHETORIC. Translated by the same. Cr. 8vo. 7s. 6d.
AN INTRODUCTION TO ARISTOTLE'S RHETORIC. With Analysis, Notes, and Appendices. By E. M. Cope, Fellow and late Tutor of Trinity College, Cambridge. 8vo. 14s.

THE ETHICS. Translated by Rev. J. E. C. WELLDON, M.A. Cr. 8vo. [In prep.
THE SOPHISTICI ELENCHI. With Translation. By E. POSTE, M.A., Fellow of Oriel College, Oxford. 8vo. 8s. 6d.
ON THE CONSTITUTION OF ATHENS. Edited by J. E. SANDYS, Litt.D.
ON THE CONSTITUTION OF ATHENS. Translated by E. POSTE, M.A. Cr. 8vo. 3s. 6d.
ON THE ART OF POETRY. A Lecture. By A. O. PRICKARD, M.A., Fellow and Tutor of New College, Oxford. Cr. 8vo. 3s. 6d.
ARISTOPHANES.—THE BIRDS. Translated into English Verse. By B. H. KENNEDY, D.D. Cr. 8vo. 6s. Help Notes to the Same, for the Use of Students. 1s. 6d.
ATTIC ORATORS.—FROM ANTIPHON TO ISAEOS. By R. C. JEBB, Litt.D., Regius Professor of Greek in the University of Cambridge. 2 vols. 8vo. 25s.
BABRIUS.—With Lexicon. By Rev. W. G..RUTHERFORD, M.A., LL.D., Head-master of Westminster. 8vo. 12s. 6d.
CICERO.—THE ACADEMICA. By J. S. REID, Litt.D., Fellow of Caius College, Cambridge. 8vo. 15s.
THE ACADEMICS. Translated by the same. 8vo. 5s. 6d.
SELECT LETTERS. After the Edition of ALBERT WATSON, M.A. Translated by G. E. JEANS, M.A., Fellow of Hertford College, Oxford. Cr. 8vo. 10s. 6d.
EURIPIDES.—MEDEA. Edited by A. W. VERRALL, Litt.D. 8vo. 7s. 6d.
IPHIGENEIA AT AULIS. Edited by E. B. ENGLAND, M.A. 8vo. 7s. 6d.
*INTRODUCTION TO THE STUDY OF EURIPIDES. By Professor J. P. MAHAFFY. Fcap. 8vo. 1s. 6d. (Classical Writers.)
HERODOTUS.—BOOKS I.-III. THE ANCIENT EMPIRES OF THE EAST. Edited by A. H. SAYCE, Deputy-Professor of Comparative Philology, Oxford. 8vo. 16s.
BOOKS IV.-IX. Edited by R. W. MACAN, M.A., Reader in Ancient History in the University of Oxford. 8vo. [In preparation.
THE HISTORY. Translated by G. C. MACAULAY, M.A. 2 vols. Cr. 8vo. 18s.
HOMER.—THE ILIAD. By WALTER LEAF, Litt.D. 8vo. Books I.-XII. 14s. Books XIII.-XXIV. 14s.
THE ILIAD. Translated into English Prose by ANDREW LANG, M.A., WALTER LEAF, Litt.D., and ERNEST MYERS, M.A. Cr. 8vo. 12s. 6d.
THE ODYSSEY. Done into English by S. H. BUTCHER, M.A., Professor of Greek in the University of Edinburgh, and ANDREW LANG, M.A. Cr. 8vo. 6s.
*INTRODUCTION TO THE STUDY OF HOMER. By the Right Hon. W. E. GLADSTONE. 18mo. 1s. (Literature Primers.)
HOMERIC DICTIONARY. Translated from the German of Dr. G. AUTENRIETH by R. P. KEEP, Ph.D. Illustrated. Cr. 8vo. 6s.
HORACE.—Translated by J. LONSDALE, M.A., and S. LEE, M.A. Gl. 8vo. 3s. 6d.
STUDIES, LITERARY AND HISTORICAL, IN THE ODES OF HORACE. By A. W. VERRALL, Litt.D. 8vo. 8s. 6d.
JUVENAL.—THIRTEEN SATIRES OF JUVENAL. By JOHN E. B. MAYOR, M.A., Professor of Latin in the University of Cambridge. Cr. 8vo. 2 vols. 10s. 6d. each. Vol. I. 10s. 6d. Vol. II. 10s. 6d.
THIRTEEN SATIRES. Translated by ALEX. LEEPER, M.A., LL.D., Warden of Trinity College, Melbourne. Cr. 8vo. 3s. 6d.
KTESIAS.—THE FRAGMENTS OF THE PERSIKA OF KTESIAS. By JOHN GILMORE, M.A. 8vo. 8s. 6d.
LIVY.—BOOKS I.-IV. Translated by Rev. H. M. STEPHENSON, M.A. [In prep.
BOOKS XXI.-XXV. Translated by A. J. CHURCH, M.A., and W. J. BRODRIBB, M.A. Cr. 8vo. 7s. 6d.
*INTRODUCTION TO THE STUDY OF LIVY. By Rev. W. W. CAPES, M.A. Fcap. 8vo. 1s. 6d. (Classical Writers.)
LONGINUS.—ON THE SUBLIME. Translated by H. L. HAVELL. B.A. With Introduction by ANDREW LANG. Cr. 8vo. 4s. 6d.

MARTIAL.—BOOKS I. AND II. OF THE EPIGRAMS. By Prof. JOHN E. B. MAYOR, M.A. 8vo. [*In the Press.*

MELEAGER.—FIFTY POEMS OF MELEAGER. Translated by WALTER HEAD-LAM. Fcap. 4to. 7s. 6d.

PAUSANIAS.—DESCRIPTION OF GREECE. Translated with Commentary by J. G. FRAZER, M.A., Fellow of Trinity College, Cambridge. [*In prep.*

PHRYNICHUS.—THE NEW PHRYNICHUS; being a Revised Text of the Ecloga of the Grammarian Phrynichus. With Introduction and Commentary by Rev. W. G. RUTHERFORD, M.A., LL.D., Headmaster of Westminster. 8vo. 18s.

PINDAR.—THE EXTANT ODES OF PINDAR. Translated by ERNEST MYERS, M.A. Cr. 8vo. 5s.

THE OLYMPIAN AND PYTHIAN ODES. Edited, with an Introductory Essay, by BASIL GILDERSLEEVE, Professor of Greek in the Johns Hopkins University, U.S.A. Cr. 8vo. 7s. 6d.

THE NEMEAN ODES. By J. B. BURY, M.A., Fellow of Trinity College, Dublin. 8vo. 12s.

THE ISTHMIAN ODES. By the same Editor. [*In the Press.*

PLATO.—PHÆDO. By R. D. ARCHER-HIND, M.A., Fellow of Trinity College, Cambridge. 8vo. 8s. 6d.

PHÆDO. By W. D. GEDDES, LL.D., Principal of the University of Aberdeen. 8vo. 8s. 6d.

TIMAEUS. With Translation. By R. D. ARCHER-HIND, M.A. 8vo. 16s.

THE REPUBLIC OF PLATO. Translated by J. LL. DAVIES, M.A., and D. J. VAUGHAN, M.A. 18mo. 4s. 6d.

EUTHYPHRO, APOLOGY, CRITO, AND PHÆDO. Translated by F. J. CHURCH. 18mo. 4s. 6d.

PHÆDRUS, LYSIS, AND PROTAGORAS. Translated by J. WRIGHT, M.A. 18mo. 4s. 6d.

PLAUTUS.—THE MOSTELLARIA. By WILLIAM RAMSAY, M.A. Edited by G. G. RAMSAY, M.A., Professor of Humanity in the University of Glasgow. 8vo. 14s.

PLINY.—CORRESPONDENCE WITH TRAJAN. C. Plinii Caecilii Secundi Epistulæ ad Traianum Imperatorem cum Eiusdem Responsis. By E. G. HARDY, M.A. 8vo. 10s. 6d.

POLYBIUS.—THE HISTORIES OF POLYBIUS. Translated by E. S. SHUCK-BURGH, M.A. 2 vols. Cr. 8vo. 24s.

SALLUST.—CATILINE AND JUGURTHA. Translated by A. W. POLLARD, B.A. Cr. 8vo. 6s. THE CATILINE (separately). 3s.

SOPHOCLES.—ŒDIPUS THE KING. Translated into English Verse by E. D. A. MORSHEAD, M.A., Assistant Master at Winchester. Fcap. 8vo. 3s. 6d.

TACITUS.—THE ANNALS. By G. O. HOLBROOKE, M.A., Professor of Latin in Trinity College, Hartford, U.S.A. With Maps. 8vo. 16s.

THE ANNALS. Translated by A. J. CHURCH, M.A., and W. J. BRODRIBB, M.A. With Maps. Cr. 8vo. 7s. 6d.

THE HISTORIES. By Rev. W. A. SPOONER, M.A., Fellow and Tutor of New College, Oxford. 8vo. 16s.

THE HISTORY. Translated by A. J. CHURCH, M.A., and W. J. BRODRIBB, M.A. With Map. Cr. 8vo. 6s.

THE AGRICOLA AND GERMANY, WITH THE DIALOGUE ON ORATORY. Translated by A. J. CHURCH, M.A., and W. J. BRODRIBB, M.A. With Maps. Cr. 8vo. 4s. 6d

*INTRODUCTION TO THE STUDY OF TACITUS. By A. J. CHURCH, M.A., and W. J. BRODRIBB, M.A. Fcap. 8vo. 1s. 6d. (*Classical Writers.*)

THEOCRITUS, BION, AND MOSCHUS. Translated by A. LANG, M.A. 18mo. 4s. 6d. Also an Edition on Large Paper. Cr. 8vo. 9s.

THUCYDIDES.—BOOK IV. A Revision of the Text, Illustrating the Principal Causes of Corruption in the Manuscripts of this Author. By Rev. W. G. RUTHERFORD, M.A., LL.D., Headmaster of Westminster. 8vo. 7s. 6d.

BOOK VIII. By H. C. GOODHART, M.A., Fellow of Trinity College, Cambridge.
[In the Press.

VIRGIL.—Translated by J. LONSDALE, M.A., and S. LEE, M.A. Gl. 8vo. 3s. 6d.
THE ÆNEID. Translated by J. W. MACKAIL, M.A., Fellow of Balliol College, Oxford. Cr. 8vo. 7s. 6d.

XENOPHON.—Translated by H. G. DAKYNS, M.A. In four vols. Cr. 8vo. Vol. I., containing "The Anabasis" and Books I. and II. of "The Hellenica." 10s. 6d. Vol. II. "Hellenica" III.-VII., and the two Polities—"Athenian" and "Laconian," the "Agesilaus," and the tract on "Revenues." With Maps and Plans. [In the Press.

GRAMMAR, COMPOSITION, & PHILOLOGY.

*BELCHER.—SHORT EXERCISES IN LATIN PROSE COMPOSITION AND EXAMINATION PAPERS IN LATIN GRAMMAR. Part I. By Rev. H. BELCHER, LL.D., Rector of the High School, Dunedin, N.Z. 18mo. 1s. 6d. KEY, for Teachers only. 18mo. 3s. 6d.
*Part II., On the Syntax of Sentences, with an Appendix, including EXERCISES IN LATIN IDIOMS, etc. 18mo. 2s. KEY, for Teachers only. 18mo. 3s.

BLACKIE.—GREEK AND ENGLISH DIALOGUES FOR USE IN SCHOOLS AND COLLEGES. By JOHN STUART BLACKIE, Emeritus Professor of Greek in the University of Edinburgh. New Edition. Fcap. 8vo. 2s. 6d.
A GREEK PRIMER, COLLOQUIAL AND CONSTRUCTIVE. Cr. 8vo. 2s. 6d.

*BRYANS.—LATIN PROSE EXERCISES BASED UPON CÆSAR'S GALLIC WAR. With a Classification of Cæsar's Chief Phrases and Grammatical Notes on Cæsar's Usages. By CLEMENT BRYANS, M.A., Assistant Master at Dulwich College. Ex. fcap. 8vo. 2s. 6d. KEY, for Teachers only. 4s. 6d.
GREEK PROSE EXERCISES based upon Thucydides. By the same.
[In preparation.

COOKSON.—A LATIN SYNTAX. By CHRISTOPHER COOKSON, M.A., Assistant Master at St. Paul's School. 8vo. [In preparation.

CORNELL UNIVERSITY STUDIES IN CLASSICAL PHILOLOGY. Edited by I. FLAGG, W. G. HALE, and B. I. WHEELER. I. The CUM-Constructions: their History and Functions. By W. G. HALE. Part 1. Critical. 1s. 8d. net. Part 2. Constructive. 3s. 4d. net. II. Analogy and the Scope of its Application in Language. By B. I. WHEELER. 1s. 3d. net.

*EICKE.—FIRST LESSONS IN LATIN. By K. M. EICKE, B.A., Assistant Master at Oundle School. Gl. 8vo. 2s. 6d.

*ENGLAND.—EXERCISES ON LATIN SYNTAX AND IDIOM. ARRANGED WITH REFERENCE TO ROBY'S SCHOOL LATIN GRAMMAR. By E. B. ENGLAND, Assistant Lecturer at the Owens College, Victoria University, Manchester. Cr. 8vo. 2s. 6d. KEY, for Teachers only. 2s. 6d.

GILES.—A SHORT MANUAL OF PHILOLOGY FOR CLASSICAL STUDENTS By P. GILES, M.A., Reader in Comparative Philology in the University of Cam bridge. Cr. 8vo. [In the Press

GOODWIN.—Works by W. W. GOODWIN, LL.D., D.C.L., Professor of Greek in Harvard University, U.S.A.
SYNTAX OF THE MOODS AND TENSES OF THE GREEK VERB. New Ed., revised and enlarged. 8vo. 14s.
*A GREEK GRAMMAR. Cr. 8vo. 6s.
*A GREEK GRAMMAR FOR SCHOOLS. Cr. 8vo. 3s. 6d.

GREENWOOD.—THE ELEMENTS OF GREEK GRAMMAR. Adapted to the System of Crude Forms. By J. G. GREENWOOD, sometime Principal of the Owens College, Manchester. Cr. 8vo. 5s. 6d.

HADLEY.—ESSAYS, PHILOLOGICAL AND CRITICAL. By JAMES HADLEY, late Professor in Yale College. 8vo. 16s.

HADLEY and ALLEN. — A GREEK GRAMMAR FOR SCHOOLS AND COLLEGES. By JAMES HADLEY, late Professor in Yale College. Revised and in part rewritten by F. DE F. ALLEN, Professor in Harvard College. Cr. 8vo. 6s.

HODGSON.—MYTHOLOGY FOR LATIN VERSIFICATION. A brief sketch of the Fables of the Ancients, prepared to be rendered into Latin Verse for Schools. By F. Hodgson, B.D., late Provost of Eton. New Ed., revised by F. C. Hodgson, M.A. 18mo. 3s.

*JACKSON.—FIRST STEPS TO GREEK PROSE COMPOSITION By Blomfield Jackson, M.A., Assistant Master at King's College School. 18mo. 1s. 6d. KEY, for Teachers only. 18mo. 3s. 6d.

*SECOND STEPS TO GREEK PROSE COMPOSITION, with Miscellaneous Idioms, Aids to Accentuation, and Examination Papers in Greek Scholarship. By the same. 18mo. 2s. 6d. KEY, for Teachers only. 18mo. 3s. 6d.

KYNASTON.—EXERCISES IN THE COMPOSITION OF GREEK IAMBIC VERSE by Translations from English Dramatists. By Rev. H. Kynaston, D.D., Professor of Classics in the University of Durham. With Vocabulary. Ex. fcap. 8vo. 5s.
KEY, for Teachers only. Ex. fcap. 8vo. 4s. 6d.

LUPTON.—*AN INTRODUCTION TO LATIN ELEGIAC VERSE COMPOSI-TION. By J. H. Lupton, Sur-Master of St. Paul's School. Gl. 8vo. 2s. 6d. KEY TO PART II. (XXV.–C.) Gl. 8vo. 3s. 6d.

*AN INTRODUCTION TO LATIN LYRIC VERSE COMPOSITION. By the same. Gl. 8vo. 3s. KEY, for Teachers only. Gl. 8vo. 4s. 6d.

MACKIE.—PARALLEL PASSAGES FOR TRANSLATION INTO GREEK AND ENGLISH. With Indexes. By Rev. Ellis C. Mackie, M.A., Classical Master at Heversham Grammar School. Gl. 8vo. 4s. 6d.

*MACMILLAN.—FIRST LATIN GRAMMAR. By M. C. Macmillan, M.A. Fcap. 8vo. 1s. 6d.

MACMILLAN'S GREEK COURSE.—Edited by Rev. W. G. Rutherford, M.A., LL.D., Headmaster of Westminster. Gl. 8vo.

*FIRST GREEK GRAMMAR—ACCIDENCE. By the Editor. 2s.

*FIRST GREEK GRAMMAR—SYNTAX. By the same. 2s.
ACCIDENCE AND SYNTAX. In one volume. 3s. 6d.

*EASY EXERCISES IN GREEK ACCIDENCE. By H. G. Underhill, M.A., Assistant Master at St. Paul's Preparatory School. 2s.

*A SECOND GREEK EXERCISE BOOK. By Rev. W. A. Heard, M.A., Headmaster of Fettes College, Edinburgh. 2s. 6d.
EASY EXERCISES IN GREEK SYNTAX. By Rev. G. H. Nall, M.A., Assistant Master at Westminster School. [In preparation.
MANUAL OF GREEK ACCIDENCE. By the Editor. [In preparation.
MANUAL OF GREEK SYNTAX. By the Editor. [In preparation.
ELEMENTARY GREEK COMPOSITION. By the Editor. [In preparation.

*MACMILLAN'S GREEK READER.—STORIES AND LEGENDS. A First Greek Reader, with Notes, Vocabulary, and Exercises. By F. H. Colson, M.A., Headmaster of Plymouth College. Gl. 8vo. 3s.

MACMILLAN'S LATIN COURSE.—By A. M. Cook, M.A., Assistant Master at St. Paul's School.

*FIRST PART. Gl. 8vo. 3s. 6d.
*SECOND PART. 2s. 6d. [Third Part in preparation.

*MACMILLAN'S SHORTER LATIN COURSE.—By A. M. Cook, M.A. Being an abridgment of "Macmillan's Latin Course," First Part. Gl. 8vo. 1s. 6d.

*MACMILLAN'S LATIN READER.—A LATIN READER FOR THE LOWER FORMS IN SCHOOLS. By H. J. Hardy, M.A., Assistant Master at Win-chester. Gl. 8vo. 2s. 6d.

*MARSHALL.—A TABLE OF IRREGULAR GREEK VERBS, classified according to the arrangement of Curtius's Greek Grammar. By J. M. Marshall, M.A., Headmaster of the Grammar School, Durham. 8vo. 1s.

MAYOR.—FIRST GREEK READER. By Prof. John E. B. Mayor, M.A., Fellow of St. John's College, Cambridge. Fcap. 8vo. 4s. 6d.

MAYOR.—GREEK FOR BEGINNERS. By Rev. J. B. MAYOR, M.A., late
   Professor of Classical Literature in King's College, London. Part I., with
   Vocabulary, 1s. 6d. Parts II. and III., with Vocabulary and Index. Fcap.
   8vo. 3s. 6d. Complete in one Vol. 4s. 6d.
NIXON.—PARALLEL EXTRACTS, Arranged for Translation into English and
   Latin, with Notes on Idioms. By J. E. NIXON, M.A., Fellow and Classical
   Lecturer, King's College, Cambridge. Part I.—Historical and Epistolary.
   Cr. 8vo. 3s. 6d.
   PROSE EXTRACTS, Arranged for Translation into English and Latin, with
   General and Special Prefaces on Style and Idiom. By the same. I. Oratorical.
   II. Historical. III. Philosophical. IV. Anecdotes and Letters. 2d Ed.,
   enlarged to 280 pp. Cr. 8vo. 4s. 6d. SELECTIONS FROM THE SAME. 3s.
   Translations of about 70 Extracts can be supplied to Schoolmasters (2s. 6d.),
   on application to the Author: and about 40 similarly of "Parallel Extracts."
   1s. 6d. post free.
*PANTIN.—A FIRST LATIN VERSE BOOK. By W. E. P. PANTIN, M.A.,
   Assistant Master at St. Paul's School. Gl. 8vo. 1s. 6d.
*PEILE.—A PRIMER OF PHILOLOGY. By J. PEILE, Litt.D., Master of Christ's
   College, Cambridge. 18mo. 1s.
*POSTGATE.—SERMO LATINUS. A short Guide to Latin Prose Composition.
   By Prof. J. P. POSTGATE, Litt.D., Fellow of Trinity College, Cambridge. Gl.
   8vo. 2s. 6d. KEY to "Selected Passages." Gl. 8vo. 3s. 6d.
POSTGATE and VINCE.—A DICTIONARY OF LATIN ETYMOLOGY. By
   J. P. POSTGATE and C. A. VINCE.   [In preparation.
POTTS.—*HINTS TOWARDS LATIN PROSE COMPOSITION. By A. W. POTTS,
   M.A., LL.D., late Fellow of St. John's College, Cambridge. Ex. fcap. 8vo. 3s.
   *PASSAGES FOR TRANSLATION INTO LATIN PROSE. Edited with Notes and
   References to the above. Ex. fcap. 8vo. 2s. 6d. KEY, for Teachers only. 2s. 6d.
*PRESTON.—EXERCISES IN LATIN VERSE OF VARIOUS KINDS. By Rev.
   G. PRESTON. Gl. 8vo. 2s. 6d. KEY, for Teachers only. Gl. 8vo. 5s.
REID.—A GRAMMAR OF TACITUS. By J. S. REID, Litt.D., Fellow of Caius
   College, Cambridge.   [In the Press.
   A GRAMMAR OF VIRGIL. By the same.   [In preparation.
ROBY.—Works by H. J. ROBY, M.A., late Fellow of St. John's College, Cambridge.
   A GRAMMAR OF THE LATIN LANGUAGE, from Plautus to Suetonius. Part
   I. Sounds, Inflexions, Word-formation, Appendices. Cr. 8vo. 9s. Part II.
   Syntax, Prepositions, etc. 10s. 6d.
   *SCHOOL LATIN GRAMMAR. Cr. 8vo. 5s.
   AN ELEMENTARY LATIN GRAMMAR.   [In the Press.
*RUSH.—SYNTHETIC LATIN DELECTUS. With Notes and Vocabulary. By E.
   RUSH, B.A. Ex. fcap. 8vo. 2s. 6d.
*RUST.—FIRST STEPS TO LATIN PROSE COMPOSITION. By Rev. G. RUST,
   M.A. 18mo. 1s. 6d. KEY, for Teachers only. By W. M. YATES. 18mo. 3s. 6d.
RUTHERFORD.—Works by the Rev. W. G. RUTHERFORD, M.A., LL.D., Head-
   master of Westminster.
   REX LEX. A Short Digest of the principal Relations between the Latin,
   Greek, and Anglo-Saxon Sounds. 8vo.   [In preparation.
   THE NEW PHRYNICHUS; being a Revised Text of the Ecloga of the Gram-
   marian Phrynichus. With Introduction and Commentary. 8vo. 18s. (See
   also Macmillan's Greek Course.)
SHUCKBURGH.—PASSAGES FROM LATIN AUTHORS FOR TRANSLATION
   INTO ENGLISH. Selected with a view to the needs of Candidates for the
   Cambridge Local, and Public Schools' Examinations. By E. S. SHUCKBURGH,
   M.A. Cr. 8vo. 2s.
*SIMPSON. — LATIN PROSE AFTER THE BEST AUTHORS: Cæsarian Prose.
   By F. P. SIMPSON, B.A. Ex. fcap. 8vo. 2s. 6d. KEY, for Teachers only.
   Ex. fcap. 8vo. 5s.
STRACHAN and WILKINS.—ANALECTA. Selected Passages for Translation.
   By J. S. STRACHAN, M.A., Professor of Greek, and A. S. WILKINS, Litt.D.,
   Professor of Latin in the Owens College, Manchester. Cr. 8vo. 5s. KEY to
   Latin Passages. Cr. 8vo. Sewed, 6d.

THRING.—Works by the Rev. E. THRING, M.A., late Headmaster of Uppingham.
  A LATIN GRADUAL. A First Latin Construing Book for Beginners. With
    Coloured Sentence Maps. Fcap. 8vo. 2s. 6d.
  A MANUAL OF MOOD CONSTRUCTIONS. Fcap. 8vo. 1s. 6d.
'WELCH and DUFFIELD.—LATIN ACCIDENCE AND EXERCISES AR-
  RANGED FOR BEGINNERS. By W. WELCH and C. G. DUFFIELD,
  Assistant Masters at Cranleigh School. 18mo. 1s. 6d.
WHITE.—FIRST LESSONS IN GREEK. Adapted to GOODWIN'S GREEK GRAM-
  MAR, and designed as an introduction to the ANABASIS OF XENOPHON. By
  JOHN WILLIAMS WHITE, Assistant Professor of Greek in Harvard University,
  U.S.A. Cr. 8vo. 3s. 6d.
WRIGHT.—Works by J. WRIGHT, M.A., late Headmaster of Sutton Coldfield School.
  A HELP TO LATIN GRAMMAR; or, the Form and Use of Words in Latin,
    with Progressive Exercises. Cr. 8vo. 4s. 6d.
  THE SEVEN KINGS OF ROME. An Easy Narrative, abridged from the First
    Book of Livy by the omission of Difficult Passages; being a First Latin Read-
    ing Book, with Grammatical Notes and Vocabulary. Fcap. 8vo. 3s. 6d.
  FIRST LATIN STEPS; OR, AN INTRODUCTION BY A SERIES OF
    EXAMPLES TO THE STUDY OF THE LATIN LANGUAGE. Cr. 8vo. 3s.
  ATTIC PRIMER. Arranged for the Use of Beginners. Ex. fcap. 8vo. 2s. 6d.
  A COMPLETE LATIN COURSE, comprising Rules with Examples, Exercises,
    both Latin and English, on each Rule, and Vocabularies. Cr. 8vo. 2s. 6d.

ANTIQUITIES, ANCIENT HISTORY, AND PHILOSOPHY.

ARNOLD.—A HISTORY OF THE EARLY ROMAN EMPIRE. By W. T. ARNOLD,
  M.A.   [In preparation.
ARNOLD.—THE SECOND PUNIC WAR. Being Chapters from THE HISTORY
  OF ROME by the late THOMAS ARNOLD, D.D., Headmaster of Rugby.
  Edited, with Notes, by W. T. ARNOLD, M.A. With 8 Maps. Cr. 8vo. 5s.
'BEESLY.—STORIES FROM THE HISTORY OF ROME. By Mrs. BEESLY.
  Fcap. 8vo. 2s. 6d.
BLACKIE.—HORÆ HELLENICÆ. By JOHN STUART BLACKIE, Emeritus Pro-
  fessor of Greek in the University of Edinburgh. 8vo. 12s.
BURN.—ROMAN LITERATURE IN RELATION TO ROMAN ART. By Rev.
  ROBERT BURN, M.A., late Fellow of Trinity College, Cambridge. Illustrated.
  Ex. cr. 8vo. 14s.
BURY.—A HISTORY OF THE LATER ROMAN EMPIRE FROM ARCADIUS
  TO IRENE, A.D. 395-800. By J. B. BURY, M.A., Fellow of Trinity College,
  Dublin. 2 vols. 8vo. 32s.
'CLASSICAL WRITERS.—Edited by JOHN RICHARD GREEN, M.A., LL.D. Fcap.
  8vo. 1s. 6d. each.
  SOPHOCLES. By Prof. L. CAMPBELL, M.A.
  EURIPIDES. By Prof. MAHAFFY, D.D.
  DEMOSTHENES. By Prof. S. H. BUTCHER, M.A.
  VIRGIL. By Prof. NETTLESHIP, M.A.
  LIVY. By Rev. W. W. CAPES, M.A.
  TACITUS. By Prof. A. J. CHURCH, M.A., and W. J. BRODRIBB, M.A.
  MILTON. By Rev. STOPFORD A. BROOKE, M.A.
DYER.—STUDIES OF THE GODS IN GREECE AT CERTAIN SANCTUARIES
  RECENTLY EXCAVATED. By LOUIS DYER, B.A. Ex. Cr. 8vo. 8s. 6d. net.
FREEMAN.—Works by EDWARD A. FREEMAN, D.C.L., LL.D., Regius Professor of
  Modern History in the University of Oxford.
  HISTORY OF ROME. (Historical Course for Schools.) 18mo. [In preparation.
  HISTORY OF GREECE. (Historical Course for Schools.) 18mo. [In preparation.
  A SCHOOL HISTORY OF ROME. Cr 8vo. [In preparation.
  HISTORICAL ESSAYS. Second Series. [Greek and Roman History.] 8vo.
  10s. 6d.

GARDNER.—SAMOS AND SAMIAN COINS. An Essay. By Percy Gardner, Litt.D., Professor of Archæology in the University of Oxford. 8vo. 7s. 6d.

GEDDES.—THE PROBLEM OF THE HOMERIC POEMS. By W. D. Geddes, Principal of the University of Aberdeen. 8vo. 14s.

GLADSTONE.—Works by the Rt. Hon. W. E. Gladstone, M.P.
THE TIME AND PLACE OF HOMER. Cr. 8vo. 6s. 6d.
LANDMARKS OF HOMERIC STUDY. Cr. 8vo. 2s. 6d.
*A PRIMER OF HOMER. 18mo. 1s.

GOW.—A COMPANION TO SCHOOL CLASSICS. By James Gow, Litt.D., Master of the High School, Nottingham. With Illustrations. 2d Ed., revised Cr. 8vo. 6s.

HARRISON and VERRALL.—MYTHOLOGY AND MONUMENTS OF ANCIENT ATHENS. Translation of a portion of the "Attica" of Pausanias. By Margaret de G. Verrall. With Introductory Essay and Archæological Commentary by Jane E. Harrison. With Illustrations and Plans. Cr. 8vo. 16s.

JEBB.—Works by R. C. Jebb, Litt.D., Professor of Greek in the University of Cambridge.
THE ATTIC ORATORS FROM ANTIPHON TO ISAEOS. 2 vols. 8vo. 25s.
*A PRIMER OF GREEK LITERATURE. 18mo. 1s.
(See also *Classical Series.*)

KIEPERT.—MANUAL OF ANCIENT GEOGRAPHY. By Dr. H. Kiepert. Cr. 8vo. 5s.

LANCIANI.—ANCIENT ROME IN THE LIGHT OF RECENT DISCOVERIES. By Rodolfo Lanciani, Professor of Archæology in the University of Rome. Illustrated. 4to. 24s.

LEAF.—INTRODUCTION TO THE ILIAD FOR ENGLISH READERS. By Walter Leaf, Litt.D. [*In preparation.*]

MAHAFFY.—Works by J. P. Mahaffy, D.D., Fellow of Trinity College, Dublin and Professor of Ancient History in the University of Dublin.
SOCIAL LIFE IN GREECE; from Homer to Menander. Cr. 8vo. 9s.
GREEK LIFE AND THOUGHT; from the Age of Alexander to the Roman Conquest. Cr. 8vo. 12s. 6d.
THE GREEK WORLD UNDER ROMAN SWAY. From Plutarch to Polybius. Cr. 8vo. 10s. 6d.
RAMBLES AND STUDIES IN GREECE. With Illustrations. With Map. Cr. 8vo. 10s. 6d.
A HISTORY OF CLASSICAL GREEK LITERATURE. Cr. 8vo. Vol. I. In two parts. Part I. The Poets, with an Appendix on Homer by Prof. Sayce. Part II. Dramatic Poets. Vol. II. The Prose Writers. In two parts. Part I. Herodotus to Plato. Part II. Isocrates to Aristotle. 4s. 6d. each.
*A PRIMER OF GREEK ANTIQUITIES. With Illustrations. 18mo. 1s.
*EURIPIDES. 18mo. 1s. 6d. (*Classical Writers.*)

MAYOR.—BIBLIOGRAPHICAL CLUE TO LATIN LITERATURE. Edited after Hübner. By Prof. John E. B. Mayor. Cr. 8vo. 10s. 6d.

NEWTON.—ESSAYS ON ART AND ARCHÆOLOGY. By Sir Charles Newton, K.C.B., D.C.L. 8vo. 12s. 6d.

PHILOLOGY.—THE JOURNAL OF PHILOLOGY. Edited by W. A. Wright, M.A., I. Bywater, M.A., and H. Jackson, Litt.D. 4s. 6d. each (half-yearly).

SAYCE.—THE ANCIENT EMPIRES OF THE EAST. By A. H. Sayce, M.A., Deputy-Professor of Comparative Philology, Oxford. Cr. 8vo. 6s.

SCHMIDT and WHITE. AN INTRODUCTION TO THE RHYTHMIC AND METRIC OF THE CLASSICAL LANGUAGES. By Dr. J. H. Heinrich Schmidt. Translated by John Williams White, Ph.D. 8vo. 10s. 6d.

SHUCHHARDT.—DR. SCHLIEMANN'S EXCAVATIONS AT TROY, TIRYNS, MYCENÆ, ORCHOMENOS, ITHACA, presented in the light of recent knowledge. By Dr. Carl Shuchhardt. Translated by Eugenie Sellers. Introduction by Walter Leaf, Litt.D. Illustrated. 8vo. 18s. net.

SHUCKBURGH.—A SCHOOL HISTORY OF ROME. By E. S. SHUCKBURGH, M.A. Cr. 8vo. [In preparation.

*STEWART.—THE TALE OF TROY. Done into English by AUBREY STEWART. Gl. 8vo. 3s. 6d.

'TOZER.—A PRIMER OF CLASSICAL GEOGRAPHY. By H. F. TOZER, M.A. 18mo. 1s.

WALDSTEIN.—CATALOGUE OF CASTS IN THE MUSEUM OF CLASSICAL ARCHÆOLOGY, CAMBRIDGE. By CHARLES WALDSTEIN, University Reader in Classical Archæology. Cr. 8vo. 1s. 6d.
₊ Also an Edition on Large Paper, small 4to. 5s.

WILKINS.—Works by Prof. WILKINS, Litt.D., LL.D.
*A PRIMER OF ROMAN ANTIQUITIES. Illustrated. 18mo. 1s.
*A PRIMER OF ROMAN LITERATURE. 18mo. 1s.

WILKINS and ARNOLD.—A MANUAL OF ROMAN ANTIQUITIES. By Prof. A. S. WILKINS, Litt.D., and W. T. ARNOLD, M.A. Cr. 8vo. Illustrated. [In preparation.

MODERN LANGUAGES AND LITERATURE.

English; French; German; Modern Greek; Italian; Spanish.

ENGLISH.

*ABBOTT.—A SHAKESPEARIAN GRAMMAR. An Attempt to Illustrate some of the Differences between Elizabethan and Modern English. By the Rev. E. A. ABBOTT, D.D., formerly Headmaster of the City of London School. Ex. fcap. 8vo. 6s.

*BACON.—ESSAYS. With Introduction and Notes, by F. G. SELBY, M.A., Professor of Logic and Moral Philosophy, Deccan College, Poona. Gl. 8vo. 3s. ; sewed, 2s. 6d.

*BURKE.—REFLECTIONS ON THE FRENCH REVOLUTION. By the same. Gl. 8vo. 5s.

BROOKE.—*PRIMER OF ENGLISH LITERATURE. By Rev. STOPFORD A. BROOKE, M.A. 18mo. 1s.
EARLY ENGLISH LITERATURE. By the same. 2 vols. 8vo. [Vol. I. In the Press.

BUTLER.—HUDIBRAS. With Introduction and Notes, by ALFRED MILNES, M.A. Ex. fcap. 8vo. Part I. 3s. 6d. Parts II. and III. 4s. 6d.

CAMPBELL.—SELECTIONS. With Introduction and Notes, by CECIL M. BARROW, M.A., Principal of Victoria College, Palghát. Gl. 8vo. [In preparation.

COWPER.—*THE TASK: an Epistle to Joseph Hill, Esq. ; TIROCINIUM, or a Review of the Schools ; and THE HISTORY OF JOHN GILPIN. Edited, with Notes, by W. BENHAM, B.D. Gl. 8vo. 1s. (Globe Readings from Standard Authors.)
THE TASK. With Introduction and Notes, by F. J. ROWE, M.A., and W. T. WEBB, M.A., Professors of English Literature, Presidency College, Calcutta. [In preparation.

'DOWDEN.—A PRIMER OF SHAKESPERE. By Prof. DOWDEN. 18mo. 1s.

DRYDEN.—SELECT PROSE WORKS. Edited, with Introduction and Notes, by Prof. C. D. YONGE. Fcap. 8vo. 2s. 6d.

*GLOBE READERS. For Standards I.-VI. Edited by A. F. MURISON. Illustrated. Gl. 8vo.

Primer I. (48 pp.)	3d.		Book III. (232 pp.)	1s. 3d.
Primer II. (48 pp.)	3d.		Book IV. (328 pp.)	1s. 9d.
Book I. (132 pp.)	6d.		Book V. (408 pp.)	2s.
Book II. (136 pp.)	9d.		Book VI. (436 pp.)	2s. 6d.

*THE SHORTER GLOBE READERS.—Illustrated. Gl. 8vo.

Primer I. (48 pp.)	3d.		Standard III. (178 pp.)	1s.
Primer II. (48 pp.)	3d.		Standard IV. (182 pp.)	1s.
Standard I. (90 pp.)	6d.		Standard V. (216 pp.)	1s. 3d.
Standard II. (124 pp.)	9d.		Standard VI. (228 pp.)	1s. 6d.

*GOLDSMITH.—THE TRAVELLER, or a Prospect of Society ; and THE DESERTED VILLAGE. With Notes, Philological and Explanatory, by J. W. HALES, M.A. Cr. 8vo. 6d.

*THE TRAVELLER AND THE DESERTED VILLAGE. With Introduction and Notes, by A. BARRETT, B.A., Professor of English Literature, Elphinstone College, Bombay. Gl. 8vo. 1s. 9d. ; sewed, 1s. 6d. The Traveller (separately), 1s., sewed.

*THE VICAR OF WAKEFIELD. With a Memoir of Goldsmith, by Prof. MASSON. Gl. 8vo. 1s. (*Globe Readings from Standard Authors.*)

SELECT ESSAYS. With Introduction and Notes, by Prof. C. D. YONGE. Fcap. 8vo. 2s. 6d.

GOSSE.—A HISTORY OF EIGHTEENTH CENTURY LITERATURE (1660-1780). By EDMUND GOSSE, M.A. Cr. 8vo. 7s. 6d.

*GRAY.—POEMS. With Introduction and Notes, by JOHN BRADSHAW, LL.D. Gl. 8vo. 1s. 9d. ; sewed, 1s. 6d.

*HALES.—LONGER ENGLISH POEMS. With Notes, Philological and Explanatory, and an Introduction on the Teaching of English, by J. W. HALES, M.A., Professor of English Literature at King's College, London. Ex. fcap. 8vo. 4s. 6d.

*HELPS.—ESSAYS WRITTEN IN THE INTERVALS OF BUSINESS. With Introduction and Notes, by F. J. ROWE, M.A., and W. T. WEBB, M.A. Gl. 8vo. 1s. 9d. ; sewed, 1s. 6d.

*JOHNSON.—LIVES OF THE POETS. The Six Chief Lives (Milton, Dryden, Swift, Addison, Pope, Gray), with Macaulay's "Life of Johnson." With Preface and Notes by MATTHEW ARNOLD. Cr. 8vo. 4s. 6d.

KELLNER.—HISTORICAL OUTLINES OF ENGLISH SYNTAX. By L. KELLNER, Ph.D. [*In the Press.*

*LAMB.—TALES FROM SHAKSPEARE. With Preface by the Rev. CANON AINGER, M.A., LL.D. Gl. 8vo. 2s. (*Globe Readings from Standard Authors.*)

*LITERATURE PRIMERS.—Edited by JOHN RICHARD GREEN, LL.D. 18mo. 1s. each.

ENGLISH GRAMMAR. By Rev. R. MORRIS, LL.D.

ENGLISH GRAMMAR EXERCISES. By R. MORRIS, LL.D., and H. C. BOWEN, M.A.

EXERCISES ON MORRIS'S PRIMER OF ENGLISH GRAMMAR. By J. WETHERELL, M.A.

ENGLISH COMPOSITION. By Professor NICHOL.

QUESTIONS AND EXERCISES ON ENGLISH COMPOSITION. By Prof. NICHOL and W. S. M'CORMICK.

ENGLISH LITERATURE. By STOPFORD BROOKE, M.A.

SHAKSPERE. By Professor DOWDEN.

THE CHILDREN'S TREASURY OF LYRICAL POETRY. Selected and arranged with Notes by FRANCIS TURNER PALGRAVE. In Two Parts. 1s. each.

PHILOLOGY. By J. PEILE, Litt.D.

ROMAN LITERATURE. By Prof. A. S. WILKINS, Litt.D.

GREEK LITERATURE. By Prof. JEBB, Litt.D.

HOMER. By the Rt. Hon. W. E. GLADSTONE, M.P.

A HISTORY OF ENGLISH LITERATURE IN FOUR VOLUMES. Cr. 8vo.

EARLY ENGLISH LITERATURE. By STOPFORD BROOKE, M.A. [*In preparation.*

ELIZABETHAN LITERATURE. (1560–1665.) By GEORGE SAINTSBURY. 7s. 6d.

EIGHTEENTH CENTURY LITERATURE. (1660–1780.) By EDMUND GOSSE, M.A. 7s. 6d.

THE MODERN PERIOD. By Prof. DOWDEN. [*In preparation.*

*MACMILLAN'S READING BOOKS.

PRIMER. 18mo. 48 pp. 2d.
BOOK I. for Standard I. 96 pp. 4d.
BOOK II. for Standard II. 144 pp. 5d.
BOOK III. for Standard III. 160 pp. 6d.
BOOK IV. for Standard IV. 176 pp. 8d.
BOOK V. for Standard V. 380 pp. 1s.
BOOK VI. for Standard VI. Cr. 8vo. 430 pp. 2s.

Book VI. is fitted for Higher Classes, and as an Introduction to English Literature.

'MACMILLAN'S COPY BOOKS.—1. Large Post 4to. Price 4d. each. 2. Post Oblong. Price 2d. each.

    1. INITIATORY EXERCISES AND SHORT LETTERS.
    2. WORDS CONSISTING OF SHORT LETTERS.
    3. LONG LETTERS. With Words containing Long Letters—Figures.
    4. WORDS CONTAINING LONG LETTERS.
    4a. PRACTISING AND REVISING COPY-BOOK. For Nos. 1 to 4.
    5. CAPITALS AND SHORT HALF-TEXT. Words beginning with a Capital.
    6. HALF-TEXT WORDS beginning with Capitals—Figures.
    7. SMALL-HAND AND HALF-TEXT. With Capitals and Figures.
    8. SMALL-HAND AND HALF-TEXT. With Capitals and Figures.
    8a. PRACTISING AND REVISING COPY-BOOK. For Nos. 5 to 8.
    9. SMALL-HAND SINGLE HEADLINES—Figures.
    10. SMALL-HAND SINGLE HEADLINES—Figures.
    11. SMALL-HAND DOUBLE HEADLINES—Figures.
    12. COMMERCIAL AND ARITHMETICAL EXAMPLES, &c.
    12a. PRACTISING AND REVISING COPY-BOOK. For Nos. 8 to 12.
    Nos. 3, 4, 5, 6, 7, 8, 9 may be had with Goodman's Patent Sliding Copies. Large Post 4to. Price 6d. each.

MARTIN.—*THE POET'S HOUR : Poetry selected and arranged for Children. By FRANCES MARTIN. 18mo. 2s. 6d.

  *SPRING-TIME WITH THE POETS. By the same. 18mo. 3s. 6d.

*MILTON.—PARADISE LOST. Books I. and II. With Introduction and Notes, by MICHAEL MACMILLAN, B.A., Professor of Logic and Moral Philosophy, Elphinstone College, Bombay. Gl. 8vo. 1s. 9d. ; sewed, 1s. 6d. Or separately, 1s. 3d. ; sewed, 1s. each.

  *L'ALLEGRO, IL PENSEROSO, LYCIDAS, ARCADES, SONNETS, &c. With Introduction and Notes, by W. BELL, M.A., Professor of Philosophy and Logic, Government College, Lahore. Gl. 8vo. 1s. 9d. ; sewed, 1s. 6d.

  *COMUS. By the same. Gl. 8vo. 1s. 3d. ; sewed, 1s.

  *SAMSON AGONISTES. By H. M. PERCIVAL, M.A., Professor of English Literature, Presidency College, Calcutta. Gl. 8vo. 2s. ; sewed, 1s. 9d.

  *INTRODUCTION TO THE STUDY OF MILTON. By STOPFORD BROOKE, M.A. Fcap. 8vo. 1s. 6d. (*Classical Writers.*)

MORRIS.—Works by the Rev. R. MORRIS, LL.D.

  *PRIMER OF ENGLISH GRAMMAR. 18mo. 1s.

  *ELEMENTARY LESSONS IN HISTORICAL ENGLISH GRAMMAR, containing Accidence and Word-Formation. 18mo. 2s. 6d.

  *HISTORICAL OUTLINES OF ENGLISH ACCIDENCE, comprising Chapters on the History and Development of the Language, and on Word-Formation. Ex. fcap. 8vo. 6s.

NICHOL and M'CORMICK.—A SHORT HISTORY OF ENGLISH LITERATURE. By Prof. JOHN NICHOL and Prof. W. S. M'CORMICK. [*In preparation.*

OLIPHANT.—THE OLD AND MIDDLE ENGLISH. By T. L. KINGTON OLIPHANT. New Ed., revised and enlarged, of "The Sources of Standard English." 2nd Ed. Gl. 8vo. 9s.

  THE NEW ENGLISH. By the same. 2 vols. Cr. 8vo. 21s.

*PALGRAVE.—THE CHILDREN'S TREASURY OF LYRICAL POETRY. Selected and arranged, with Notes, by FRANCIS T. PALGRAVE. 18mo. 2s. 6d. Also in Two Parts. 1s. each.

PATMORE.—THE CHILDREN'S GARLAND FROM THE BEST POETS. Selected and arranged by COVENTRY PATMORE. Gl. 8vo. 2s. (*Globe Readings from Standard Authors.*)

PLUTARCH.—Being a Selection from the Lives which illustrate Shakespeare. North's Translation. Edited, with Introductions, Notes, Index of Names, and Glossarial Index, by Prof. W. W. SKEAT, Litt.D. Cr. 8vo. 6s.

*RANSOME.—SHORT STUDIES OF SHAKESPEARE'S PLOTS. By Cyril Ransome, Professor of Modern History and Literature, Yorkshire College, Leeds. Cr. 8vo. 3s. 6d.

*RYLAND.—CHRONOLOGICAL OUTLINES OF ENGLISH LITERATURE. By F. Ryland, M.A. Cr. 8vo. 6s.

SAINTSBURY.—A HISTORY OF ELIZABETHAN LITERATURE. 1560-1665. By George Saintsbury. Cr. 8vo. 7s. 6d.

SCOTT.—*LAY OF THE LAST MINSTREL, and THE LADY OF THE LAKE. Edited, with Introduction and Notes, by Francis Turner Palgrave. Gl. 8vo. 1s. (*Globe Readings from Standard Authors.*)

 *THE LAY OF THE LAST MINSTREL. With Introduction and Notes, by G. H. Stuart, M.A., and E. H. Elliot, B.A. Gl. 8vo. 2s.; sewed, 1s. 9d. Introduction and Canto I. 9d. sewed. Cantos I. to III. 1s. 3d.; sewed, 1s. Cantos IV. to VI. 1s. 3d.; sewed, 1s.

 *MARMION, and THE LORD OF THE ISLES. By F. T. Palgrave. Gl. 8vo. 1s. (*Globe Readings from Standard Authors.*)

 *MARMION. With Introduction and Notes, by Michael Macmillan, B.A. Gl. 8vo. 3s.; sewed, 2s. 6d.

 *THE LADY OF THE LAKE. By G. H. Stuart, M.A. Gl. 8vo. 2s. 6d.; sewed, 2s.

 *ROKEBY. With Introduction and Notes, by Michael Macmillan, B.A. Gl. 8vo. 3s.; sewed, 2s. 6d.

SHAKESPEARE.—*A SHAKESPEARIAN GRAMMAR. By Rev. E. A. Abbott, D.D. Gl. 8vo. 6s.

 A SHAKESPEARE MANUAL. By F. G. Fleay, M.A. 2d Ed. Ex. fcap. 8vo. 4s. 6d.

 *A PRIMER OF SHAKESPERE. By Prof. Dowden. 18mo. 1s.

 *SHORT STUDIES OF SHAKESPEARE'S PLOTS. By Cyril Ransome, M.A. Cr. 8vo. 3s. 6d.

 *THE TEMPEST. With Introduction and Notes, by K. Deighton, late Principal of Agra College. Gl. 8vo. 1s. 9d.; sewed, 1s. 6d.

 *MUCH ADO ABOUT NOTHING. By the same. Gl. 8vo. 1s. 9d.; sewed, 1s. 6d.

 *A MIDSUMMER NIGHT'S DREAM. By the same. Gl. 8vo. 1s. 9d.; sewed, 1s. 6d.

 *THE MERCHANT OF VENICE. By the same. Gl. 8vo. 1s. 9d.; sewed, 1s. 6d.

 *AS YOU LIKE IT. By the same. Gl. 8vo. 1s. 9d.; sewed, 1s. 6d.

 *TWELFTH NIGHT. By the same. Gl. 8vo. 1s. 9d.; sewed, 1s. 6d.

 *THE WINTER'S TALE. By the same. Gl. 8vo. 2s.; sewed, 1s. 9d.

 *KING JOHN. By the same. Gl. 8vo. 1s. 9d.; sewed, 1s. 6d.

 *RICHARD II. By the same. Gl. 8vo. 1s. 9d.; sewed, 1s. 6d.

 *HENRY V. By the same. Gl. 8vo. 1s. 9d.; sewed, 1s. 6d.

 *RICHARD III. By C. H. Tawney, M.A., Principal and Professor of English Literature, Presidency College, Calcutta. Gl. 8vo. 2s. 6d.; sewed, 2s.

 *CORIOLANUS. By K. Deighton. Gl. 8vo. 2s. 6d.; sewed, 2s.

 *JULIUS CÆSAR. By the same. Gl. 8vo. 1s. 9d.; sewed, 1s. 6d

 *MACBETH. By the same. Gl. 8vo. 1s. 9d.; sewed, 1s. 6d.

 *HAMLET. By the same. Gl. 8vo. 2s. 6d.; sewed, 2s.

 *KING LEAR. By the same. Gl. 8vo. 1s. 9d.; sewed, 1s. 6d.

 *OTHELLO. By the same. Gl. 8vo. 2s.; sewed, 1s. 9d.

 *ANTONY AND CLEOPATRA. By the same. Gl. 8vo. 2s. 6d.; sewed, 2s.

 *CYMBELINE. By the same. Gl. 8vo. 2s. 6d.; sewed, 2s.

*SONNENSCHEIN and MEIKLEJOHN.—THE ENGLISH METHOD OF TEACHING TO READ. By A. Sonnenschein and J. M. D. Meiklejohn, M.A. Fcap. 8vo.

 THE NURSERY BOOK, containing all the Two-Letter Words in the Language. 1d. (Also in Large Type on Sheets for School Walls. 5s.)

THE FIRST COURSE, consisting of Short Vowels with Single Consonants. 7d.
THE SECOND COURSE, with Combinations and Bridges, consisting of Short
Vowels with Double Consonants. 7d.
THE THIRD AND FOURTH COURSES, consisting of Long Vowels, and all
the Double Vowels in the Language. 7d.

*SOUTHEY.—LIFE OF NELSON. With Introduction and Notes, by MICHAEL
MACMILLAN, B.A. Gl. 8vo. 3s. ; sewed, 2s. 6d.

SPENSER.—FAIRY QUEEN. Book L. With Introduction and Notes, by H. M.
PERCIVAL, M.A.   [In the Press.

TAYLOR.—WORDS AND PLACES; or, Etymological Illustrations of History,
Ethnology, and Geography. By Rev. ISAAC TAYLOR, Litt.D. With Maps.
Gl. 8vo. 6s.

TENNYSON.—THE COLLECTED WORKS OF LORD TENNYSON. An Edition
for Schools. In Four Parts. Cr. 8vo. 2s. 6d. each.
TENNYSON FOR THE YOUNG. Edited, with Notes for the Use of Schools,
by the Rev. ALFRED AINGER, LL.D., Canon of Bristol. 18mo. 1s. net.
      [In the Press.

*SELECTIONS FROM TENNYSON. With Introduction and Notes, by F. J.
ROWE, M.A., and W. T. WEBB, M.A. Gl. 8vo. 3s. 6d.
This selection contains :—Recollections of the Arabian Nights, The Lady of
Shalott, Œnone, The Lotos Eaters, Ulysses, Tithonus, Morte d'Arthur, Sir
Galahad, Dora, Ode on the Death of the Duke of Wellington, and The Revenge.
*ENOCH ARDEN. By W. T. WEBB, M.A. Gl. 8vo. 2s.
AYLMER'S FIELD. By W. T. WEBB, M.A.   [In the Press.
THE PRINCESS ; A MEDLEY. By P. M. WALLACE, B.A.   [In the Press.
*THE COMING OF ARTHUR, AND THE PASSING OF ARTHUR. By F. J.
ROWE, M.A. Gl. 8vo. 2s.

THRING.—THE ELEMENTS OF GRAMMAR TAUGHT IN ENGLISH. By
EDWARD THRING, M.A. With Questions. 4th Ed. 18mo. 2s.

*VAUGHAN.—WORDS FROM THE POETS. By C. M. VAUGHAN. 18mo. 1s.

WARD.—THE ENGLISH POETS. Selections, with Critical Introductions by
various Writers and a General Introduction by MATTHEW ARNOLD. Edited
by T. H. WARD, M.A. 4 Vols. Vol. I. CHAUCER TO DONNE.—Vol. II. BEN
JONSON TO DRYDEN.—Vol. III. ADDISON TO BLAKE.—Vol. IV. WORDSWORTH
TO ROSSETTI. 2d Ed. Cr. 8vo. 7s. 6d. each.

*WETHERELL.—EXERCISES ON MORRIS'S PRIMER OF ENGLISH GRAM-
MAR. By JOHN WETHERELL, M.A., Headmaster of Towcester Grammar
School. 18mo. 1s.

WOODS.—*A FIRST POETRY BOOK. By M. A. WOODS, Head Mistress of the
Clifton High School for Girls. Fcap. 8vo. 2s. 6d.
*A SECOND POETRY BOOK. By the same. In Two Parts. 2s. 6d. each.
*A THIRD POETRY BOOK. By the same. 4s. 6d.
HYMNS FOR SCHOOL WORSHIP. By the same. 18mo. 1s. 6d.

WORDSWORTH.—SELECTIONS. With Introduction and Notes, by F. J. ROWE,
M.A., and W. T. WEBB, M.A. Gl. 8vo.   [In preparation.

YONGE.—*A BOOK OF GOLDEN DEEDS. By CHARLOTTE M. YONGE. Gl. 8vo. 2s.
*THE ABRIDGED BOOK OF GOLDEN DEEDS. 18mo. 1s.

FRENCH.

BEAUMARCHAIS.—LE BARBIER DE SEVILLE. With Introduction and
Notes. By L. P. BLOUET. Fcap. 8vo. 3s. 6d.

*BOWEN.—FIRST LESSONS IN FRENCH. By H. COURTHOPE BOWEN, M.A.
Ex. fcap. 8vo. 1s.

BREYMANN.—Works by HERMANN BREYMANN, Ph.D., Professor of Philology in
the University of Munich.
FIRST FRENCH EXERCISE BOOK. Ex. fcap. 8vo. 4s. 6d.
SECOND FRENCH EXERCISE BOOK. Ex. fcap. 8vo. 2s. 6d.

FASNACHT.—Works by G. E. FASNACHT, late Assistant Master at Westminster.
THE ORGANIC METHOD OF STUDYING LANGUAGES. Ex. fcap. 8vo. I.
French. 3s. 6d.

A SYNTHETIC FRENCH GRAMMAR FOR SCHOOLS. Cr. 8vo. **3s. 6d.**

GRAMMAR AND GLOSSARY OF THE FRENCH LANGUAGE OF THE SEVENTEENTH CENTURY. Cr. 8vo. [In preparation.

MACMILLAN'S PRIMARY SERIES OF FRENCH READING BOOKS.—Edited by G. E. FASNACHT. With Illustrations, Notes, Vocabularies, and Exercises. Gl. 8vo.

*FRENCH READINGS FOR CHILDREN. By G. E. FASNACHT. 1s. 6d.

'CORNAZ—NOS ENFANTS ET LEURS AMIS. By EDITH HARVEY. 1s. 6d.

*DE MAISTRE—LA JEUNE SIBÉRIENNE ET LE LÉPREUX DE LA CITÉ D'AOSTE. By STEPHANE BARLET, B.Sc. etc. 1s. 6d.

*FLORIAN—FABLES. By Rev. CHARLES YELD, M.A., Headmaster of University School, Nottingham. 1s. 6d.

*LA FONTAINE—A SELECTION OF FABLES. By L. M. MORIARTY, B.A., Assistant Master at Harrow. 2s. 6d.

*MOLESWORTH—FRENCH LIFE IN LETTERS. By Mrs. MOLESWORTH. 1s. 6d.

*PERRAULT—CONTES DE FÉES. By G. E. FASNACHT. 1s. 6d.

MACMILLAN'S PROGRESSIVE FRENCH COURSE.—By G. E. FASNACHT. Ex. fcap. 8vo.

*FIRST YEAR, containing Easy Lessons on the Regular Accidence. 1s.

*SECOND YEAR, containing an Elementary Grammar with copious Exercises, Notes, and Vocabularies. 2s.

*THIRD YEAR, containing a Systematic Syntax, and Lessons in Composition. 2s. 6d.

THE TEACHER'S COMPANION TO MACMILLAN'S PROGRESSIVE FRENCH COURSE. With Copious Notes, Hints for Different Renderings, Synonyms, Philological Remarks, etc. By G. E. FASNACHT. Ex. fcap. 8vo. Each Year 4s. 6d.

*MACMILLAN'S FRENCH COMPOSITION.—By G. E. FASNACHT. Ex. fcap. 8vo. Part I. Elementary. 2s. 6d. Part II. Advanced. [In the Press.

THE TEACHER'S COMPANION TO MACMILLAN'S COURSE OF FRENCH COMPOSITION. By G. E. FASNACHT. Part I. Ex. fcap. 8vo. 4s. 6d.

MACMILLAN'S PROGRESSIVE FRENCH READERS. By G. E. FASNACHT. Ex. fcap. 8vo.

*FIRST YEAR, containing Tales, Historical Extracts, Letters, Dialogues, Ballads, Nursery Songs, etc., with Two Vocabularies: (1) in the order of subjects; (2) in alphabetical order. With Imitative Exercises. 2s. 6d.

*SECOND YEAR, containing Fiction in Prose and Verse, Historical and Descriptive Extracts, Essays, Letters, Dialogues, etc. With Imitative Exercises. 2s. 6d.

MACMILLAN'S FOREIGN SCHOOL CLASSICS. Edited by G. E. FASNACHT. 18mo.

*CORNEILLE—LE CID. By G. E. FASNACHT. 1s.

*DUMAS—LES DEMOISELLES DE ST. CYR. By VICTOR OGER, Lecturer at University College, Liverpool. 1s. 6d.

LA FONTAINE'S FABLES. Books I.-VI. By L. M. MORIARTY, B.A., Assistant Master at Harrow. [In preparation.

*MOLIÈRE—L'AVARE. By the same. 1s.

*MOLIÈRE—LE BOURGEOIS GENTILHOMME. By the same. 1s. 6d.

*MOLIÈRE—LES FEMMES SAVANTES. By G. E. FASNACHT. 1s.

*MOLIÈRE—LE MISANTHROPE. By the same. 1s.

*MOLIÈRE—LE MÉDECIN MALGRÉ LUI. By the same. 1s.

*MOLIÈRE—LES PRÉCIEUSES RIDICULES. By the same. 1s.

*RACINE—BRITANNICUS. By E. PELLISSIER, M.A. 2s.

*FRENCH READINGS FROM ROMAN HISTORY. Selected from various Authors, by C. COLBECK, M.A., Assistant Master at Harrow. 4s. 6d.

*SAND, GEORGE—LA MARE AU DIABLE. By W. E. RUSSELL, M.A., Assistant Master at Haileybury. 1s.

*SANDEAU, JULES—MADEMOISELLE DE LA SEIGLIÈRE. By H. C. STEEL, Assistant Master at Winchester. 1s. 6d.

*VOLTAIRE—CHARLES XII. By G. E. FASNACHT. 3s. 6d.

*MASSON.—A COMPENDIOUS DICTIONARY OF THE FRENCH LANGUAGE. Adapted from the Dictionaries of Professor A. ELWALL. By GUSTAVE MASSON. Cr. 8vo. 3s. 6d.

MOLIÈRE.—LE MALADE IMAGINAIRE. With Introduction and Notes, by F. TARVER, M.A., Assistant Master at Eton. Fcap. 8vo. 2s. 6d.

*PELLISSIER.—FRENCH ROOTS AND THEIR FAMILIES. A Synthetic Vocabulary, based upon Derivations. By E. PELLISSIER, M.A., Assistant Master at Clifton College. Gl. 8vo. 6s.

GERMAN.

BEHAGEL.—THE GERMAN LANGUAGE. By Dr. OTTO BEHAGEL. Translated by EMIL TRECHMANN, B.A., Ph.D., Lecturer in Modern Literature in the University of Sydney, N.S.W. Gl. 8vo. [Nearly Ready.

HUSS.—A SYSTEM OF ORAL INSTRUCTION IN GERMAN, by means of Progressive Illustrations and Applications of the leading Rules of Grammar. By H. C. O. Huss, Ph.D. Cr. 8vo. 5s.

MACMILLAN'S PRIMARY SERIES OF GERMAN READING BOOKS. Edited by G. E. FASNACHT. With Notes, Vocabularies, and Exercises. Gl. 8vo.

*GRIMM—KINDER UND HAUSMÄRCHEN. By G. E. FASNACHT. 2s. 6d.

*HAUFF—DIE KARAVANE. By HERMAN HAGER, Ph.D., Lecturer in the Owens College, Manchester. 3s.

*SCHMID, CHR. VON—H. VON EICHENFELS. By G. E. FASNACHT. 2s. 6d.

MACMILLAN'S PROGRESSIVE GERMAN COURSE. By G. E. FASNACHT. Ex. fcap. 8vo.

*FIRST YEAR. Easy lessons and Rules on the Regular Accidence. 1s. 6d.

*SECOND YEAR. Conversational Lessons in Systematic Accidence and Elementary Syntax. With Philological Illustrations and Etymological Vocabulary. 3s. 6d.

THIRD YEAR. [In the Press.

TEACHER'S COMPANION TO MACMILLAN'S PROGRESSIVE GERMAN COURSE. With copious Notes, Hints for Different Renderings, Synonyms, Philological Remarks, etc. By G. E. FASNACHT. Ex. fcap. 8vo. FIRST YEAR. 4s. 6d. SECOND YEAR. 4s. 6d.

MACMILLAN'S GERMAN COMPOSITION. By G. E. FASNACHT. Ex. fcap. 8vo.
*I. FIRST COURSE. Parallel German-English Extracts and Parallel English-German Syntax. 2s. 6d.

TEACHER'S COMPANION TO MACMILLAN'S GERMAN COMPOSITION. By G. E. FASNACHT. FIRST COURSE. Gl. 8vo. 4s. 6d.

MACMILLAN'S PROGRESSIVE GERMAN READERS. By G. E. FASNACHT. Ex. fcap. 8vo.
*FIRST YEAR, containing an Introduction to the German order of Words, with Copious Examples, extracts from German Authors in Prose and Poetry; Notes, and Vocabularies. 2s. 6d.

MACMILLAN'S FOREIGN SCHOOL CLASSICS.—Edited by G. E. FASNACHT. 18mo.
FREYTAG (G.)—DOKTOR LUTHER. By F. STORR, M.A., Headmaster of the Modern Side, Merchant Taylors' School. [In preparation.
*GOETHE—GÖTZ VON BERLICHINGEN. By H. A. BULL, M.A., Assistant Master at Wellington. 2s.
*GOETHE—FAUST. PART I., followed by an Appendix on PART II. By JANE LEE, Lecturer in German Literature at Newnham College, Cambridge. 4s. 6d.
*HEINE—SELECTIONS FROM THE REISEBILDER AND OTHER PROSE WORKS. By C. COLBECK, M.A., Assistant Master at Harrow. 2s. 6d.
LESSING—MINNA VON BARNHELM. By JAMES SIME, M.A. [In preparation.

*SCHILLER—SELECTIONS FROM SCHILLER'S LYRICAL POEMS. With a Memoir of Schiller. By E. J. TURNER, B.A., and E. D. A. MORSHEAD, M.A., Assistant Masters at Winchester. 2s. 6d.

*SCHILLER—DIE JUNGFRAU VON ORLEANS. By JOSEPH GOSTWICK. 2s. 6d.

*SCHILLER—MARIA STUART. By C. SHELDON, D.Litt., of the Royal Academical Institution, Belfast. 2s. 6d.

*SCHILLER—WILHELM TELL. By G. E. FASNACHT. 2s. 6d.

*SCHILLER—WALLENSTEIN. Part I. DAS LAGER. By H. B. COTTERILL, M.A. 2s.

*UHLAND—SELECT BALLADS. Adapted as a First Easy Reading Book for Beginners. With Vocabulary. By G. E. FASNACHT. 1s.

*PYLODET.—NEW GUIDE TO GERMAN CONVERSATION; containing an Alphabetical List of nearly 800 Familiar Words; followed by Exercises, Vocabulary of Words in frequent use, Familiar Phrases and Dialogues, a Sketch of German Literature, Idiomatic Expressions, etc. By L. PYLODET. 18mo. 2s. 6d.

SMITH.—COMMERCIAL GERMAN. By F. C. SMITH, M.A. [In the Press.

WHITNEY.—A COMPENDIOUS GERMAN GRAMMAR. By W. D. WHITNEY, Professor of Sanskrit and Instructor in Modern Languages in Yale College. Cr. 8vo. 4s. 6d.

A GERMAN READER IN PROSE AND VERSE. By the same. With Notes and Vocabulary. Cr. 8vo. 5s.

*WHITNEY and EDGREN.—A COMPENDIOUS GERMAN AND ENGLISH DICTIONARY, with Notation of Correspondences and Brief Etymologies. By Prof. W. D. WHITNEY, assisted by A. H. EDGREN. Cr. 8vo. 7s. 6d.

THE GERMAN-ENGLISH PART, separately, 5s.

MODERN GREEK.

VINCENT and DICKSON.—HANDBOOK TO MODERN GREEK. By Sir EDGAR VINCENT, K.C.M.G., and T. G. DICKSON, M.A. With Appendix on the relation of Modern and Classical Greek by Prof. JEBB. Cr. 8vo. 6s.

ITALIAN.

DANTE.—THE INFERNO OF DANTE. With Translation and Notes, by A. J. BUTLER, M.A. Cr. 8vo. [In the Press.

THE PURGATORIO OF DANTE. With Translations and Notes, by the same. Cr. 8vo. 12s. 6d.

THE PARADISO OF DANTE. With Translation and Notes, by the same. 2d. Ed. Cr. 8vo. 12s. 6d.

READINGS ON THE PURGATORIO OF DANTE. Chiefly based on the Commentary of Benvenuto Da Imola. By the Hon. W. WARREN VERNON, M.A. With an Introduction by the Very Rev. the DEAN OF ST. PAUL'S. 2 vols. Cr. 8vo. 24s.

SPANISH.

CALDERON.—FOUR PLAYS OF CALDERON. With Introduction and Notes. By NORMAN MacCOLL, M.A. Cr. 8vo. 14s.

The four plays here given are *El Principe Constante, La Vida es Sueno, El Alcalde de Zalamea,* and *El Escondido y La Tapada.*

MATHEMATICS.

Arithmetic, Book-keeping, Algebra, Euclid and Pure Geometry, Geometrical
Drawing, Mensuration, Trigonometry, Analytical Geometry (Plane and
Solid), Problems and Questions in Mathematics, Higher Pure Mathematics, Mechanics (Statics, Dynamics, Hydrostatics, Hydrodynamics: see
also Physics), Physics (Sound, Light, Heat, Electricity, Elasticity, Attrac-
tions, &c.), Astronomy, Historical.

ARITHMETIC.

*ALDIS.—THE GREAT GIANT ARITHMOS.   A most Elementary Arithmetic
for Children.   By MARY STEADMAN ALDIS.   Illustrated.   Gl. 8vo.   2s. 6d.

ARMY PRELIMINARY EXAMINATION, SPECIMENS OF PAPERS SET AT
THE, 1882-89.—With Answers to the Mathematical Questions.   Subjects:
Arithmetic, Algebra, Euclid, Geometrical Drawing, Geography, French,
English Dictation.   Cr. 8vo.   3s. 6d.

*BRADSHAW.—A COURSE OF EASY ARITHMETICAL EXAMPLES FOR
BEGINNERS.   By J. G. BRADSHAW, B.A., Assistant Master at Clifton College.
Gl. 8vo.   2s.   With Answers, 2s. 6d.

*BROOKSMITH.—ARITHMETIC IN THEORY AND PRACTICE. By J. BROOK-
SMITH, M.A.   Cr. 8vo.   4s. 6d.   KEY.   Crown 8vo.   10s. 6d.

*BROOKSMITH.—ARITHMETIC FOR BEGINNERS.   By J. and E. J. BROOK-
SMITH.   Gl. 8vo.   1s. 6d.

CANDLER.—HELP TO ARITHMETIC.   Designed for the use of Schools.   By H.
CANDLER, Mathematical Master of Uppingham School.   2d Ed.   Ex. fcap. 8vo.
2s. 6d.

*DALTON.—RULES AND EXAMPLES IN ARITHMETIC.   By the Rev. T. DAL-
TON, M.A., Senior Mathematical Master at Eton.   New Ed., with Answers.
18mo.   2s. 6d.

*GOYEN.—HIGHER ARITHMETIC AND ELEMENTARY MENSURATION.
By P. GOYEN, Inspector of Schools, Dunedin, New Zealand.   Cr. 8vo.   5s.

*HALL and KNIGHT.—ARITHMETICAL EXERCISES AND EXAMINATION
PAPERS.   With an Appendix containing Questions in LOGARITHMS and
MENSURATION.   By H. S. HALL, M.A., Master of the Military and Engineering
Side, Clifton College, and S. R. KNIGHT, B.A.   Gl. 8vo.   2s. 6d.

LOCK.—Works by Rev. J. B. LOCK, M.A., Senior Fellow and Bursar of Gonville
and Caius College, Cambridge.

*ARITHMETIC FOR SCHOOLS.   With Answers and 1000 additional Examples
for Exercise.   3d Ed., revised.   Gl. 8vo.   4s. 6d.   Or, Part I.   2s.   Part II.   3s.
KEY.   Cr. 8vo.   10s. 6d.

*ARITHMETIC FOR BEGINNERS.   A School Class-Book of Commercial Arith-
metic.   Gl. 8vo.   2s. 6d.   KEY.   Cr. 8vo.   8s. 6d.

*A SHILLING BOOK OF ARITHMETIC, FOR ELEMENTARY SCHOOLS.
18mo.   1s.   With Answers.   1s. 6d.

*PEDLEY.—EXERCISES IN ARITHMETIC for the Use of Schools.   Containing
more than 7000 original Examples.   By SAMUEL PEDLEY.   Cr. 8vo.   5s.
Also in Two Parts, 2s. 6d. each.

SMITH.—Works by Rev. BARNARD SMITH, M.A., late Fellow and Senior Bursar of
St. Peter's College, Cambridge.

ARITHMETIC AND ALGEBRA, in their Principles and Application; with
numerous systematically arranged Examples taken from the Cambridge Examination Papers, with especial reference to the Ordinary Examination for the
B.A. Degree.   New Ed., carefully revised.   Cr. 8vo.   10s. 6d.

*ARITHMETIC FOR SCHOOLS.   Cr. 8vo.   4s. 6d.   KEY.   Cr. 8vo.   8s. 6d.
New Edition.   Revised by Prof W. H. HUDSON.   [In preparation.

EXERCISES IN ARITHMETIC. Cr. 8vo. 2s. With Answers, 2s. 6d. Answers separately, 6d.

SCHOOL CLASS-BOOK OF ARITHMETIC. 18mo. 3s. Or separately, in Three Parts, 1s. each. KEYS. Parts I., II., and III., 2s. 6d. each.

SHILLING BOOK OF ARITHMETIC. 18mo. Or separately, Part I., 2d.; Part II., 3d.; Part III., 7d. Answers, 6d. KEY. 18mo. 4s. 6d.

*THE SAME, with Answers. 18mo, cloth. 1s. 6d.

EXAMINATION PAPERS IN ARITHMETIC. 18mo. 1s. 6d. The Same, with Answers. 18mo. 2s. Answers, 6d. KEY. 18mo. 4s. 6d.

THE METRIC SYSTEM OF ARITHMETIC, ITS PRINCIPLES AND APPLICATIONS, with Numerous Examples. 18mo. 3d.

A CHART OF THE METRIC SYSTEM, on a Sheet, size 42 in. by 34 in. on Roller. 3s. 6d. Also a Small Chart on a Card. Price 1d.

EASY LESSONS IN ARITHMETIC, combining Exercises in Reading, Writing, Spelling, and Dictation. Part I. Cr. 8vo. 9d.

EXAMINATION CARDS IN ARITHMETIC. With Answers and Hints. Standards I. and II., in box, 1s. Standards III., IV., and V., in boxes, 1s. each. Standard VI. in Two Parts, in boxes, 1s. each.

A and B papers, of nearly the same difficulty, are given so as to prevent copying, and the colours of the A and B papers differ in each Standard, and from those of every other Standard, so that a master or mistress can see at a glance whether the children have the proper papers.

BOOK-KEEPING.

*THORNTON.—FIRST LESSONS IN BOOK-KEEPING. By J. THORNTON. Cr. 8vo. 2s. 6d. KEY. Oblong 4to. 10s. 6d.

*PRIMER OF BOOK-KEEPING. 18mo. 1s. KEY. Demy 8vo. 2s. 6d.

ALGEBRA.

*DALTON.—RULES AND EXAMPLES IN ALGEBRA. By Rev. T. DALTON, Senior Mathematical Master at Eton. Part I. 18mo. 2s. KEY. Cr. 8vo. 7s. 6d. Part II. 18mo. 2s. 6d.

HALL and KNIGHT.—Works by H. S. HALL, M.A., Master of the Military and Engineering Side, Clifton College, and S. R. KNIGHT, B.A.

*ELEMENTARY ALGEBRA FOR SCHOOLS. 6th Ed., revised and corrected. Gl. 8vo, bound in maroon coloured cloth, 3s. 6d.; with Answers, bound in green coloured cloth, 4s. 6d. KEY. 8s. 6d.

*ALGEBRAICAL EXERCISES AND EXAMINATION PAPERS. To accompany ELEMENTARY ALGEBRA. 2d Ed., revised. Gl. 8vo. 2s. 6d.

*HIGHER ALGEBRA. 3d Ed. Cr. 8vo. 7s. 6d. KEY. Cr. 8vo. 10s. 6d.

*JONES and CHEYNE.—ALGEBRAICAL EXERCISES. Progressively Arranged. By Rev. C. A. JONES and C. H. CHEYNE, M.A., late Mathematical Masters at Westminster School. 18mo. 2s. 6d.

KEY. By Rev. W. FAILES, M.A., Mathematical Master at Westminster School. Cr. 8vo. 7s. 6d.

SMITH.—ARITHMETIC AND ALGEBRA, in their Principles and Application: with numerous systematically arranged Examples taken from the Cambridge Examination Papers, with especial reference to the Ordinary Examination for the B.A. Degree. By Rev. BARNARD SMITH, M.A. New Edition, carefully revised. Cr. 8vo. 10s. 6d.

SMITH.—Works by CHARLES SMITH, M.A., Master of Sidney Sussex College, Cambridge.

*ELEMENTARY ALGEBRA. 2d Ed., revised. Gl. 8vo. 4s. 6d. KEY. By A. G. CRACKNELL, B.A. Cr. 8vo. 10. 6d.

*A TREATISE ON ALGEBRA. 2d Ed. Cr. 8vo. 7s. 6d. KEY. Cr. 8vo. 10s. 6d.

TODHUNTER.—Works by ISAAC TODHUNTER, F.R.S.

*ALGEBRA FOR BEGINNERS. 18mo. 2s. 6d. KEY. Cr. 8vo. 6s. 6d.

*ALGEBRA FOR COLLEGES AND SCHOOLS. By Isaac Todhunter, F.R.S. Cr. 8vo. 7s. 6d. KEY. Cr. 8vo. 10s. 6d.

EUCLID AND PURE GEOMETRY.

COCKSHOTT and WALTERS.—A TREATISE ON GEOMETRICAL CONICS. In accordance with the Syllabus of the Association for the Improvement of Geometrical Teaching. By A. Cockshott, M.A., Assistant Master at Eton, and Rev. F. B. Walters, M.A., Principal of King William's College, Isle of Man. Cr. 8vo. 5s.

CONSTABLE.—GEOMETRICAL EXERCISES FOR BEGINNERS. By Samuel Constable. Cr. 8vo. 3s. 6d.

CUTHBERTSON.—EUCLIDIAN GEOMETRY. By Francis Cuthbertson, M.A., LL.D. Ex. feap. 8vo. 4s. 6d.

DAY.—PROPERTIES OF CONIC SECTIONS PROVED GEOMETRICALLY. By Rev. H. G. Day, M.A. Part I. The Ellipse, with an ample collection of Problems. Cr. 8vo. 3s. 6d.

*DEAKIN.—RIDER PAPERS ON EUCLID. BOOKS I. and II. By Rupert Deakin, M.A. 18mo. 1s.

DODGSON.—Works by Charles L. Dodgson, M.A., Student and late Mathematical Lecturer, Christ Church, Oxford.
EUCLID, BOOKS I. and II. 6th Ed., with words substituted for the Algebraical Symbols used in the 1st Ed. Cr. 8vo. 2s.
EUCLID AND HIS MODERN RIVALS. 2d Ed. Cr. 8vo. 6s.
CURIOSA MATHEMATICA. Part I. A New Theory of Parallels. 3d Ed. Cr. 8vo. 2s.

DREW.—GEOMETRICAL TREATISE ON CONIC SECTIONS. By W. H. Drew, M.A. New Ed., enlarged. Cr. 8vo. 5s.

DUPUIS.—ELEMENTARY SYNTHETIC GEOMETRY OF THE POINT, LINE AND CIRCLE IN THE PLANE. By N. F. Dupuis, M.A., Professor of Pure Mathematics in the University of Queen's College, Kingston, Canada. Gl. 8vo. 4s. 6d.

*HALL and STEVENS.—A TEXT-BOOK OF EUCLID'S ELEMENTS. Including Alternative Proofs, together with additional Theorems and Exercises, classified and arranged. By H. S. Hall, M.A., and F. H. Stevens, M.A., Masters of the Military and Engineering Side, Clifton College. Gl. 8vo. Book I., 1s.; Books I. and II., 1s. 6d.; Books I.–IV., 3s.; Books III.–IV., 2s.; Books III.–VI., 3s.; Books V.–VI. and XI., 2s. 6d.; Books I.–VI. and XI., 4s. 6d.; Book XI., 1s. [KEY. *In preparation.*

HALSTED.—THE ELEMENTS OF GEOMETRY. By G. B. Halsted, Professor of Pure and Applied Mathematics in the University of Texas. 8vo. 12s. 6d.

HAYWARD.—THE ELEMENTS OF SOLID GEOMETRY. By R. B. Hayward, M.A., F.R.S. Gl. 8vo. 3s.

LOCK.—EUCLID FOR BEGINNERS. Being an Introduction to existing Text-Books. By Rev. J. B. Lock, M.A. [*In the Press.*

MILNE and DAVIS.—GEOMETRICAL CONICS. Part I. The Parabola. By Rev. J. J. Milne, M.A., and R. F. Davis, M.A. Cr. 8vo. 2s.

*RICHARDSON.—THE PROGRESSIVE EUCLID. Books I. and II. With Notes, Exercises, and Deductions. Edited by A. T. Richardson, M.A., Senior Mathematical Master at the Isle of Wight College. Gl. 8vo. 2s. 6d.

SYLLABUS OF PLANE GEOMETRY (corresponding to Euclid, Books I.–VI.)— Prepared by the Association for the Improvement of Geometrical Teaching. Cr. 8vo. Sewed, 1s.

SYLLABUS OF MODERN PLANE GEOMETRY.—Prepared by the Association for the Improvement of Geometrical Teaching. Cr. 8vo. Sewed. 1s.

*TODHUNTER.—THE ELEMENTS OF EUCLID. By I. Todhunter, F.R.S. 18mo. 3s. 6d. *Books I. and II. 1s. KEY. Cr. 8vo. 6s. 6d.

WILSON.—Works by Ven. Archdeacon Wilson, M.A., formerly Headmaster of Clifton College.
ELEMENTARY GEOMETRY. BOOKS I.–V. Containing the Subjects of Euclid's first Six Books. Following the Syllabus of the Geometrical Association. Ex. feap. 8vo. 4s. 6d.

WILSON.—Works by Ven. Archdeacon WILSON—*continued*.
SOLID GEOMETRY AND CONIC SECTIONS. With Appendices on Transversals and Harmonic Division. Ex. fcap. 8vo. 3s. 6d.

GEOMETRICAL DRAWING.

EAGLES.—CONSTRUCTIVE GEOMETRY OF PLANE CURVES. By T. H. EAGLES, M.A., Instructor in Geometrical Drawing and Lecturer in Architecture at the Royal Indian Engineering College, Cooper's Hill. Cr. 8vo. 12s.

EDGAR and PRITCHARD. — NOTE-BOOK ON PRACTICAL SOLID OR DESCRIPTIVE GEOMETRY. Containing Problems with help for Solutions. By J. H. EDGAR and G. S. PRITCHARD. 4th Ed., revised by A. MEEZE. Gl. 8vo. 4s. 6d.

*KITCHENER.—A GEOMETRICAL NOTE-BOOK. Containing Easy Problems in Geometrical Drawing preparatory to the Study of Geometry. For the Use of Schools. By F. E. KITCHENER, M.A., Headmaster of the Newcastle-under-Lyme High School. 4to. 2s.

MILLAR.—ELEMENTS OF DESCRIPTIVE GEOMETRY. By J. B. MILLAR, Civil Engineer, Lecturer on Engineering in the Victoria University, Manchester. 2d Ed. Cr. 8vo. 6s.

PLANT.—PRACTICAL PLANE AND DESCRIPTIVE GEOMETRY. By E. C. PLANT. Globe 8vo.   [*In preparation*.

MENSURATION.

STEVENS.—ELEMENTARY MENSURATION. With Exercises on the Mensuration of Plane and Solid Figures. By F. H. STEVENS, M.A. Gl. 8vo.
[*In preparation*.

TEBAY.—ELEMENTARY MENSURATION FOR SCHOOLS. By S. TEBAY. Ex. fcap. 8vo. 3s. 6d.

*TODHUNTER.—MENSURATION FOR BEGINNERS. By ISAAC TODHUNTER, F.R.S. 18mo. 2s. 6d. KEY. By Rev. FR. L. McCARTHY. Cr. 8vo. 7s. 6d.

TRIGONOMETRY.

BEASLEY.—AN ELEMENTARY TREATISE ON PLANE TRIGONOMETRY. With Examples. By R. D. BEASLEY, M.A. 9th Ed., revised and enlarged. Cr. 8vo. 3s. 6d.

BOTTOMLEY.—FOUR-FIGURE MATHEMATICAL TABLES. Comprising Logarithmic and Trigonometrical Tables, and Tables of Squares, Square Roots, and Reciprocals. By J. T. BOTTOMLEY, M.A., Lecturer in Natural Philosophy in the University of Glasgow. 8vo. 2s. 6d.

HAYWARD.—THE ALGEBRA OF CO-PLANAR VECTORS AND TRIGONOMETRY. By R. B. HAYWARD, M.A., F.R.S., Assistant Master at Harrow.
[*In preparation*.

JOHNSON.—A TREATISE ON TRIGONOMETRY. By W. E. JOHNSON, M.A., late Scholar and Assistant Mathematical Lecturer at King's College, Cambridge. Cr. 8vo. 8s. 6d.

LEVETT and DAVISON.—ELEMENTS OF TRIGONOMETRY. By RAWDON LEVETT and A. F. DAVISON, Assistant Masters at King Edward's School, Birmingham.   [*In the Press*.

LOCK.—Works by Rev. J. B. LOCK, M.A., Senior Fellow and Bursar of Gonville and Caius College, Cambridge.
*THE TRIGONOMETRY OF ONE ANGLE. Gl. 8vo. 2s. 6d.
*TRIGONOMETRY FOR BEGINNERS, as far as the Solution of Triangles. 3d Ed. Gl. 8vo. 2s. 6d. KEY. Cr. 8vo. 6s. 6d.
*ELEMENTARY TRIGONOMETRY. 6th Ed. (in this edition the chapter on logarithms has been carefully revised). Gl. 8vo. 4s. 6d. KEY. Cr. 8vo. 8s. 6d.

HIGHER TRIGONOMETRY. 5th Ed. Gl. 8vo. 4s. 6d. Both Parts complete in One Volume. Gl. 8vo. 7s. 6d.

M'CLELLAND and PRESTON.—A TREATISE ON SPHERICAL TRIGONO-METRY. With applications to Spherical Geometry and numerous Examples. By W. J. M'CLELLAND, M.A., Principal of the Incorporated Society's School, Santry, Dublin, and T. PRESTON, M.A. Cr. 8vo. 8s. 6d., or: Part I. To the End of Solution of Triangles, 4s. 6d. Part II., 5s.

MATTHEWS.—MANUAL OF LOGARITHMS. By G. F. MATTHEWS, B.A. 8vo. 5s. net.

PALMER.—TEXT-BOOK OF PRACTICAL LOGARITHMS AND TRIGONO-METRY. By J. H. PALMER, Headmaster, R.N., H.M.S. Cambridge, Devonport. Gl. 8vo. 4s. 6d.

SNOWBALL.—THE ELEMENTS OF PLANE AND SPHERICAL TRIGONO-METRY. By J. C. SNOWBALL. 14th Ed. Cr. 8vo. 7s. 6d.

TODHUNTER.—Works by ISAAC TODHUNTER, F.R.S.
*TRIGONOMETRY FOR BEGINNERS. 18mo. 2s. 6d. KEY. Cr. 8vo. 8s. 6d.
PLANE TRIGONOMETRY. Cr. 8vo. 5s. A New Edition, revised by R. W. HOGG, M.A. Cr. 8vo. 5s. KEY. Cr. 8vo. 10s. 6d.
A TREATISE ON SPHERICAL TRIGONOMETRY. Cr. 8vo. 4s. 6d.

WOLSTENHOLME.—EXAMPLES FOR PRACTICE IN THE USE OF SEVEN-FIGURE LOGARITHMS. By JOSEPH WOLSTENHOLME, D.Sc., late Professor of Mathematics in the Royal Indian Engineering Coll., Cooper's Hill. 8vo. 5s.

ANALYTICAL GEOMETRY (Plane and Solid).

DYER.—EXERCISES IN ANALYTICAL GEOMETRY. By J. M. DYER, M.A., Assistant Master at Eton. Illustrated. Cr. 8vo. 4s. 6d.

FERRERS.—AN ELEMENTARY TREATISE ON TRILINEAR CO-ORDIN-ATES, the Method of Reciprocal Polars, and the Theory of Projectors. By the Rev. N. M. FERRERS, D.D., F.R.S., Master of Gonville and Caius College, Cambridge. 4th Ed., revised. Cr. 8vo. 6s. 6d.

FROST.—Works by PERCIVAL FROST, D.Sc., F.R.S., Fellow and Mathematical Lecturer at King's College, Cambridge.
AN ELEMENTARY TREATISE ON CURVE TRACING. 8vo. 12s.
SOLID GEOMETRY. 3d Ed. Demy 8vo. 16s.
HINTS FOR THE SOLUTION OF PROBLEMS in the Third Edition of SOLID GEOMETRY. 8vo. 8s. 6d.

JOHNSON.—CURVE TRACING IN CARTESIAN CO-ORDINATES. By W. WOOLSEY JOHNSON, Professor of Mathematics at the U.S. Naval Academy, Annapolis, Maryland. Cr. 8vo. 4s. 6d.

M'CLELLAND.—THE GEOMETRY OF THE CIRCLE. By W. J. M'CLELLAND, M.A. Cr. 8vo. [In the Press.

PUCKLE.—AN ELEMENTARY TREATISE ON CONIC SECTIONS AND AL-GEBRAIC GEOMETRY. With Numerous Examples and Hints for their Solution. By G. H. PUCKLE, M.A. 5th Ed., revised and enlarged. Cr. 8vo. 7s. 6d.

SMITH.—Works by CHARLES SMITH, M.A., Master of Sidney Sussex College, Cambridge.
CONIC SECTIONS. 7th Ed. Cr. 8vo. 7s. 6d.
SOLUTIONS TO CONIC SECTIONS. Cr. 8vo. 10s. 6d.
AN ELEMENTARY TREATISE ON SOLID GEOMETRY. 2d Ed. Cr. 8vo. 9s. 6d.

TODHUNTER.—Works by ISAAC TODHUNTER, F.R.S.
PLANE CO-ORDINATE GEOMETRY, as applied to the Straight Line and the Conic Sections. Cr. 8vo. 7s. 6d.
KEY. By C. W. BOURNE, M.A., Headmaster of King's College School. Cr. 8vo. 10s. 6d.

TODHUNTER.—Works by Isaac Todhunter, F.R.S.—*continued.*
EXAMPLES OF ANALYTICAL GEOMETRY OF THREE DIMENSIONS. New Ed., revised. Cr. 8vo. 4s.

PROBLEMS AND QUESTIONS IN MATHEMATICS.

ARMY PRELIMINARY EXAMINATION, 1882-1890, Specimens of Papers set at the. With Answers to the Mathematical Questions. Subjects: Arithmetic, Algebra, Euclid, Geometrical Drawing, Geography, French, English Dictation. Cr. 8vo. 3s. 6d.

CAMBRIDGE SENATE-HOUSE **PROBLEMS AND RIDERS, WITH SOLU-**TIONS:—
1875—PROBLEMS AND RIDERS. By A. G. Greenhill, F.R.S. Cr. 8vo. 8s. 6d.
1878—SOLUTIONS OF SENATE-HOUSE PROBLEMS. By the Mathematical Moderators and Examiners. Edited by J. W. L. Glaisher, F.R.S., Fellow of Trinity College, Cambridge. 12s.

CHRISTIE.—A COLLECTION OF ELEMENTARY TEST-QUESTIONS IN PURE AND MIXED MATHEMATICS; with Answers and Appendices on Synthetic Division, and on the Solution of Numerical Equations by Horner's Method. By James R. Christie, F.R.S. Cr. 8vo. 8s. 6d.

CLIFFORD.—MATHEMATICAL PAPERS. By W. K. Clifford. Edited by R. Tucker. With an Introduction by H. J. Stephen Smith, M.A. 8vo. 30s.

MILNE.—Works by Rev. John J. Milne, Private Tutor.
WEEKLY PROBLEM PAPERS. With Notes intended for the use of Students preparing for Mathematical Scholarships, and for Junior Members of the Universities who are reading for Mathematical Honours. Pott 8vo. 4s. 6d.
SOLUTIONS TO WEEKLY PROBLEM PAPERS. Cr. 8vo. 10s. 6d.
COMPANION TO WEEKLY PROBLEM PAPERS. Cr. 8vo. 10s. 6d.

RICHARDSON.—MISCELLANEOUS MATHEMATICAL PAPERS. Elementary and Advanced. By A. T. Richardson, M.A., Senior Mathematical Master at the Isle of Wight College. [*In the Press.*

SANDHURST MATHEMATICAL PAPERS, for admission into the Royal Military College, 1881-1889. Edited by E. J. Brooksmith, B.A., Instructor in Mathematics at the Royal Military Academy, Woolwich. Cr. 8vo. 3s. 6d.

WOOLWICH MATHEMATICAL PAPERS, for Admission into the Royal Military Academy, Woolwich, 1880-1888 inclusive. By the same Editor. Cr. 8vo. 6s.

WOLSTENHOLME.—Works by Joseph Wolstenholme, D.Sc., late Professor of Mathematics in the Royal Engineering Coll., Cooper's Hill.
MATHEMATICAL PROBLEMS, on Subjects included in the First and Second Divisions of the Schedule of Subjects for the Cambridge Mathematical Tripos Examination. New Ed., greatly enlarged. 8vo. 18s.
EXAMPLES FOR PRACTICE IN THE USE OF SEVEN-FIGURE LOGARITHMS. 8vo. 5s.

HIGHER PURE MATHEMATICS.

AIRY.—Works by Sir G. B. Airy, K.C.B., formerly Astronomer-Royal.
ELEMENTARY TREATISE ON PARTIAL DIFFERENTIAL EQUATIONS. With Diagrams. 2d Ed. Cr. 8vo. 5s. 6d.
ON THE ALGEBRAICAL AND NUMERICAL THEORY OF ERRORS OF OBSERVATIONS AND THE COMBINATION OF OBSERVATIONS. 2d Ed., revised. Cr. 8vo. 6s. 6d.

BOOLE.—THE CALCULUS OF FINITE DIFFERENCES. By G. Boole. 3d Ed., revised by J. F. Moulton, Q.C. Cr. 8vo. 10s. 6d.

EDWARDS.—THE DIFFERENTIAL CALCULUS. By Joseph Edwards, M.A. With Applications and numerous Examples. Cr. 8vo. 10s. 6d.

FERRERS.—AN ELEMENTARY TREATISE ON SPHERICAL HARMONICS, AND SUBJECTS CONNECTED WITH THEM. By Rev. N. M. Ferrers, D.D F.R.S., Master of Gonville and Caius College, Cambridge. Cr. 8vo. 7s. 6d.

FORSYTH.—A TREATISE ON DIFFERENTIAL EQUATIONS. By ANDREW RUSSELL FORSYTH, F.R.S., Fellow and Assistant Tutor of Trinity College, Cambridge. 2d Ed. 8vo. 14s.

FROST.—AN ELEMENTARY TREATISE ON CURVE TRACING. By PERCIVAL FROST, M.A., D.Sc. 8vo. 12s.

GRAHAM.—GEOMETRY OF POSITION. By R. H. GRAHAM. Cr. 8vo. 7s. 6d.

GREENHILL.—DIFFERENTIAL AND INTEGRAL CALCULUS. By A. G. GREENHILL, Professor of Mathematics to the Senior Class of Artillery Officers, Woolwich. New Ed. Cr. 8vo. 10s. 6d.

APPLICATIONS OF ELLIPTIC FUNCTIONS. By the same. [*In the Press.*

JOHNSON.—Works by WILLIAM WOOLSEY JOHNSON, Professor of Mathematics at the U.S. Naval Academy, Annapolis, Maryland.

INTEGRAL CALCULUS, an Elementary Treatise on the. Founded on the Method of Rates or Fluxions. 8vo. 9s.

CURVE TRACING IN CARTESIAN CO-ORDINATES. Cr. 8vo. 4s. 6d.

A TREATISE ON ORDINARY AND DIFFERENTIAL EQUATIONS. Ex. cr. 8vo. 15s.

KELLAND and TAIT.—INTRODUCTION TO QUATERNIONS, with numerous examples. By P. KELLAND and P. G. TAIT, Professors in the Department of Mathematics in the University of Edinburgh. 2d Ed. Cr. 8vo. 7s. 6d.

KEMPE.—HOW TO DRAW A STRAIGHT LINE: a Lecture on Linkages. By A. B. KEMPE. Illustrated. Cr. 8vo. 1s. 6d.

KNOX.—DIFFERENTIAL CALCULUS FOR BEGINNERS. By ALEXANDER KNOX. Fcap. 8vo. 3s. 6d.

MUIR.—THE THEORY OF DETERMINANTS IN THE HISTORICAL ORDER OF ITS DEVELOPMENT. Part I. Determinants in General. Leibnitz (1693) to Cayley (1841). By THOS. MUIR, Mathematical Master in the High School of Glasgow. 8vo. 10s. 6d.

RICE and JOHNSON.—AN ELEMENTARY TREATISE ON THE DIFFEREN-TIAL CALCULUS. Founded on the Method of Rates or Fluxions. By J. M. RICE, Professor of Mathematics in the United States Navy, and W. W. JOHN-SON, Professor of Mathematics at the United States Naval Academy. 3d Ed., revised and corrected. 8vo. 18s. Abridged Ed. 9s.

TODHUNTER.—Works by ISAAC TODHUNTER, F.R.S.

AN ELEMENTARY TREATISE ON THE THEORY OF EQUATIONS. Cr. 8vo. 7s. 6d.

A TREATISE ON THE DIFFERENTIAL CALCULUS. Cr. 8vo. 10s. 6d. KEY. Cr. 8vo. 10s. 6d.

A TREATISE ON THE INTEGRAL CALCULUS AND ITS APPLICATIONS. Cr. 8vo. 10s. 6d. KEY. Cr. 8vo. 10s. 6d.

A HISTORY OF THE MATHEMATICAL THEORY OF PROBABILITY, from the time of Pascal to that of Laplace. 8vo. 18s.

AN ELEMENTARY TREATISE ON LAPLACE'S, LAME'S, AND BESSEL'S FUNCTIONS. Cr. 8vo. 10s. 6d.

MECHANICS: Statics, Dynamics, Hydrostatics, Hydrodynamics. (See also Physics.)

ALEXANDER and THOMSON.—ELEMENTARY APPLIED MECHANICS. By Prof. T. ALEXANDER and A. W. THOMSON. Part II. Transverse Stress. Cr. 8vo. 10s. 6d.

BALL.—EXPERIMENTAL MECHANICS. A Course of Lectures delivered at the Royal College of Science for Ireland. By Sir R. S. BALL, F.R.S. 2d Ed. Illustrated. Cr. 8vo. 6s.

CLIFFORD.—THE ELEMENTS OF DYNAMIC. An Introduction to the Study of Motion and Rest in Solid and Fluid Bodies. By W. K. CLIFFORD. Part I.—Kinematic. Cr. 8vo. Books I.–III. 7s. 6d.; Book IV. and Appendix, 6s.

COTTERILL.—APPLIED MECHANICS: An Elementary General Introduction to the Theory of Structures and Machines. By J. H. COTTERILL, F.R.S., Professor of Applied Mechanics in the Royal Naval College, Greenwich. 8vo. 18s.

COTTERILL and SLADE.—LESSONS IN APPLIED MECHANICS. By Prof. J. H. COTTERILL and J. H. SLADE. Fcap. 8vo. 5s. 6d.

DYNAMICS, SYLLABUS OF ELEMENTARY. Part I. Linear Dynamics. With an Appendix on the Meanings of the Symbols in Physical Equations. Prepared by the Association for the Improvement of Geometrical Teaching. 4to. 1s.

GANGUILLET and KUTTER.—A GENERAL FORMULA FOR THE UNIFORM FLOW OF WATER IN RIVERS AND OTHER CHANNELS. By E. GANGUILLET and W. R. KUTTER. Translated, with Additions, including Tables and Diagrams, and the Elements of over 1200 Gaugings of Rivers, Small Channels, and Pipes in English Measure, by R. HERING, Assoc. Am. Soc., C.E., M. Inst. C.E., and J. C. TRAUTWINE Jun., Assoc. Am. Soc. C.E., Assoc. Inst. C.E. 8vo. 17s.

GRAHAM.—GEOMETRY OF POSITION. By R. H. GRAHAM. Cr. 8vo. 7s. 6d.

GREAVES.—Works by JOHN GREAVES, M.A., Fellow and Mathematical Lecturer at Christ's College, Cambridge.

*STATICS FOR BEGINNERS. Gl. 8vo. 3s. 6d.

A TREATISE ON ELEMENTARY STATICS. 2d Ed. Cr. 8vo. 6s. 6d.

GREENHILL.—HYDROSTATICS. By A. G. GREENHILL, Professor of Mathematics to the Senior Class of Artillery Officers, Woolwich. Cr. 8vo. [In preparation.

*HICKS.—ELEMENTARY DYNAMICS OF PARTICLES AND SOLIDS. By W. M. HICKS, D.Sc., Principal and Professor of Mathematics and Physics, Firth College, Sheffield. Cr. 8vo. 6s. 6d.

JELLETT.—A TREATISE ON THE THEORY OF FRICTION. By JOHN H. JELLETT, B.D., late Provost of Trinity College, Dublin. 8vo. 8s. 6d.

KENNEDY.—THE MECHANICS OF MACHINERY. By A. B. W. KENNEDY, F.R.S. Illustrated. Cr. 8vo. 12s. 6d.

LOCK.—Works by Rev. J. B. LOCK, M.A.

*ELEMENTARY STATICS. 2d Ed. Gl. 8vo. 4s. 6d.

*ELEMENTARY DYNAMICS. 3d Ed. Gl. 8vo. 4s. 6d.

ELEMENTARY HYDROSTATICS. Gl. 8vo. [In preparation.

MECHANICS FOR BEGINNERS. Gl. 8vo. Part I. MECHANICS OF SOLIDS. [In the Press. Part II. MECHANICS OF FLUIDS. [In preparation.

MACGREGOR.—KINEMATICS AND DYNAMICS. An Elementary Treatise. By J G. MACGREGOR, D.Sc., Munro Professor of Physics in Dalhousie College, Halifax, Nova Scotia. Illustrated. Cr. 8vo. 10s. 6d.

PARKINSON.—AN ELEMENTARY TREATISE ON MECHANICS. By S. PARKINSON, D.D., F.R.S., late Tutor and Prælector of St. John's College, Cambridge. 6th Ed., revised. Cr. 8vo. 9s. 6d.

PIRIE.—LESSONS ON RIGID DYNAMICS. By Rev. G. PIRIE, M.A., Professor of Mathematics in the University of Aberdeen. Cr. 8vo. 6s.

ROUTH.—Works by EDWARD JOHN ROUTH, D.Sc., LL.D., F.R.S., Hon. Fellow of St. Peter's College, Cambridge.

A TREATISE ON THE DYNAMICS OF THE SYSTEM OF RIGID BODIES. With numerous Examples. Two Vols. 8vo. Vol. I.—Elementary Parts. 5th Ed. 14s. Vol. II.—The Advanced Parts. 4th Ed. 14s.

STABILITY OF A GIVEN STATE OF MOTION, PARTICULARLY STEADY MOTION. Adams Prize Essay for 1877. 8vo. 8s. 6d.

*SANDERSON.—HYDROSTATICS FOR BEGINNERS. By F. W. SANDERSON, M.A., Assistant Master at Dulwich College. Gl. 8vo. 4s. 6d.

TAIT and STEELE.—A TREATISE ON DYNAMICS OF A PARTICLE. By Professor TAIT, M.A., and W. J. STEELE, B.A. 6th Ed., revised. Cr. 8vo. 12s.

TODHUNTER.—Works by ISAAC TODHUNTER, F.R.S.

*MECHANICS FOR BEGINNERS. 18mo. 4s. 6d. KEY. Cr. 8vo. 6s. 6d.

A TREATISE ON ANALYTICAL STATICS. 5th Ed. Edited by Prof. J. D. EVERETT, F.R.S. Cr. 8vo. 10s. 6d.

PHYSICS: Sound, Light, Heat, Electricity, Elasticity, Attractions, etc. (See also Mechanics.)

AIRY.—Works by Sir G. B. AIRY, K.C.B., formerly Astronomer-Royal.

ON SOUND AND ATMOSPHERIC VIBRATIONS. With the Mathematical Elements of Music. 2d Ed., revised and enlarged. Cr. 8vo. 9s.

GRAVITATION: An Elementary Explanation of the Principal Perturbations in the Solar System. 2d Ed. Cr. 8vo. 7s. 6d.

CLAUSIUS.—MECHANICAL THEORY OF HEAT. By R. CLAUSIUS. Translated by W. R. BROWNE, M.A. Cr. 8vo. 10s. 6d.

CUMMING.—AN INTRODUCTION TO THE THEORY OF ELECTRICITY. By LINNÆUS CUMMING, M.A., Assistant Master at Rugby. Illustrated. Cr. 8vo. 8s. 6d.

DANIELL.—A TEXT-BOOK OF THE PRINCIPLES OF PHYSICS. By ALFRED DANIELL, D.Sc. Illustrated. 2d Ed., revised and enlarged. 8vo. 21s.

DAY.—ELECTRIC LIGHT ARITHMETIC. By R. E. DAY, Evening Lecturer in Experimental Physics at King's College, London. Pott 8vo. 2s.

EVERETT.—ILLUSTRATIONS OF THE C. G. S. SYSTEM OF UNITS WITH TABLES OF PHYSICAL CONSTANTS. By J D. EVERETT, F.R.S., Professor of Natural Philosophy, Queen's College, Belfast. New Ed. Ex. fcap. 8vo. 5s.

FERRERS.—AN ELEMENTARY TREATISE ON SPHERICAL HARMONICS, and Subjects connected with them. By Rev. N. M. FERRERS, D.D., F.R.S., Master of Gonville and Caius College, Cambridge. Cr. 8vo. 7s. 6d.

FESSENDEN.—PHYSICS FOR PUBLIC SCHOOLS. By C. FESSENDEN. Illustrated. Fcap. 8vo. [In the Press.

GRAY.—THE THEORY AND PRACTICE OF ABSOLUTE MEASUREMENTS IN ELECTRICITY AND MAGNETISM. By A. GRAY, F.R.S.E., Professor of Physics in the University College of North Wales. Two Vols. Cr. 8vo. Vol. I. 12s. 6d. [Vol. II. In the Press.

ABSOLUTE MEASUREMENTS IN ELECTRICITY AND MAGNETISM. 2d Ed., revised and greatly enlarged. Fcap. 8vo. 5s. 6d.

IBBETSON.—THE MATHEMATICAL THEORY OF PERFECTLY ELASTIC SOLIDS, with a Short Account of Viscous Fluids. By W. J. IBBETSON, late Senior Scholar of Clare College, Cambridge. 8vo. 21s.

`JONES.`—EXAMPLES IN PHYSICS. Containing over 1000 Problems with Answers and numerous solved Examples. Suitable for candidates preparing for the Intermediate, Science, Preliminary, Scientific, and other Examinations of the University of London. By D. E. JONES, B.Sc., Professor of Physics in the University College of Wales, Aberystwyth. Fcap. 8vo. 3s. 6d.

*ELEMENTARY LESSONS IN HEAT, LIGHT, AND SOUND. By the same. Gl. 8vo. 2s. 6d.

LOCKYER.—CONTRIBUTIONS TO SOLAR PHYSICS. By J. NORMAN LOCKYER, F.R.S. With Illustrations. Royal 8vo. 31s. 6d.

LODGE.—MODERN VIEWS OF ELECTRICITY. By OLIVER J. LODGE, F.R.S., Professor of Experimental Physics in University College, Liverpool. Illustrated. Cr. 8vo. 6s. 6d.

LOEWY.—*QUESTIONS AND EXAMPLES ON EXPERIMENTAL PHYSICS: Sound, Light, Heat, Electricity, and Magnetism. By B. LOEWY, Examiner in Experimental Physics to the College of Preceptors. Fcap. 8vo. 2s.

`A GRADUATED COURSE OF NATURAL SCIENCE FOR ELEMENTARY AND TECHNICAL SCHOOLS AND COLLEGES. By the same. In Three Parts. Part I. FIRST YEAR'S COURSE. Gl. 8vo. 2s. Part II. [In preparation.

LUPTON.—NUMERICAL TABLES AND CONSTANTS IN ELEMENTARY SCIENCE. By S. LUPTON, M.A., late Assistant Master at Harrow. Ex. fcap. 8vo. 2s. 6d.

MACFARLANE.—PHYSICAL ARITHMETIC. By A. MACFARLANE, D.Sc., late Examiner in Mathematics at the University of Edinburgh. Cr. 8vo. 7s. 6d.

*MAYER.—SOUND: A Series of Simple, Entertaining, and Inexpensive Experiments in the Phenomena of Sound. By A. M. MAYER, Professor of Physics in the Stevens Institute of Technology. Illustrated. Cr. 8vo. 3s. 6d.

*MAYER and BARNARD.—LIGHT: A Series of Simple, Entertaining, and Inexpensive Experiments in the Phenomena of Light. By A. M. MAYER and C. BARNARD. Illustrated. Cr. 8vo. 2s. 6d.

MOLLOY.—GLEANINGS IN SCIENCE: Popular Lectures on Scientific Subjects. By the Rev. GERALD MOLLOY, D.Sc., Rector of the Catholic University of Ireland. 8vo. 7s. 6d.

NEWTON.—PRINCIPIA. Edited by Prof. Sir W. THOMSON, P.R.S., and Prof. BLACKBURNE. 4to. 31s. 6d.

THE FIRST THREE SECTIONS OF NEWTON'S PRINCIPIA. With Notes and Illustrations. Also a Collection of Problems, principally intended as Examples of Newton's Methods. By P. FROST, M.A., D.Sc. 3d Ed. 8vo. 12s.

PARKINSON.—A TREATISE ON OPTICS. By S. PARKINSON, D.D., F.R.S., late Tutor and Prælector of St. John's College, Cambridge. 4th Ed., revised and enlarged. Cr. 8vo. 10s. 6d.

PEABODY.—THERMODYNAMICS OF THE STEAM-ENGINE AND OTHER HEAT-ENGINES. By CECIL H. PEABODY, Associate Professor of Steam Engineering, Massachusetts Institute of Technology. 8vo. 21s.

PERRY.—STEAM: An Elementary Treatise. By JOHN PERRY, Professor of Mechanical Engineering and Applied Mechanics at the Technical College, Finsbury. 18mo. 4s. 6d.

PICKERING.—ELEMENTS OF PHYSICAL MANIPULATION. By Prof. EDWARD C. PICKERING. Medium 8vo. Part I., 12s. 6d. Part II., 14s.

PRESTON.—THE THEORY OF LIGHT. By THOMAS PRESTON, M.A. Illustrated. 8vo. 12s. 6d.

THE THEORY OF HEAT. By the same Author. 8vo. [In preparation.

RAYLEIGH.—THE THEORY OF SOUND. By LORD RAYLEIGH, F.R.S. 8vo. Vol. I., 12s. 6d. Vol. II., 12s. 6d. [Vol. III. In the Press.

SHANN.—AN ELEMENTARY TREATISE ON HEAT, IN RELATION TO STEAM AND THE STEAM-ENGINE. By G. SHANN, M.A. Illustrated. Cr. 8vo. 4s. 6d.

SPOTTISWOODE.—POLARISATION OF LIGHT. By the late W. SPOTTISWOODE, F.R.S. Illustrated. Cr. 8vo. 3s. 6d.

STEWART.—Works by BALFOUR STEWART, F.R.S., late Langworthy Professor of Physics in the Owens College, Victoria University, Manchester.

*PRIMER OF PHYSICS. Illustrated. With Questions. 18mo. 1s.

*LESSONS IN ELEMENTARY PHYSICS. Illustrated. Fcap. 8vo. 4s. 6d.

*QUESTIONS. By Prof. T. H. CORE. Fcap. 8vo. 2s.

STEWART and GEE.—LESSONS IN ELEMENTARY PRACTICAL PHYSICS. By BALFOUR STEWART, F.R.S., and W. W. HALDANE GEE, B.Sc. Cr. 8vo. Vol. I. GENERAL PHYSICAL PROCESSES. 6s. Vol. II. ELECTRICITY AND MAGNETISM. 7s. 6d. [Vol. III. OPTICS, HEAT, AND SOUND. In the Press.

*PRACTICAL PHYSICS FOR SCHOOLS AND THE JUNIOR STUDENTS OF COLLEGES. Gl. 8vo. Vol. I. ELECTRICITY AND MAGNETISM. 2s. 6d. [Vol. II. OPTICS, HEAT, AND SOUND. In the Press.

STOKES.—ON LIGHT. Burnett Lectures, delivered in Aberdeen in 1883-4-5. By Sir G. G. STOKES, F.R.S., Lucasian Professor of Mathematics in the University of Cambridge. First Course: ON THE NATURE OF LIGHT. Second Course: ON LIGHT AS A MEANS OF INVESTIGATION. Third Course: ON THE BENEFICIAL EFFECTS OF LIGHT. Cr. 8vo. 7s. 6d.

⁎ The 2d and 3d Courses may be had separately. Cr. 8vo. 2s. 6d. each.

STONE.—AN ELEMENTARY TREATISE ON SOUND. By W. H. STONE. Illustrated. Fcap. 8vo. 3s. 6d.

TAIT.—HEAT. By P. G. TAIT, Professor of Natural Philosophy in the University of Edinburgh. Cr. 8vo. 6s.

LECTURES ON SOME RECENT ADVANCES IN PHYSICAL SCIENCE. By the same. 3d Edition. Crown 8vo. 9s.

TAYLOR.—SOUND AND MUSIC. An Elementary Treatise on the Physical Constitution of Musical Sounds and Harmony, including the Chief Acoustical Discoveries of Professor Helmholtz. By SEDLEY TAYLOR, M.A. Illustrated. 2d Ed. Ex. cr. 8vo. 8s. 6d.

*THOMPSON. — ELEMENTARY LESSONS IN ELECTRICITY AND MAGNETISM. By SILVANUS P. THOMPSON, Principal and Professor of Physics in the Technical College, Finsbury. Illustrated. New Ed., revised. Fcap. 8vo. 4s. 6d.

THOMSON.—Works by J. J. THOMSON, Professor of Experimental Physics in the University of Cambridge.
A TREATISE ON THE MOTION OF VORTEX RINGS. Adams Prize Essay 1882. 8vo. 6s.
APPLICATIONS OF DYNAMICS TO PHYSICS AND CHEMISTRY. Cr. 8vo. 7s. 6d.

THOMSON.—Works by Sir W. THOMSON, P.R.S., Professor of Natural Philosophy in the University of Glasgow.
ELECTROSTATICS AND MAGNETISM, REPRINTS OF PAPERS ON. 2d Ed. 8vo. 18s.
POPULAR LECTURES AND ADDRESSES. 3 Vols. Illustrated. Cr. 8vo. Vol. I. CONSTITUTION OF MATTER. 7s. 6d. Vol. III. NAVIGATION. 7s. 6d.

TODHUNTER.—Works by ISAAC TODHUNTER, F.R.S.
AN ELEMENTARY TREATISE ON LAPLACE'S, LAMÉ'S, AND BESSEL'S FUNCTIONS. Crown 8vo. 10s. 6d.
A HISTORY OF THE MATHEMATICAL THEORIES OF ATTRACTION, AND THE FIGURE OF THE EARTH, from the time of Newton to that of Laplace. 2 vols. 8vo. 24s.

TURNER.—A COLLECTION OF EXAMPLES ON HEAT AND ELECTRICITY. By H. H. TURNER, Fellow of Trinity College, Cambridge. Cr. 8vo. 2s. 6d.

WRIGHT.—LIGHT: A Course of Experimental Optics, chiefly with the Lantern. By LEWIS WRIGHT. Illustrated. Cr. 8vo. 7s. 6d.

ASTRONOMY.

AIRY.—Works by Sir G. B. AIRY, K.C.B., formerly Astronomer-Royal.
*POPULAR ASTRONOMY. 7th Ed. Revised by H. H. TURNER, M.A. 18mo. 4s. 6d.
GRAVITATION: An Elementary Explanation of the Principal Perturbations in the Solar System. 2d Ed. Cr. 8vo. 7s. 6d.

CHEYNE.—AN ELEMENTARY TREATISE ON THE PLANETARY THEORY. By C. H. H. CHEYNE. With Problems. 3d Ed. Edited by Rev. A. FREEMAN, M.A., F.R.A.S. Cr. 8vo. 7s. 6d.

CLARK and SADLER.—THE STAR GUIDE. By L. CLARK and H. SADLER. Roy. 8vo. 5s.

CROSSLEY, GLEDHILL, and WILSON.—A HANDBOOK OF DOUBLE STARS. By E. CROSSLEY, J. GLEDHILL, and J. M. WILSON. 8vo. 21s.
CORRECTIONS TO THE HANDBOOK OF DOUBLE STARS. 8vo. 1s.

FORBES.—TRANSIT OF VENUS. By G. FORBES, Professor of Natural Philosophy in the Andersonian University, Glasgow. Illustrated. Cr. 8vo. 3s. 6d.

GODFRAY.—Works by HUGH GODFRAY, M.A., Mathematical Lecturer at Pembroke College, Cambridge.
A TREATISE ON ASTRONOMY. 4th Ed. 8vo. 12s. 6d.
AN ELEMENTARY TREATISE ON THE LUNAR THEORY, with a brief Sketch of the Problem up to the time of Newton. 2d Ed., revised. Cr. 8vo. 5s. 6d.

LOCKYER.—Works by J. NORMAN LOCKYER, F.R.S.
*PRIMER OF ASTRONOMY. Illustrated. 18mo. 1s.
*ELEMENTARY LESSONS IN ASTRONOMY. With Spectra of the Sun, Stars, and Nebulæ, and numerous Illustrations. 36th Thousand. Revised throughout. Fcap. 8vo. 5s. 6d.

*QUESTIONS ON LOCKYER'S ELEMENTARY LESSONS IN ASTRONOMY. By J. FORBES ROBERTSON. 18mo. 1s. 6d.

THE CHEMISTRY OF THE SUN. Illustrated. 8vo. 14s.

THE METEORITIC HYPOTHESIS OF THE ORIGIN OF COSMICAL SYSTEMS. Illustrated. 8vo. 17s. net.

THE EVOLUTION OF THE HEAVENS AND THE EARTH. Cr. 8vo. Illustrated. [In the Press.

LOCKYER and SEABROKE.—STAR-GAZING PAST AND PRESENT. By J. NORMAN LOCKYER, F.R.S. Expanded from Shorthand Notes with the assistance of G. M. SEABROKE, F.R.A.S. Royal 8vo. 21s.

NEWCOMB.—POPULAR ASTRONOMY. By S. NEWCOMB, LL.D., Professor U.S. Naval Observatory. Illustrated. 2d Ed., revised. 8vo. 18s.

HISTORICAL.

BALL.—A SHORT ACCOUNT OF THE HISTORY OF MATHEMATICS. By W. W. R. BALL, M.A. Cr. 8vo. 10s. 6d.

NATURAL SCIENCES.

Chemistry; Physical Geography, Geology, and Mineralogy; Biology; Medicine.

(For MECHANICS, PHYSICS, and ASTRONOMY, see *MATHEMATICS*.)

CHEMISTRY.

ARMSTRONG.—A MANUAL OF INORGANIC CHEMISTRY. By HENRY ARMSTRONG, F.R.S., Professor of Chemistry in the City and Guilds of London Technical Institute. Cr. 8vo. [In preparation.

*COHEN.—THE OWENS COLLEGE COURSE OF PRACTICAL ORGANIC CHEMISTRY. By JULIUS B. COHEN, Ph.D., Assistant Lecturer on Chemistry in the Owens College, Manchester. With a Preface by Sir HENRY ROSCOE, F.R.S., and C. SCHORLEMMER, F.R.S. Fcap. 8vo. 2s. 6d.

COOKE.—ELEMENTS OF CHEMICAL PHYSICS. By JOSIAH P. COOKE, Jun., Erving Professor of Chemistry and Mineralogy in Harvard University. 4th Ed. 8vo. 21s.

FLEISCHER.—A SYSTEM OF VOLUMETRIC ANALYSIS. By EMIL FLEISCHER. Translated, with Notes and Additions, by M. M. P. MUIR, F.R.S.E. Illustrated. Cr. 8vo. 7s. 6d.

FRANKLAND.—A HANDBOOK OF AGRICULTURAL CHEMICAL ANALYSIS. By P. F. FRANKLAND, F.R.S., Professor of Chemistry in University College, Dundee. Cr. 8vo. 7s. 6d.

HARTLEY.—A COURSE OF QUANTITATIVE ANALYSIS FOR STUDENTS. By W. NOEL HARTLEY, F.R.S., Professor of Chemistry and of Applied Chemistry, Science and Art Department, Royal College of Science, Dublin. Gl. 8vo. 5s.

HEMPEL.—METHODS OF GAS ANALYSIS. By Dr. WALTHER HEMPEL. Translated by Dr. L. M. DENNIS. [In preparation.

HIORNS.—PRACTICAL METALLURGY AND ASSAYING. A Text-Book for the use of Teachers, Students, and Assayers. By ARTHUR H. HIORNS, Principal of the School of Metallurgy, Birmingham and Midland Institute. Illustrated. Gl. 8vo. 6s.

A TEXT-BOOK OF ELEMENTARY METALLURGY FOR THE USE OF STUDENTS. To which is added an Appendix of Examination Questions, embracing the whole of the Questions set in the three stages of the subject by the Science and Art Department for the past twenty years. By the same. Gl. 8vo. 4s.

IRON AND STEEL MANUFACTURE. A Text-Book for Beginners. By the same. Illustrated. Gl. 8vo. 3s. 6d.

MIXED METALS OR METALLIC ALLOYS. By the same. Gl. 8vo. 6s.

JONES.—*THE OWENS COLLEGE JUNIOR COURSE OF PRACTICAL CHEM-ISTRY. By FRANCIS JONES, F.R.S.E., Chemical Master at the Grammar School, Manchester. With Preface by Sir HENRY ROSCOE, F.R.S. Illustrated. Fcap. 8vo. 2s. 6d.

*QUESTIONS ON CHEMISTRY. A Series of Problems and Exercises in Inorganic and Organic Chemistry. By the same. Fcap. 8vo. 3s.

LANDAUER.—BLOWPIPE ANALYSIS. By J. LANDAUER. Authorised English Edition by J. TAYLOR and W. E. KAY, of Owens College, Manchester.
[*New Edition in the Press.*

LOCKYER.—THE CHEMISTRY OF THE SUN. By J. NORMAN LOCKYER, F.R.S. Illustrated. 8vo. 14s.

LUPTON.—CHEMICAL ARITHMETIC. With 1200 Problems. By S. LUPTON, M.A. 2d Ed., revised and abridged. Fcap. 8vo. 4s. 6d.

MANSFIELD.—A THEORY OF SALTS. By C. B. MANSFIELD. Crown 8vo. 14s.

MELDOLA.—THE CHEMISTRY OF PHOTOGRAPHY. By RAPHAEL MELDOLA, F.R.S., Professor of Chemistry in the Technical College, Finsbury. Cr. 8vo. 6s.

MEYER.— HISTORY OF CHEMISTRY FROM THE EARLIEST TIMES TO THE PRESENT DAY. By ERNST VON MEYER, Ph.D. Translated by GEORGE McGOWAN, Ph.D. 8vo. 14s. net.

MIXTER.—AN ELEMENTARY TEXT-BOOK OF CHEMISTRY. By WILLIAM G. MIXTER, Professor of Chemistry in the Sheffield Scientific School of Yale College. 2d and revised Ed. Cr. 8vo. 7s. 6d.

MUIR.—PRACTICAL CHEMISTRY FOR MEDICAL STUDENTS. Specially ar-ranged for the first M.B. Course. By M. M. P. MUIR, F.R.S.E., Fellow and Præ-lector in Chemistry at Gonville and Caius College, Cambridge. Fcap. 8vo. 1s. 6d.

MUIR and WILSON.—THE ELEMENTS OF THERMAL CHEMISTRY. By M. M. P. MUIR, F.R.S.E.; assisted by D. M. WILSON. 8vo. 12s. 6d.

OSTWALD. —OUTLINES OF GENERAL CHEMISTRY (PHYSICAL AND THEORETICAL). By Prof. W. OSTWALD. Translated by JAMES WALKER, D.Sc., Ph.D. 8vo. 10s. net.

RAMSAY.—EXPERIMENTAL PROOFS OF CHEMICAL THEORY FOR BE-GINNERS. By WILLIAM RAMSAY, F.R.S., Professor of Chemistry in Univer-sity College, London. Pott 8vo. 2s. 6d.

REMSEN.—Works by IRA REMSEN, Professor of Chemistry in the Johns Hopkins University, U.S.A.

COMPOUNDS OF CARBON : or, Organic Chemistry, an Introduction to the Study of. Cr. 8vo. 6s. 6d.

AN INTRODUCTION TO THE STUDY OF CHEMISTRY (INORGANIC CHEMISTRY). Cr. 8vo. 6s. 6d.

*THE ELEMENTS OF CHEMISTRY. A Text-Book for Beginners. Fcap. 8vo. 2s. 6d.

A TEXT-BOOK OF INORGANIC CHEMISTRY. 8vo. 16s.

ROSCOE.— Works by Sir HENRY E. ROSCOE, F.R.S., formerly Professor of Chemistry in the Owens College, Victoria University, Manchester.

*PRIMER OF CHEMISTRY. Illustrated. With Questions. 18mo. 1s.

*LESSONS IN ELEMENTARY CHEMISTRY, INORGANIC AND ORGANIC. With Illustrations and Chromolitho of the Solar Spectrum, and of the Alkalies and Alkaline Earths. Fcap. 8vo. 4s. 6d.

ROSCOE and SCHORLEMMER.—INORGANIC AND ORGANIC CHEMISTRY. A Complete Treatise on Inorganic and Organic Chemistry. By Sir HENRY E. ROSCOE, F.R.S., and Prof. C. SCHORLEMMER, F.R.S. Illustrated. 8vo.

Vols. I. and II. INORGANIC CHEMISTRY. Vol. I.—The Non-Metallic Ele-ments. 2d Ed. 21s. Vol. II. Part I.—Metals. 18s. Part II.—Metals. 18s.

Vol. III.—ORGANIC CHEMISTRY. THE CHEMISTRY OF THE HYDRO-CARBONS and their Derivatives. Five Parts. Parts I., II., and IV. 21s. Parts III. and V. 18s. each.

ROSCOE and SCHUSTER.—SPECTRUM ANALYSIS. Lectures delivered in 1868. By Sir HENRY ROSCOE, F.R.S. 4th Ed., revised and considerably enlarged by the Author and by A. SCHUSTER, F.R.S., Ph.D., Professor of Applied Mathematics in the Owens College, Victoria University. With Appendices, Illustrations, and Plates. 8vo. 21s.

*THORPE.—A SERIES OF CHEMICAL PROBLEMS. With Key. For use in Colleges and Schools. By T. E. THORPE, B.Sc. (Vic.), Ph.D., F.R.S. Revised and Enlarged by W. TATE, Assoc.N.S.S. With Preface by Sir H. E. ROSCOE, F.R.S. New Ed. Fcap. 8vo. 2s.

THORPE and RÜCKER.—A TREATISE ON CHEMICAL PHYSICS. By Prof. T. E. THORPE, F.R.S., and Prof. A. W. RÜCKER, F.R.S. Illustrated. 8vo. *[In preparation.*

WURTZ.—A HISTORY OF CHEMICAL THEORY. By AD. WURTZ. Translated by HENRY WATTS, F.R.S. Crown 8vo. 6s.

PHYSICAL GEOGRAPHY, GEOLOGY, AND MINERALOGY.

BLANFORD.—THE RUDIMENTS OF PHYSICAL GEOGRAPHY FOR THE USE OF INDIAN SCHOOLS; with a Glossary of Technical Terms employed. By H. F. BLANFORD, F.G.S. Illustrated. Cr. 8vo. 2s. 6d.

FERREL.—A POPULAR TREATISE ON THE WINDS. Comprising the General Motions of the Atmosphere, Monsoons, Cyclones, Tornadoes, Waterspouts, Hailstorms, etc. By WILLIAM FERREL, M.A., Member of the American National Academy of Sciences. 8vo. 18s.

FISHER.—PHYSICS OF THE EARTH'S CRUST. By the Rev. OSMOND FISHER, M.A., F.G.S., Hon. Fellow of King's College, London. 2d Ed., altered and enlarged. 8vo. 12s.

GEIKIE.—Works by Sir ARCHIBALD GEIKIE, F.R.S., Director-General of the Geological Survey of the United Kingdom.

*PRIMER OF PHYSICAL GEOGRAPHY. Illustrated. With Questions. 18mo. 1s.

*ELEMENTARY LESSONS IN PHYSICAL GEOGRAPHY. Illustrated. Fcap. 8vo. 4s. 6d. *QUESTIONS ON THE SAME. 1s. 6d.

*PRIMER OF GEOLOGY. Illustrated. 18mo. 1s.

*CLASS-BOOK OF GEOLOGY. Illustrated. New and Cheaper Ed. Cr. 8vo. 4s. 6d.

TEXT-BOOK OF GEOLOGY. Illustrated. 2d Ed., 7th Thousand, revised and enlarged. 8vo. 28s.

OUTLINES OF FIELD GEOLOGY. Illustrated. New Ed., revised and enlarged. Gl. 8vo. 3s. 6d.

THE SCENERY AND GEOLOGY OF SCOTLAND, VIEWED IN CONNEXION WITH ITS PHYSICAL GEOLOGY. Illustrated. Cr. 8vo. 12s. 6d.

HUXLEY.—PHYSIOGRAPHY. An Introduction to the Study of Nature. By T. H. HUXLEY, F.R.S. Illustrated. New and Cheaper Edition. Cr. 8vo. 6s.

LOCKYER.—OUTLINES OF PHYSIOGRAPHY—THE MOVEMENTS OF THE EARTH. By J. NORMAN LOCKYER, F.R.S., Examiner in Physiography for the Science and Art Department. Illustrated. Cr. 8vo. Sewed, 1s. 6d.

MIERS.—A TREATISE ON MINERALOGY. By H. A. MIERS, of the British Museum. 8vo. *[In preparation.*

PHILLIPS. A TREATISE ON ORE DEPOSITS. By J. ARTHUR PHILLIPS, F.R.S. Illustrated. 8vo. 25s.

ROSENBUSCH and IDDINGS.—MICROSCOPICAL PHYSIOGRAPHY OF THE ROCK-MAKING MINERALS: AN AID TO THE MICROSCOPICAL STUDY OF ROCKS. By H. ROSENBUSCH. Translated and Abridged by J. P. IDDINGS. Illustrated. 8vo. 24s.

WILLIAMS.—ELEMENTS OF CRYSTALLOGRAPHY FOR STUDENTS OF CHEMISTRY, PHYSICS, AND MINERALOGY. By G. H. WILLIAMS, Ph.D., Cr. 8vo. 6s.

BIOLOGY.

ALLEN.—ON THE COLOURS OF FLOWERS, as Illustrated in the British Flora. By GRANT ALLEN. Illustrated. Cr. 8vo. 3s. 6d.

BALFOUR.—A TREATISE ON COMPARATIVE EMBRYOLOGY. By F. M. BALFOUR, F.R.S., Fellow and Lecturer of Trinity College, Cambridge. Illustrated. 2d Ed., reprinted without alteration from the 1st Ed. 2 vols. 8vo. Vol. I. 18s. Vol. II. 21s.

BALFOUR and WARD.—A GENERAL TEXT-BOOK OF BOTANY. By ISAAC BAYLEY BALFOUR, F.R.S., Professor of Botany in the University of Edinburgh, and H. MARSHALL WARD, F.R.S., Professor of Botany in the Royal Indian Engineering College, Cooper's Hill. 8vo. [In preparation.

*BETTANY.—FIRST LESSONS IN PRACTICAL BOTANY. By G. T. BETTANY. 18mo. 1s.

*BOWER.—A COURSE OF PRACTICAL INSTRUCTION IN BOTANY. By F. O. BOWER, D.Sc., F.R.S., Regius Professor of Botany in the University of Glasgow. New Ed., revised. Cr. 8vo. 10s. 6d. Abridged Ed. [In preparation.

BUCKTON.—MONOGRAPH OF THE BRITISH CICADÆ, OR TETTIGIDÆ. By G. B. BUCKTON. In 8 parts, Quarterly. Part I. January, 1890. 8vo. Parts I.-VI. ready. 8s. each, net. Vol. I. 33s. 6d. net.

CHURCH and SCOTT.—MANUAL OF VEGETABLE PHYSIOLOGY. By Professor A. H. CHURCH, and D. H. SCOTT, D.Sc., Lecturer in the Normal School of Science. Illustrated. Cr. 8vo. [In preparation.

COPE.—THE ORIGIN OF THE FITTEST. Essays on Evolution. By E. D COPE, M.A., Ph.D. 8vo. 12s. 6d.

COUES.—HANDBOOK OF FIELD AND GENERAL ORNITHOLOGY. By Prof. ELLIOTT COUES, M.A. Illustrated. 8vo. 10s. net.

DARWIN.—MEMORIAL NOTICES OF CHARLES DARWIN, F.R.S., etc. By T. H. HUXLEY, F.R.S., G. J. ROMANES, F.R.S., ARCHIBALD GEIKIE, F.R.S., and W. THISELTON DYER, F.R.S. Reprinted from *Nature*. With a Portrait. Cr. 8vo. 2s. 6d.

EIMER.—ORGANIC EVOLUTION AS THE RESULT OF THE INHERITANCE OF ACQUIRED CHARACTERS ACCORDING TO THE LAWS OF ORGANIC GROWTH. By Dr. G. H. THEODOR EIMER. Translated by J. T. CUNNINGHAM, F.R.S.E., late Fellow of University College, Oxford. 8vo. 12s. 6d.

FEARNLEY.—A MANUAL OF ELEMENTARY PRACTICAL HISTOLOGY. By WILLIAM FEARNLEY. Illustrated. Cr. 8vo. 7s. 6d.

FLOWER and GADOW.—AN INTRODUCTION TO THE OSTEOLOGY OF THE MAMMALIA. By W. H. FLOWER, F.R.S., Director of the Natural History Departments of the British Museum. Illustrated. 3d Ed. Revised with the assistance of HANS GADOW, Ph.D., Lecturer on the Advanced Morphology of Vertebrates in the University of Cambridge. Cr. 8vo. 10s. 6d.

FOSTER.—Works by MICHAEL FOSTER, M.D., F.R.S., Professor of Physiology in the University of Cambridge.

   *PRIMER OF PHYSIOLOGY. Illustrated. 18mo. 1s.

   A TEXT-BOOK OF PHYSIOLOGY. Illustrated. 5th Ed., largely revised. 8vo. Part I., comprising Book I. Blood—The Tissues of Movement, The Vascular Mechanism. 10s. 6d. Part II., comprising Book II. The Tissues of Chemical Action, with their Respective Mechanisms—Nutrition. 10s. 6d. Part III. The Central Nervous System. 7s. 6d.

FOSTER and BALFOUR.—THE ELEMENTS OF EMBRYOLOGY. By Prof. MICHAEL FOSTER, M.D., F.R.S., and the late F. M. BALFOUR, F.R.S., Professor of Animal Morphology in the University of Cambridge. 2d Ed., revised. Edited by A. SEDGWICK, M.A., Fellow and Assistant Lecturer of Trinity College, Cambridge, and W. HEAPE, M.A., late Demonstrator in the Morphological Laboratory of the University of Cambridge. Illustrated. Cr. 8vo. 10s. 6d.

FOSTER and LANGLEY.—A COURSE OF ELEMENTARY PRACTICAL PHYSIOLOGY AND HISTOLOGY. By Prof. MICHAEL FOSTER, M.D., F.R.S., and J. N. LANGLEY, F.R.S., Fellow of Trinity College, Cambridge. 6th Ed. Cr. 8vo. 7s. 6d.

GAMGEE.—A TEXT-BOOK OF THE PHYSIOLOGICAL CHEMISTRY OF THE ANIMAL BODY. Including an Account of the Chemical Changes occurring in Disease. By A. GAMGEE, M.D., F.R.S. Illustrated. 8vo. Vol. I. 18s.

GOODALE.—PHYSIOLOGICAL BOTANY. I. Outlines of the Histology of Phænogamous Plants. II. Vegetable Physiology. By GEORGE LINCOLN GOODALE, M.A., M.D., Professor of Botany in Harvard University. 8vo. 10s. 6d.

GRAY.—STRUCTURAL BOTANY, OR ORGANOGRAPHY ON THE BASIS OF MORPHOLOGY. To which are added the Principles of Taxonomy and Phytography, and a Glossary of Botanical Terms. By Prof. ASA GRAY, LL.D. 8vo. 10s. 6d.

THE SCIENTIFIC PAPERS OF ASA GRAY. Selected by C. SPRAGUE SARGENT. 2 vols. Vol. I. Reviews of Works on Botany and Related Subjects, 1834-1887. Vol. II. Essays, Biographical Sketches, 1841-1886. 8vo. 21s.

HAMILTON.—A SYSTEMATIC AND PRACTICAL TEXT-BOOK OF PATHOLOGY. By D. J. HAMILTON, F.R.S.E., Professor of Pathological Anatomy in the University of Aberdeen. Illustrated. 8vo. Vol. I. 25s.

HARTIG.—TEXT-BOOK OF THE DISEASES OF TREES. By Dr. ROBERT HARTIG. Translated by WM. SOMERVILLE, B.Sc., D.Œ., Professor of Agriculture and Forestry, Durham College of Science, Newcastle-on-Tyne. Edited, with Introduction, by Prof. H. MARSHALL WARD. 8vo. [In preparation.

HOOKER.—Works by Sir JOSEPH HOOKER, F.R.S., &c.
*PRIMER OF BOTANY. Illustrated. 18mo. 1s.
THE STUDENT'S FLORA OF THE BRITISH ISLANDS. 3d Ed., revised. Gl. 8vo. 10s. 6d.

HOWES.—AN ATLAS OF PRACTICAL ELEMENTARY BIOLOGY. By G. B. Howes, Assistant Professor of Zoology, Normal School of Science and Royal School of Mines. With a Preface by Prof. T. H. HUXLEY, F.R.S. 4to. 14s.

HUXLEY.—Works by Prof. T. H. HUXLEY, F.R.S.
*INTRODUCTORY PRIMER OF SCIENCE. 18mo. 1s.
*LESSONS IN ELEMENTARY PHYSIOLOGY. Illustrated. Fcap. 8vo. 4s. 6d.
*QUESTIONS ON HUXLEY'S PHYSIOLOGY. By T. ALCOCK, M.D. 18mo. 1s. 6d.

HUXLEY and MARTIN.—A COURSE OF PRACTICAL INSTRUCTION IN ELEMENTARY BIOLOGY. By Prof. T. H. HUXLEY, F.R.S., assisted by H. N. MARTIN, F.R.S., Professor of Biology in the Johns Hopkins University, U.S.A. New Ed., revised and extended by G. B. Howes and D. H. Scott, Ph.D., Assistant Professors, Normal School of Science and Royal School of Mines. With a Preface by T. H. HUXLEY, F.R.S. Cr. 8vo. 10s. 6d.

KLEIN.—Works by E. KLEIN, F.R.S., Lecturer on General Anatomy and Physiology in the Medical School of St. Bartholomew's Hospital, Professor of Bacteriology at the College of State Medicine, London.
MICRO-ORGANISMS AND DISEASE. An Introduction into the Study of Specific Micro-Organisms. Illustrated. 3d Ed., revised. Cr. 8vo. 6s.
THE BACTERIA IN ASIATIC CHOLERA. Cr. 8vo. 5s.

LANG.—TEXT-BOOK OF COMPARATIVE ANATOMY. By Dr. ARNOLD LANG. Professor of Zoology in the University of Zurich. Translated by H. M. BERNARD, M.A., and M BERNARD. Introduction by Prof. E. HAECKEL. 2 vols. Illustrated. 8vo. [In the Press.

LANKESTER.—Works by E. RAY LANKESTER, F.R.S., Linacre Professor of Human and Comparative Anatomy in the University of Oxford.
A TEXT-BOOK OF ZOOLOGY. 8vo. [In preparation.
THE ADVANCEMENT OF SCIENCE. Occasional Essays and Addresses. 8vo. 10s. 6d.

LUBBOCK.—Works by the Right Hon. Sir JOHN LUBBOCK, F R.S., D.C.L.
THE ORIGIN AND METAMORPHOSES OF INSECTS. Illustrated. Cr. 8vo. 3s. 6d.
ON BRITISH WILD FLOWERS CONSIDERED IN RELATION TO INSECTS. Illustrated. Cr. 8vo. 4s. 6d.

LUBBOCK.—Works by the Right Hon. Sir JOHN LUBBOCK, F.R.S., D.C.L.—*cont.*
FLOWERS, FRUITS, AND LEAVES. Illustrated. 2d Ed. Cr. 8vo. 4s. 6d.
SCIENTIFIC LECTURES. 2d Ed. 8vo. 8s. 6d.
FIFTY YEARS OF SCIENCE. Being the Address delivered at York to the British Association, August 1881. 5th Ed. Cr. 8vo. 2s. 6d.

MARTIN and MOALE.—ON THE DISSECTION OF VERTEBRATE ANIMALS. By Prof. H. N. MARTIN and W. A. MOALE. Cr. 8vo. [*In preparation.*

MIVART.—LESSONS IN ELEMENTARY ANATOMY. By ST. GEORGE MIVART, F.R.S., Lecturer on Comparative Anatomy at St. Mary's Hospital. Illustrated. Fcap. 8vo. 6s. 6d.

MÜLLER.—THE FERTILISATION OF FLOWERS. By HERMANN MÜLLER. Translated and Edited by D'ARCY W. THOMPSON, B.A., Professor of Biology in University College, Dundee. With a Preface by C. DARWIN, F.R.S. Illustrated. 8vo. 21s.

OLIVER.—Works by DANIEL OLIVER, F.R.S., late Professor of Botany in University College, London.
*LESSONS IN ELEMENTARY BOTANY. Illustrated. Fcap. 8vo. 4s. 6d.
FIRST BOOK OF INDIAN BOTANY. Illustrated. Ex. fcap. 8vo. 6s. 6d.

PARKER.—Works by T. JEFFERY PARKER, F.R.S., Professor of Biology in the University of Otago, New Zealand.
A COURSE OF INSTRUCTION IN ZOOTOMY (VERTEBRATA). Illustrated. Cr. 8vo. 8s. 6d.
LESSONS IN ELEMENTARY BIOLOGY. Illustrated. Cr. 8vo. 10s. 6d.

PARKER and BETTANY.—THE MORPHOLOGY OF THE SKULL. By Prof. W. K. PARKER, F.R.S., and G. T. BETTANY. Illustrated. Cr. 8vo. 10s. 6d.

ROMANES.—THE SCIENTIFIC EVIDENCES OF ORGANIC EVOLUTION. By GEORGE J. ROMANES, F.R.S., Zoological Secretary of the Linnean Society. Cr. 8vo. 2s. 6d.

SEDGWICK.—TREATISE ON EMBRYOLOGY. By ADAM SEDGWICK, F.R.S., Fellow and Lecturer of Trinity College, Cambridge. Illustrated. 8vo.
[*In preparation.*

SHUFELDT.—THE MYOLOGY OF THE RAVEN (*Corvus corax sinuatus*). A Guide to the Study of the Muscular System in Birds. By R. W. SHUFELDT. Illustrated. 8vo. 13s. net.

SMITH.—DISEASES OF FIELD AND GARDEN CROPS, CHIEFLY SUCH AS ARE CAUSED BY FUNGI. By W. G. SMITH, F.L.S. Illustrated. Fcap. 8vo. 4s. 6d.

STEWART and CORRY.—A FLORA OF THE NORTH-EAST OF IRELAND. Including the Phanerogamia, the Cryptogamia Vascularia, and the Muscineæ. By S. A. STEWART, Curator of the Collections in the Belfast Museum, and the late T. H. CORRY, M.A., Lecturer on Botany in the University Medical and Science Schools, Cambridge. Cr. 8vo. 5s. 6d.

WALLACE.—DARWINISM: An Exposition of the Theory of Natural Selection, with some of its Applications. By ALFRED RUSSEL WALLACE, LL.D., F.R.S. 3d Ed. Cr. 8vo. 9s.
NATURAL SELECTION: AND TROPICAL NATURE. By the same. New Ed. Cr. 8vo. 6s.
ISLAND LIFE. By the same. New Ed. Cr. 8vo. 6s.

WARD.—TIMBER AND SOME OF ITS DISEASES. By H. MARSHALL WARD, F.R.S., Professor of Botany in the Royal Indian Engineering College, Cooper's Hill. Illustrated. Cr. 8vo. 6s.

WIEDERSHEIM.—ELEMENTS OF THE COMPARATIVE ANATOMY OF VERTEBRATES. By Prof. R. WIEDERSHEIM. Adapted by W. NEWTON PARKER, Professor of Biology in the University College of South Wales and Monmouthshire. With Additions. Illustrated. 8vo. 12s. 6d.

MEDICINE.

BLYTH.—A MANUAL OF PUBLIC HEALTH. By A. WYNTER BLYTH, M.R.C.S. 8vo. 17s. net.

BRUNTON.—Works by T. LAUDER BRUNTON, M.D., F.R.S., Examiner in Materia Medica in the University of London, in the Victoria University, and in the Royal College of Physicians, London.

A TEXT-BOOK OF PHARMACOLOGY, THERAPEUTICS, AND MATERIA MEDICA. Adapted to the United States Pharmacopœia by F. H. WILLIAMS, M.D., Boston, Mass. 3d Ed. Adapted to the New British Pharmacopœia, 1885, and additions, 1891. 8vo. 21s. Or in 2 Vols. 22s. 6d.

TABLES OF MATERIA MEDICA: A Companion to the Materia Medica Museum. Illustrated. Cheaper Issue. 8vo. 5s.

ON THE CONNECTION BETWEEN CHEMICAL CONSTITUTION AND PHYSIOLOGICAL ACTION, BEING AN INTRODUCTION TO MODERN THERAPEUTICS. Croonian Lectures. 8vo. [In the Press.

GRIFFITHS.—LESSONS ON PRESCRIPTIONS AND THE ART OF PRE-SCRIBING. By W. HANDSEL GRIFFITHS. Adapted to the Pharmacopœia, 1885. 18mo. 3s. 6d.

HAMILTON.—A TEXT-BOOK OF PATHOLOGY, SYSTEMATIC AND PRAC-TICAL. By D. J. HAMILTON, F.R.S.E., Professor of Pathological Anatomy, University of Aberdeen. Illustrated. Vol. I. 8vo. 25s.

KLEIN.—Works by E. KLEIN, F.R.S., Lecturer on General Anatomy and Physio-logy in the Medical School of St. Bartholomew's Hospital, London.

MICRO-ORGANISMS AND DISEASE. An Introduction into the Study of Specific Micro-Organisms. Illustrated. 3d Ed., revised. Cr. 8vo. 6s.

THE BACTERIA IN ASIATIC CHOLERA. Cr. 8vo. 5s.

WHITE.—A TEXT-BOOK OF GENERAL THERAPEUTICS. By W. HALE WHITE, M.D., Senior Assistant Physician to and Lecturer in Materia Medica at Guy's Hospital. Illustrated. Cr. 8vo. 8s. 6d.

ZIEGLER—MACALISTER.—TEXT-BOOK OF PATHOLOGICAL ANATOMY AND PATHOGENESIS. By Prof. E. ZIEGLER. Translated and Edited by DONALD MACALISTER, M.A., M.D., Fellow and Medical Lecturer of St. John's College, Cambridge. Illustrated. 8vo.

Part I.—GENERAL PATHOLOGICAL ANATOMY. 2d Ed. 12s. 6d.

Part II.—SPECIAL PATHOLOGICAL ANATOMY. Sections I.-VIII. 2d Ed. 12s. 6d. Sections IX.-XII. 12s. 6d.

HUMAN SCIENCES.

Mental and Moral Philosophy ; Political Economy ; Law and Politics ; Anthropology ; Education.

MENTAL AND MORAL PHILOSOPHY.

BALDWIN.—HANDBOOK OF PSYCHOLOGY: SENSES AND INTELLECT. By Prof. J. M. BALDWIN, M.A., LL.D. 2d Ed., revised. 8vo. 12s. 6d.

BOOLE.—THE MATHEMATICAL ANALYSIS OF LOGIC. Being an Essay towards a Calculus of Deductive Reasoning. By GEORGE BOOLE. 8vo. 5s.

CALDERWOOD.—HANDBOOK OF MORAL PHILOSOPHY. By Rev. HENRY CALDERWOOD, LL.D., Professor of Moral Philosophy in the University of Edinburgh. 14th Ed., largely rewritten. Cr. 8vo. 6s.

CLIFFORD.—SEEING AND THINKING. By the late Prof. W. K. CLIFFORD, F.R.S. With Diagrams. Cr. 8vo. 3s. 6d.

HÖFFDING.—OUTLINES OF PSYCHOLOGY. By Prof. H. HÖFFDING. Trans-lated by M. E. LOWNDES. Cr. 8vo. 6s.

JAMES.—THE PRINCIPLES OF PSYCHOLOGY. By WM. JAMES, Professor of Psychology in Harvard University. 2 vols. 8vo. 25s. net.

JARDINE.—THE ELEMENTS OF THE PSYCHOLOGY OF COGNITION. By Rev. ROBERT JARDINE, D.Sc. 3d Ed., revised. Cr. 8vo. 6s. 6d.

JEVONS.—Works by W. STANLEY JEVONS, F.R.S.

*PRIMER OF LOGIC. 18mo. 1s.

*ELEMENTARY LESSONS IN LOGIC, Deductive and Inductive, with Copious Questions and Examples, and a Vocabulary of Logical Terms. Fcap. 8vo. 3s. 6d.

THE PRINCIPLES OF SCIENCE. A Treatise on Logic and Scientific Method. New and revised Ed. Cr. 8vo. 12s. 6d.

STUDIES IN DEDUCTIVE LOGIC. 2d Ed. Cr. 8vo. 6s.

PURE LOGIC: AND OTHER MINOR WORKS. Edited by R. ADAMSON, M.A., LL.D., Professor of Logic at Owens College, Manchester, and HARRIET A. JEVONS. With a Preface by Prof. ADAMSON. 8vo. 10s. 6d.

KANT—MAX MÜLLER.—CRITIQUE OF PURE REASON. By IMMANUEL KANT. 2 vols. 8vo. 16s. each. Vol. I. HISTORICAL INTRODUCTION, by LUDWIG NOIRÉ; Vol. II. CRITIQUE OF PURE REASON, translated by F. MAX MÜLLER.

KANT—MAHAFFY and BERNARD.—KANT'S CRITICAL PHILOSOPHY FOR ENGLISH READERS. By J. P. MAHAFFY, D.D., Professor of Ancient History in the University of Dublin, and JOHN H. BERNARD, B.D., Fellow of Trinity College, Dublin. A new and complete Edition in 2 vols. Cr. 8vo.

Vol. I. THE KRITIK OF PURE REASON EXPLAINED AND DEFENDED. 7s. 6d.

Vol. II. THE PROLEGOMENA. Translated with Notes and Appendices. 6s.

KEYNES.—FORMAL LOGIC, Studies and Exercises in. Including a Generalisation of Logical Processes in their application to Complex Inferences. By JOHN NEVILLE KEYNES, D.Sc. 2d Ed., revised and enlarged. Cr. 8vo. 10s. 6d.

McCOSH.—Works by JAMES McCOSH, D.D., President of Princeton College.

PSYCHOLOGY. Cr. 8vo.

I. THE COGNITIVE POWERS. 6s. 6d.

II. THE MOTIVE POWERS. 6s. 6d.

FIRST AND FUNDAMENTAL TRUTHS: being a Treatise on Metaphysics. Ex. cr. 8vo. 9s.

THE PREVAILING TYPES OF PHILOSOPHY. CAN THEY LOGICALLY REACH REALITY? 8vo. 3s. 6d.

MAURICE.—MORAL AND METAPHYSICAL PHILOSOPHY. By F. D. MAURICE, M.A., late Professor of Moral Philosophy in the University of Cambridge. Vol. I.—Ancient Philosophy and the First to the Thirteenth Centuries. Vol. II.—Fourteenth Century and the French Revolution, with a glimpse into the Nineteenth Century. 4th Ed. 2 vols. 8vo. 16s.

*RAY.—A TEXT-BOOK OF DEDUCTIVE LOGIC FOR THE USE OF STUDENTS. By P. K. RAY, D.Sc., Professor of Logic and Philosophy, Presidency College, Calcutta. 4th Ed. Globe 8vo. 4s. 6d.

SIDGWICK.—Works by HENRY SIDGWICK, LL.D., D.C.L., Knightbridge Professor of Moral Philosophy in the University of Cambridge.

THE METHODS OF ETHICS. 4th Ed. 8vo. 14s. A Supplement to the 2d Ed., containing all the important Additions and Alterations in the 3d Ed. 8vo. 6s.

OUTLINES OF THE HISTORY OF ETHICS, for English Readers. 2d Ed., revised. Cr. 8vo. 3s. 6d.

VENN.—Works by JOHN VENN, F.R.S., Examiner in Moral Philosophy in the University of London.

THE LOGIC OF CHANCE. An Essay on the Foundations and Province of the Theory of Probability, with special Reference to its Logical Bearings and its Application to Moral and Social Science. 3d Ed., rewritten and greatly enlarged. Cr. 8vo. 10s. 6d.

SYMBOLIC LOGIC. Cr. 8vo. 10s. 6d.

THE PRINCIPLES OF EMPIRICAL OR INDUCTIVE LOGIC. 8vo. 18s.

POLITICAL ECONOMY.

BÖHM-BAWERK.—CAPITAL AND INTEREST. Translated by WILLIAM SMART, M.A. 8vo. 12s. net.

THE POSITIVE THEORY OF CAPITAL. By the same Author and Translator. 8vo. 12s. net.

CAIRNES.—THE CHARACTER AND LOGICAL METHOD OF POLITICAL ECONOMY. By J. E. CAIRNES. Cr. 8vo. 6s.

SOME LEADING PRINCIPLES OF POLITICAL ECONOMY NEWLY EXPOUNDED. By the same. 8vo. 14s.

COSSA.—GUIDE TO THE STUDY OF POLITICAL ECONOMY. By Dr. L. COSSA. Translated. With a Preface by W. S. JEVONS, F.R.S. Cr. 8vo. 4s. 6d.

'FAWCETT.—POLITICAL ECONOMY FOR BEGINNERS, WITH QUESTIONS. By Mrs. HENRY FAWCETT. 7th Ed. 18mo. 2s. 6d.

FAWCETT.—A MANUAL OF POLITICAL ECONOMY. By the Right Hon. HENRY FAWCETT, F.R.S. 7th Ed., revised. With a Chapter on "State Socialism and the Nationalisation of the Land," and an Index. Cr. 8vo. 12s. 6d.

AN EXPLANATORY DIGEST of the above. By C. A. WATERS, B.A. Cr. 8vo. 2s. 6d.

GILMAN.—PROFIT-SHARING BETWEEN EMPLOYER AND EMPLOYEE. A Study in the Evolution of the Wages System. By N. P. GILMAN. Cr. 8vo. 7s. 6d.

GUNTON.—WEALTH AND PROGRESS: A Critical Examination of the Wages Question and its Economic Relation to Social Reform. By GEORGE GUNTON. Cr. 8vo. 6s.

HOWELL.—THE CONFLICTS OF CAPITAL AND LABOUR HISTORICALLY AND ECONOMICALLY CONSIDERED. Being a History and Review of the Trade Unions of Great Britain, showing their Origin, Progress, Constitution, and Objects, in their varied Political, Social, Economical, and Industrial Aspects. By GEORGE HOWELL, M.P. 2d Ed., revised. Cr. 8vo. 7s. 6d.

JEVONS.— Works by W. STANLEY JEVONS, F.R.S.

*PRIMER OF POLITICAL ECONOMY. 18mo. 1s.

THE THEORY OF POLITICAL ECONOMY. 3d Ed., revised. 8vo. 10s. 6d.

KEYNES.—THE SCOPE AND METHOD OF POLITICAL ECONOMY. By J. N. KEYNES, D.Sc. 7s. net.

MARSHALL.—PRINCIPLES OF ECONOMICS. By ALFRED MARSHALL, M.A. 2 vols. 8vo. Vol. I. 2d Ed. 12s. 6d. net.

MARSHALL.—THE ECONOMICS OF INDUSTRY. By A. MARSHALL, M.A., Professor of Political Economy in the University of Cambridge, and MARY P. MARSHALL. Ex. fcap. 8vo. 2s. 6d.

PALGRAVE.—A DICTIONARY OF POLITICAL ECONOMY. By various Writers. Edited by R. H. INGLIS PALGRAVE, F.R.S. 3s. 6d. each, net. No. 1. July 1891.

PANTALEONI.—MANUAL OF POLITICAL ECONOMY. By Prof. M. PANTALEONI. Translated by T. BOSTON BRUCE. [In preparation.

SIDGWICK.—THE PRINCIPLES OF POLITICAL ECONOMY. By HENRY SIDGWICK, LL.D., D.C.L., Knightbridge Professor of Moral Philosophy in the University of Cambridge. 2d Ed., revised. 8vo. 16s.

SMART.—AN INTRODUCTION TO THE THEORY OF POLITICAL ECONOMY. By WILLIAM SMART, M.A. Crown 8vo.

WALKER.—Works by FRANCIS A. WALKER, M.A.

FIRST LESSONS IN POLITICAL ECONOMY. Cr. 8vo. 5s.

A BRIEF TEXT-BOOK OF POLITICAL ECONOMY. Cr. 8vo. 6s. 6d.

POLITICAL ECONOMY. 2d Ed., revised and enlarged. 8vo. 12s. 6d.

THE WAGES QUESTION. Ex. Cr. 8vo. 8s. 6d. net.

MONEY. Ex. Cr. 8vo. 8s. 6d. net.

'WICKSTEED.—ALPHABET OF ECONOMIC SCIENCE. By PHILIP H. WICKSTEED, M.A. Part I. Elements of the Theory of Value or Worth. Gl. 8vo. 2s. 6d.

LAW AND POLITICS.

ADAMS and CUNNINGHAM.—THE SWISS CONFEDERATION. By Sir F. O. ADAMS and C. CUNNINGHAM. 8vo. 14s.

ANGLO-SAXON LAW, ESSAYS ON.—Contents: Anglo-Saxon Law Courts, Land and Family Law, and Legal Procedure. 8vo. 18s.

BALL.—THE STUDENT'S GUIDE TO THE BAR. By WALTER W. R. BALL, M.A., Fellow and Assistant Tutor of Trinity College, Cambridge. 4th Ed., revised. Cr. 8vo. 2s. 6d.

BIGELOW.—HISTORY OF PROCEDURE IN ENGLAND FROM THE NORMAN CONQUEST. The Norman Period, 1066-1204. By MELVILLE M. BIGELOW, Ph.D., Harvard University. 8vo. 16s.

BOUTMY.— STUDIES IN CONSTITUTIONAL LAW. By EMILE BOUTMY. Translated by Mrs. DICEY, with Preface by Prof. A. V. DICEY. Cr. 8vo. 6s.

THE ENGLISH CONSTITUTION. By the same. Translated by Mrs. EADEN, with Introduction by Sir F. POLLOCK, Bart. Cr. 8vo. 6s.

BRYCE.—THE AMERICAN COMMONWEALTH. By JAMES BRYCE, M.P., D.C.L., Regius Professor of Civil Law in the University of Oxford. Two Volumes. Ex. cr. 8vo. 25s. Part I. The National Government. Part II. The State Governments. Part III. The Party System. Part IV. Public Opinion. Part V. Illustrations and Reflections. Part VI. Social Institutions.

BUCKLAND.—OUR NATIONAL INSTITUTIONS. A Short Sketch for Schools. By ANNA BUCKLAND. With Glossary. 18mo. 1s.

CHERRY.—LECTURES ON THE GROWTH OF CRIMINAL LAW IN ANCIENT COMMUNITIES. By R. R. CHERRY, LL.D., Reid Professor of Constitutional and Criminal Law in the University of Dublin. 8vo. 5s. net.

DICEY.—INTRODUCTION TO THE STUDY OF THE LAW OF THE CONSTITUTION. By A. V. DICEY, B.C.L., Vinerian Professor of English Law in the University of Oxford. 3d Ed. 8vo. 12s. 6d.

DILKE.—PROBLEMS OF GREATER BRITAIN. By the Right Hon. Sir CHARLES WENTWORTH DILKE. With Maps. 4th Ed. Ex. cr. 8vo. 12s. 6d.

DONISTHORPE.—INDIVIDUALISM: A System of Politics. By WORDSWORTH DONISTHORPE. 8vo. 14s.

ENGLISH CITIZEN, THE.—A Series of Short Books on his Rights and Responsibilities. Edited by HENRY CRAIK, LL.D. Cr. 8vo. 3s. 6d. each.
CENTRAL GOVERNMENT. By H. D. TRAILL, D.C.L.
THE ELECTORATE AND THE LEGISLATURE. By SPENCER WALPOLE.
THE POOR LAW. By Rev. T. W. FOWLE, M.A. New Ed. With Appendix.
THE NATIONAL BUDGET; THE NATIONAL DEBT; TAXES AND RATES. By A. J. WILSON.
THE STATE IN RELATION TO LABOUR. By W. STANLEY JEVONS, LL.D.
THE STATE AND THE CHURCH. By the Hon. ARTHUR ELLIOT.
FOREIGN RELATIONS. By SPENCER WALPOLE.
THE STATE IN ITS RELATION TO TRADE. By Sir T. H. FARRER, Bart.
LOCAL GOVERNMENT. By M. D. CHALMERS, M.A.
THE STATE IN ITS RELATION TO EDUCATION. By HENRY CRAIK, LL.D.
THE LAND LAWS. By Sir F. POLLOCK, Bart., Professor of Jurisprudence in the University of Oxford.
COLONIES AND DEPENDENCIES. Part I. INDIA. By J. S. COTTON, M.A. II. THE COLONIES. By E. J. PAYNE, M.A.
JUSTICE AND POLICE. By F. W. MAITLAND.
THE PUNISHMENT AND PREVENTION OF CRIME. By Colonel Sir EDMUND DU CANE, K.C.B., Chairman of Commissioners of Prisons.

FISKE.—CIVIL GOVERNMENT IN THE UNITED STATES CONSIDERED WITH SOME REFERENCE TO ITS ORIGINS. By JOHN FISKE, formerly Lecturer on Philosophy at Harvard University. Cr. 8vo. 6s. 6d.

HOLMES.—THE COMMON LAW. By O. W. HOLMES, Jun. Demy 8vo. 12s.

JENKS.—THE GOVERNMENT OF VICTORIA. By EDWARD JENKS, B.A., LL.B., Professor of Law in the University of Melbourne. [In preparation.

MAITLAND.—PLEAS OF THE CROWN FOR THE COUNTY OF GLOUCESTER BEFORE THE ABBOT OF READING AND HIS FELLOW JUSTICES ITINERANT, IN THE FIFTH YEAR OF THE REIGN OF KING HENRY THE THIRD, AND THE YEAR OF GRACE 1221. By F. W. MAITLAND. 8vo. 7s. 6d.

MUNRO.—COMMERCIAL LAW. By J. E. C. MUNRO, LL.D., Professor of Law and Political Economy in the Owens College, Manchester. [In preparation.

PATERSON.—Works by JAMES PATERSON, Barrister-at-Law.

COMMENTARIES ON THE LIBERTY OF THE SUBJECT, AND THE LAWS OF ENGLAND RELATING TO THE SECURITY OF THE PERSON. Cheaper Issue. Two Vols. Cr. 8vo. 21s.

THE LIBERTY OF THE PRESS, SPEECH, AND PUBLIC WORSHIP. Being Commentaries on the Liberty of the Subject and the Laws of England. Cr. 8vo. 12s.

PHILLIMORE.—PRIVATE LAW AMONG THE ROMANS. From the Pandects. By J. G. PHILLIMORE, Q.C. 8vo. 16s.

POLLOCK.—ESSAYS IN JURISPRUDENCE AND ETHICS. By Sir FREDERICK POLLOCK, Bart., Corpus Christi Professor of Jurisprudence in the University of Oxford. 8vo. 10s. 6d.

INTRODUCTION TO THE HISTORY OF THE SCIENCE OF POLITICS. By the same. Cr. 8vo. 2s. 6d.

RICHEY.—THE IRISH LAND LAWS. By ALEX. G. RICHEY, Q.C., Deputy Regius Professor of Feudal English Law in the University of Dublin. Cr. 8vo. 3s. 6d.

SIDGWICK.—THE ELEMENTS OF POLITICS. By HENRY SIDGWICK, LL.D. 8vo. 14s. net.

STEPHEN.—Works by Sir J. FITZJAMES STEPHEN, Bart.

A DIGEST OF THE LAW OF EVIDENCE. 5th Ed., revised and enlarged. Cr. 8vo. 6s.

A DIGEST OF THE CRIMINAL LAW: CRIMES AND PUNISHMENTS. 4th Ed., revised. 8vo. 16s.

A DIGEST OF THE LAW OF CRIMINAL PROCEDURE IN INDICTABLE OFFENCES. By Sir J. F. STEPHEN, Bart., and H. STEPHEN, LL.M., of the Inner Temple, Barrister-at-Law. 8vo. 12s. 6d.

A HISTORY OF THE CRIMINAL LAW OF ENGLAND. Three Vols. 8vo. 48s.

GENERAL VIEW OF THE CRIMINAL LAW OF ENGLAND. 8vo. 14s.

ANTHROPOLOGY.

DAWKINS.—EARLY MAN IN BRITAIN AND HIS PLACE IN THE TERTIARY PERIOD. By Prof. W. BOYD DAWKINS. Medium 8vo. 25s.

FRAZER.—THE GOLDEN BOUGH. A Study in Comparative Religion. By J. G. FRAZER, M.A., Fellow of Trinity College, Cambridge. 2 vols. 8vo. 28s.

M'LENNAN.—THE PATRIARCHAL THEORY. Based on the papers of the late JOHN F. M'LENNAN. Edited by DONALD M'LENNAN, M.A., Barrister-at-Law. 8vo. 14s.

STUDIES IN ANCIENT HISTORY. By the same. Comprising a Reprint of "Primitive Marriage." An inquiry into the origin of the form of capture in Marriage Ceremonies. 8vo. 16s.

TYLOR.—ANTHROPOLOGY. An Introduction to the Study of Man and Civilisation. By E. B. TYLOR, F.R.S. Illustrated. Cr. 8vo. 7s. 6d.

WESTERMARCK.—THE HISTORY OF HUMAN MARRIAGE. By Dr. EDWARD WESTERMARCK. With Preface by A. R. WALLACE. 8vo. 14s. net.

WILSON.—THE RIGHT HAND: LEFT-HANDEDNESS. By Sir D. WILSON. Cr. 8vo. 4s. 6d.

EDUCATION.

ARNOLD.—REPORTS ON ELEMENTARY SCHOOLS. 1852-1882. By MATTHEW ARNOLD, D.C.L. Edited by the Right Hon. Sir FRANCIS SANDFORD, K.C.B. Cheaper Issue. Cr. 8vo. 3s. 6d.

HIGHER SCHOOLS AND UNIVERSITIES IN GERMANY. By the same. Crown 8vo. 6s.

BALL.—THE STUDENT'S GUIDE TO THE BAR. By WALTER W. R. BALL, M.A., Fellow and Assistant Tutor of Trinity College, Cambridge. 4th Ed., revised. Cr. 8vo. 2s. 6d.

0

44 TECHNICAL KNOWLEDGE

"BLAKISTON.—THE TEACHER. Hints on School Management. A Handbook for Managers, Teachers' Assistants, and Pupil Teachers. By J. R. BLAKISTON. Cr. 8vo. 2s. 6d. (Recommended by the London, Birmingham, and Leicester School Boards.)

CALDERWOOD.—ON TEACHING. By Prof. HENRY CALDERWOOD. New Ed. Ex. fcap. 8vo. 2s. 6d.

FEARON.—SCHOOL INSPECTION. By D. R. FEARON. 6th Ed. Cr. 8vo. 2s. 6d.

FITCH.—NOTES ON AMERICAN SCHOOLS AND TRAINING COLLEGES. Reprinted from the Report of the English Education Department for 1888-89, with permission of the Controller of H.M.'s Stationery Office. By J. G. FITCH, M.A. Gl. 8vo. 2s. 6d.

GEIKIE.—THE TEACHING OF GEOGRAPHY. A Practical Handbook for the use of Teachers. By Sir ARCHIBALD GEIKIE, F R.S., Director-General of the Geological Survey of the United Kingdom. Cr. 8vo. 2s.

GLADSTONE.—SPELLING REFORM FROM A NATIONAL POINT OF VIEW. By J. H. GLADSTONE. Cr. 8vo. 1s. 6d.

HERTEL.—OVERPRESSURE IN HIGH SCHOOLS IN DENMARK. By Dr. HERTEL. Translated by C. G. SÖRENSEN. With Introduction by Sir J. CRICHTON-BROWNE, F.R.S. Cr. 8vo. 3s. 6d.

TODHUNTER.—THE CONFLICT OF STUDIES. By ISAAC TODHUNTER, F.R.S. 8vo. 10s. 6d.

TECHNICAL KNOWLEDGE.

(See also MECHANICS, LAW, and MEDICINE.)

Civil and Mechanical Engineering; Military and Naval Science; Agriculture; Domestic Economy; Book-Keeping; Commerce.

CIVIL AND MECHANICAL ENGINEERING.

ALEXANDER and THOMSON.—ELEMENTARY APPLIED MECHANICS. By T. ALEXANDER, Professor of Civil Engineering, Trinity College, Dublin, and A. W. THOMSON, Professor at College of Science, Poona, India. Part II. TRANSVERSE STRESS. Cr. 8vo. 10s. 6d.

CHALMERS.—GRAPHICAL DETERMINATION OF FORCES IN ENGINEER-ING STRUCTURES. By J. B. CHALMERS, C.E. Illustrated. 8vo. 24s.

COTTERILL.—APPLIED MECHANICS: An Elementary General Introduction to the Theory of Structures and Machines. By J. H. COTTERILL, F.R.S., Professor of Applied Mechanics in the Royal Naval College, Greenwich. 2d Ed. 8vo. 18s.

COTTERILL and SLADE.—LESSONS IN APPLIED MECHANICS. By Prof. J. H. COTTERILL and J. H. SLADE. Fcap. 8vo. 5s. 6d.

GRAHAM.—GEOMETRY OF POSITION. By R. H. GRAHAM. Cr. 8vo. 7s. 6d.

KENNEDY.—THE MECHANICS OF MACHINERY. By A. B. W. KENNEDY, F.R.S. Illustrated. Cr. 8vo. 12s. 6d.

WHITHAM.—STEAM-ENGINE DESIGN. For the Use of Mechanical Engineers, Students, and Draughtsmen. By J. M. WHITHAM, Professor of Engineering, Arkansas Industrial University. Illustrated. 8vo. 25s.

YOUNG.—SIMPLE PRACTICAL METHODS OF CALCULATING STRAINS ON GIRDERS, ARCHES, AND TRUSSES. With a Supplementary Essay on Economy in Suspension Bridges. By E. W. YOUNG, C.E. With Diagrams. 8vo. 7s. 6d.

MILITARY AND NAVAL SCIENCE.

AITKEN.—THE GROWTH OF THE RECRUIT AND YOUNG SOLDIER. With a view to the selection of "Growing Lads" for the Army, and a Regulated System of Training for Recruits. By Sir W. AITKEN, F.R.S., Professor of Pathology in the Army Medical School. Cr. 8vo. 8s. 6d.

ARMY PRELIMINARY EXAMINATION, 1882-1890, Specimens of Papers set at the. With Answers to the Mathematical Questions. Subjects: Arithmetic, Algebra, Euclid, Geometrical Drawing, Geography, French, English Dictation. Cr. 8vo. 3s. 6d.

MATTHEWS.—MANUAL OF LOGARITHMS. By G. F. MATTHEWS, B.A. 8vo. 5s. net.

MAURICE.—WAR. By FREDERICK MAURICE, Colonel C.B., R.A. 8vo. 5s. net.

MERCUR.—ELEMENTS OF THE ART OF WAR. Prepared for the use of Cadets of the United States Military Academy. By JAMES MERCUR, Professor of Civil Engineering at the United States Academy, West Point, New York. 2d Ed., revised and corrected. 8vo. 17s.

PALMER.—TEXT-BOOK OF PRACTICAL LOGARITHMS AND TRIGONO-METRY. By J. H. PALMER, Head Schoolmaster, R.N., H.M.S. *Cambridge*, Devonport. Gl. 8vo. 4s. 6d.

ROBINSON.—TREATISE ON MARINE SURVEYING. Prepared for the use of younger Naval Officers. With Questions for Examinations and Exercises principally from the Papers of the Royal Naval College. With the results. By Rev. JOHN L. ROBINSON, Chaplain and Instructor in the Royal Naval College, Greenwich. Illustrated. Cr. 8vo. 7s. 6d.

SANDHURST MATHEMATICAL PAPERS, for Admission into the Royal Military College, 1881-1889. Edited by E. J. BROOKSMITH, B.A., Instructor in Mathematics at the Royal Military Academy, Woolwich. Cr. 8vo. 3s. 6d.

SHORTLAND.—NAUTICAL SURVEYING. By the late Vice-Admiral SHORTLAND, LL.D. 8vo. 21s.

THOMSON.—POPULAR LECTURES AND ADDRESSES. By Sir WILLIAM THOMSON, LL.D., P.R.S. In 3 vols. Illustrated. Cr. 8vo. Vol. III. Navigation. 7s. 6d.

WILKINSON.—THE BRAIN OF AN ARMY. A Popular Account of the German General Staff. By SPENSER WILKINSON. Cr. 8vo. 2s. 6d.

WOLSELEY.—Works by General Viscount WOLSELEY, G.C.M.G.
THE SOLDIER'S POCKET-BOOK FOR FIELD SERVICE. 5th Ed., revised and enlarged. 16mo. Roan. 5s.
FIELD POCKET-BOOK FOR THE AUXILIARY FORCES. 16mo. 1s. 6d.

WOOLWICH MATHEMATICAL PAPERS, for Admission into the Royal Military Academy, Woolwich, 1880-1888 inclusive. Edited by E. J. BROOKSMITH, B.A., Instructor in Mathematics at the Royal Military Academy, Woolwich. Cr. 8vo. 6s.

AGRICULTURE.

FRANKLAND.—AGRICULTURAL CHEMICAL ANALYSIS, A Handbook of. By PERCY F. FRANKLAND, F.R.S., Professor of Chemistry, University College, Dundee. Founded upon *Leitfaden für die Agriculture Chemiche Analyse*, von Dr. F. KROCKER. Cr. 8vo. 7s. 6d.

HARTIG.—TEXT-BOOK OF THE DISEASES OF TREES. By Dr. ROBERT HARTIG. Translated by WM. SOMERVILLE, B.Sc., D.Œ., Professor of Agriculture and Forestry, Durham College of Science, Newcastle-on-Tyne. Edited, with Introduction, by Prof. H. MARSHALL WARD. 8vo. [*In preparation.*]

LASLETT.—TIMBER AND TIMBER TREES, NATIVE AND FOREIGN. By THOMAS LASLETT. Cr. 8vo. 8s. 6d.

SMITH.—DISEASES OF FIELD AND GARDEN CROPS, CHIEFLY SUCH AS ARE CAUSED BY FUNGI. By WORTHINGTON G. SMITH, F.L.S. Illustrated. Fcap. 8vo. 4s. 6d.

TANNER.—*ELEMENTARY LESSONS IN THE SCIENCE OF AGRICULTURAL PRACTICE. By HENRY TANNER, F.C.S., M.R.A.C., Examiner in the Principles of Agriculture under the Government Department of Science. Fcap. 8vo. 3s. 6d.
*FIRST PRINCIPLES OF AGRICULTURE. By the same. 18mo. 1s.
THE PRINCIPLES OF AGRICULTURE. By the same. A Series of Reading Books for use in Elementary Schools. Ex. fcap. 8vo.
  *I. The Alphabet of the Principles of Agriculture. 6d.
  *II. Further Steps in the Principles of Agriculture. 1s.
  *III. Elementary School Readings on the Principles of Agriculture for the third stage. 1s.

WARD.—TIMBER AND SOME OF ITS DISEASES. By H. Marshall Ward, M.A., F.L.S., F.R.S., Fellow of Christ's College, Cambridge, Professor of Botany at the Royal Indian Engineering College, Cooper's Hill. With Illustrations. Cr. 8vo. 6s.

DOMESTIC ECONOMY.

*BARKER.—FIRST LESSONS IN THE PRINCIPLES OF COOKING. By Lady Barker. 18mo. 1s.

*BERNERS.—FIRST LESSONS ON HEALTH. By J. Berners. 18mo. 1s.

*COOKERY BOOK.—THE MIDDLE CLASS COOKERY BOOK. Edited by the Manchester School of Domestic Cookery. Fcap. 8vo. 1s. 6d.

CRAVEN.—A GUIDE TO DISTRICT NURSES. By Mrs. Dacre Craven (née Florence Sarah Lees), Hon. Associate of the Order of St. John of Jerusalem, etc. Cr. 8vo. 2s. 6d.

FREDERICK.—HINTS TO HOUSEWIVES ON SEVERAL POINTS, PARTICULARLY ON THE PREPARATION OF ECONOMICAL AND TASTEFUL DISHES. By Mrs. Frederick. Cr. 8vo. 1s.

*GRAND'HOMME.—CUTTING-OUT AND DRESSMAKING. From the French of Mdlle. E. Grand'homme. With Diagrams. 18mo. 1s.

JEX-BLAKE.—THE CARE OF INFANTS. A Manual for Mothers and Nurses. By Sophia Jex-Blake, M.D., Lecturer on Hygiene at the London School of Medicine for Women. 18mo. 1s.

RATHBONE.—THE HISTORY AND PROGRESS OF DISTRICT NURSING FROM ITS COMMENCEMENT IN THE YEAR 1859 TO THE PRESENT DATE, including the foundation by the Queen of the Queen Victoria Jubilee Institute for Nursing the Poor in their own Homes. By William Rathbone, M.P. Cr. 8vo. 2s. 6d.

*TEGETMEIER.—HOUSEHOLD MANAGEMENT AND COOKERY. With an Appendix of Recipes used by the Teachers of the National School of Cookery. By W. B. Tegetmeier. Compiled at the request of the School Board for London. 18mo. 1s.

'WRIGHT.—THE SCHOOL COOKERY-BOOK. Compiled and Edited by C. E. Guthrie Wright, Hon. Sec. to the Edinburgh School of Cookery. 18mo. 1s.

BOOK-KEEPING.

*THORNTON.—FIRST LESSONS IN BOOK-KEEPING. By J. Thornton. Cr. 8vo. 2s. 6d. KEY. Oblong 4to. 10s. 6d.

*PRIMER OF BOOK-KEEPING. By the same. 18mo. 1s. KEY. 8vo. 2s. 6d.

COMMERCE.

MACMILLAN'S ELEMENTARY COMMERCIAL CLASS BOOKS. Edited by James Gow, Litt.D., Headmaster of Nottingham School. Globe 8vo.

The following volumes are arranged for :—

*THE HISTORY OF COMMERCE IN EUROPE. By H. de B. Gibbins, M.A. 3s. 6d. [Ready.

COMMERCIAL GERMAN. By F. C. Smith, B.A., formerly scholar of Magdalene College, Cambridge. [In the Press.

COMMERCIAL GEOGRAPHY. By E. C. K. Gonner, M.A., Professor of Political Economy in University College, Liverpool. [In preparation.

COMMERCIAL FRENCH.

COMMERCIAL ARITHMETIC. By A. W. Sunderland, M.A., late Scholar of Trinity College, Cambridge; Fellow of the Institute of Actuaries. [In prep.

COMMERCIAL LAW. By J. E. C. Munro, LL.D., Professor of Law and Political Economy in the Owens College, Manchester.

GEOGRAPHY.

(See also PHYSICAL GEOGRAPHY.)

BARTHOLOMEW.—*THE ELEMENTARY SCHOOL ATLAS. By JOHN BARTHOLOMEW, F.R.G.S. 4to. 1s.

*MACMILLAN'S SCHOOL ATLAS, PHYSICAL AND POLITICAL. Consisting of 80 Maps and complete Index. By the same. Prepared for the use of Senior Pupils. Royal 4to. 8s. 6d. Half-morocco. 10s. 6d.

THE LIBRARY REFERENCE ATLAS OF THE WORLD. By the same. A Complete Series of 84 Modern Maps. With Geographical Index to 100,000 places. Half-morocco. Gilt edges. Folio. £2:12:6 net. Also issued in parts, 5s. each net. Geographical Index, 7s. 6d. net. Part I., April 1891.

'**CLARKE.**—CLASS-BOOK OF GEOGRAPHY. By C. B. CLARKE, F.R.S. New Ed., revised 1889, with 18 Maps. Fcap. 8vo. 3s. Sewed, 2s. 6d.

GEIKIE.—Works by Sir ARCHIBALD GEIKIE, F.R.S., Director-General of the Geological Survey of the United Kingdom.

*THE TEACHING OF GEOGRAPHY. A Practical Handbook for the use of Teachers. Cr. 8vo. 2s.

'GEOGRAPHY OF THE BRITISH ISLES. 18mo. 1s.

'**GREEN.**—A SHORT GEOGRAPHY OF THE BRITISH ISLANDS. By JOHN RICHARD GREEN and A. S. GREEN. With Maps. Fcap. 8vo. 3s. 6d.

*'**GROVE.**—A PRIMER OF GEOGRAPHY. By Sir GEORGE GROVE, D.C.L. Illustrated. 18mo. 1s.

KIEPERT.—A MANUAL OF ANCIENT GEOGRAPHY. By Dr. H. KIEPERT. Cr. 8vo. 5s.

MACMILLAN'S GEOGRAPHICAL SERIES.—Edited by Sir ARCHIBALD GEIKIE, F.R.S., Director-General of the Geological Survey of the United Kingdom.

*THE TEACHING OF GEOGRAPHY. A Practical Handbook for the Use of Teachers. By Sir ARCHIBALD GEIKIE, F.R.S. Cr. 8vo. 2s.

*MAPS AND MAP-DRAWING. By W. A. ELDERTON. 18mo. 1s.

*GEOGRAPHY OF THE BRITISH ISLES. By Sir A. GEIKIE, F.R.S. 18mo. 1s.

*AN ELEMENTARY CLASS-BOOK OF GENERAL GEOGRAPHY. By H. R. MILL, D.Sc., Lecturer on Physiography and on Commercial Geography in the Heriot-Watt College, Edinburgh. Illustrated. Cr. 8vo. 3s. 6d.

'GEOGRAPHY OF EUROPE. By J. SIME, M.A. Illustrated. Gl. 8vo. 3s.

*ELEMENTARY GEOGRAPHY OF INDIA, BURMA, AND CEYLON. By H. F. BLANFORD, F.G.S. Gl. 8vo. 2s. 6d.

GEOGRAPHY OF NORTH AMERICA. By Prof. N. S. SHALER. [In preparation.

GEOGRAPHY OF THE BRITISH COLONIES. By G. M. DAWSON and A. SUTHERLAND. [In the Press.

STRACHEY.—LECTURES ON GEOGRAPHY. By General RICHARD STRACHEY, R.E. Cr. 8vo. 4s. 6d.

'**TOZER.**—A PRIMER OF CLASSICAL GEOGRAPHY. By H. F. TOZER, M.A. 18mo. 1s.

HISTORY.

ARNOLD.—THE SECOND PUNIC WAR. Being Chapters from THE HISTORY OF ROME, by the late THOMAS ARNOLD, D.D., Headmaster of Rugby. Edited, with Notes, by W. T. ARNOLD, M.A. With 8 Maps. Cr. 8vo. 5s.

ARNOLD.—A HISTORY OF THE EARLY ROMAN EMPIRE. By W. T. ARNOLD, M.A. Cr. 8vo. [In preparation.

***BEESLY.**—STORIES FROM THE HISTORY OF ROME. By Mrs. BEESLY. Fcap. 8vo. 2s. 6d.

BRYCE.—Works by JAMES BRYCE, M.P., D.C.L., Regius Professor of Civil Law in the University of Oxford.

THE HOLY ROMAN EMPIRE. 9th Ed. Cr. 8vo. **7s. 6d.**
  ₊ Also a *Library Edition*. Demy 8vo. **14s.**
THE AMERICAN COMMONWEALTH. 2 vols. Ex. cr. 8vo. **25s.** Part I.
  The National Government. Part II. The State Governments. Part III.
  The Party System. Part IV. Public Opinion. Part V. Illustrations and
  Reflections. Part VI. Social Institutions.
*BUCKLEY.—A HISTORY OF ENGLAND FOR BEGINNERS. By ARABELLA
  B. BUCKLEY. With Maps and Tables. Gl. 8vo. **3s.**
BURY.—A HISTORY OF THE LATER ROMAN EMPIRE FROM ARCADIUS
  TO IRENE, A.D. 395-800. By JOHN B. BURY, M.A., Fellow of Trinity College,
  Dublin. 2 vols. 8vo. **32s.**
CASSEL.—MANUAL OF JEWISH HISTORY AND LITERATURE. By Dr. D.
  CASSEL. Translated by Mrs. HENRY LUCAS. Fcap. 8vo. **2s. 6d.**
ENGLISH STATESMEN, TWELVE. Cr. 8vo. **2s. 6d.** each.
  WILLIAM THE CONQUEROR. By EDWARD A. FREEMAN, D.C.L., LL.D.
  HENRY II. By Mrs. J. R. GREEN.
  EDWARD I. By F. YORK POWELL.   [*In preparation.*
  HENRY VII. By JAMES GAIRDNER.
  CARDINAL WOLSEY. By Bishop CREIGHTON.
  ELIZABETH. By E. S. BEESLY.   [*In preparation.*
  OLIVER CROMWELL. By FREDERIC HARRISON.
  WILLIAM III. By H. D. TRAILL.
  WALPOLE. By JOHN MORLEY.
  CHATHAM. By JOHN MORLEY.   [*In preparation.*
  PITT. By JOHN MORLEY.   [*In preparation.*
  PEEL. By J. R. THURSFIELD.
FISKE.—Works by JOHN FISKE, formerly Lecturer on Philosophy at Harvard
  University.
  THE CRITICAL PERIOD IN AMERICAN HISTORY, 1783-1789. Ex. cr.
  8vo. **10s. 6d.**
  THE BEGINNINGS OF NEW ENGLAND; or, The Puritan Theocracy in its
  Relations to Civil and Religious Liberty. Cr. 8vo. **7s. 6d.**
  THE AMERICAN REVOLUTION. 2 vols. Cr. 8vo. **18s.**
FREEMAN.—Works by EDWARD A. FREEMAN, D.C.L., Regius Professor of Modern
  History in the University of Oxford, etc.
  *OLD ENGLISH HISTORY. With Maps. Ex. fcap. 8vo. **6s.**
  A SCHOOL HISTORY OF ROME. Cr. 8vo.   [*In preparation.*
  METHODS OF HISTORICAL STUDY. 8vo. **10s. 6d.**
  THE CHIEF PERIODS OF EUROPEAN HISTORY. Six Lectures. With an
  Essay on Greek Cities under Roman Rule. 8vo. **10s. 6d.**
  HISTORICAL ESSAYS. First Series. 4th Ed. 8vo. **10s. 6d.**
  HISTORICAL ESSAYS. Second Series. 3d Ed., with additional Essays. 8vo.
  10s. 6d.
  HISTORICAL ESSAYS. Third Series. 8vo. **12s.**
  THE GROWTH OF THE ENGLISH CONSTITUTION FROM THE EARLIEST
  TIMES. 4th Ed. Cr. 8vo. **5s.**
  *GENERAL SKETCH OF EUROPEAN HISTORY. Enlarged, with Maps, etc.
  18mo. **3s. 6d.**
  *PRIMER OF EUROPEAN HISTORY. 18mo. **1s.** (*History Primers.*)
FRIEDMANN.—ANNE BOLEYN. A Chapter of English History, 1527-1536. By
  PAUL FRIEDMANN. 2 vols. 8vo. **28s.**
*GIBBINS.—THE HISTORY OF COMMERCE IN EUROPE. By H. de B.
  GIBBINS, M.A. With Maps. Globe 8vo. **3s. 6d.**
GREEN.—Works by JOHN RICHARD GREEN, LL.D., late Honorary Fellow of
  Jesus College, Oxford.
  *A SHORT HISTORY OF THE ENGLISH PEOPLE. New and Revised Ed.
  With Maps, Genealogical Tables, and Chronological Annals. Cr. 8vo. **8s. 6d.**
  159th Thousand.
  *Also the same in Four Parts. With the corresponding portion of Mr. Tait's
  "Analysis." Crown 8vo. **3s.** each. Part I. 607-1265. Part II. 1204-1553.
  Part III. 1540-1689 Part IV. 1660-1873.

HISTORY OF THE ENGLISH PEOPLE. In four vols. 8vo. 16s. each.
  Vol. I.—Early England, 449-1071; Foreign Kings, 1071-1214; The Charter,
    1214-1291; The Parliament, 1307-1461. With 8 Maps.
  Vol. II.—The Monarchy, 1461-1540; The Reformation, 1540-1603.
  Vol. III.—Puritan England, 1603-1660; The Revolution, 1660-1688. With four
    Maps.
  Vol. IV.—The Revolution, 1688-1760; Modern England, 1760-1815. With
    Maps and Index.
  THE MAKING OF ENGLAND. With Maps. 8vo. 16s.
  THE CONQUEST OF ENGLAND. With Maps and Portrait. 8vo. 18s.
*ANALYSIS OF ENGLISH HISTORY, based on Green's "Short History of the
  English People." By C. W. A. TAIT, M.A., Assistant Master at Clifton College.
  Revised and Enlarged Ed. Crown 8vo. 4s. 6d.
*READINGS FROM ENGLISH HISTORY. Selected and Edited by JOHN
  RICHARD GREEN. Three Parts. Gl. 8vo. 1s. 6d. each. I. Hengist to Cressy.
  II. Cressy to Cromwell. III. Cromwell to Balaklava.
GUEST.—LECTURES ON THE HISTORY OF ENGLAND. By M. J. GUEST.
  With Maps. Cr. 8vo. 6s.
*HISTORICAL COURSE FOR SCHOOLS.—Edited by E. A. FREEMAN, D.C.L.,
  Regius Professor of Modern History in the University of Oxford. 18mo.
  GENERAL SKETCH OF EUROPEAN HISTORY. By E. A. FREEMAN,
    D.C.L. New Ed., revised and enlarged. With Chronological Table, Maps, and
    Index. 3s. 6d.
  HISTORY OF ENGLAND. By EDITH THOMPSON. New Ed., revised and
    enlarged. With Coloured Maps. 2s. 6d.
  HISTORY OF SCOTLAND. By MARGARET MACARTHUR. 2s.
  HISTORY OF ITALY. By Rev. W. HUNT, M.A. New Ed. With Coloured
    Maps. 3s. 6d.
  HISTORY OF GERMANY. By J. SIME, M.A. New Ed., revised. 3s.
  HISTORY OF AMERICA. By JOHN A. DOYLE. With Maps. 4s. 6d.
  HISTORY OF EUROPEAN COLONIES. By E. J. PAYNE, M.A. With Maps.
    4s. 6d.
  HISTORY OF FRANCE. By CHARLOTTE M. YONGE. With Maps. 3s. 6d.
  HISTORY OF GREECE. By EDWARD A. FREEMAN, D.C.L. [In preparation.
  HISTORY OF ROME. By EDWARD A. FREEMAN, D.C.L. [In preparation.
*HISTORY PRIMERS.—Edited by JOHN RICHARD GREEN, LL.D. 18mo. 1s. each.
  ROME. By Bishop CREIGHTON. Maps.
  GREECE. By C. A. FYFFE, M.A., late Fellow of University College, Oxford.
    Maps.
  EUROPE. By E. A. FREEMAN, D.C.L. Maps.
  FRANCE. By CHARLOTTE M. YONGE.
  GREEK ANTIQUITIES. By Rev. J. P. MAHAFFY, D.D. Illustrated.
  CLASSICAL GEOGRAPHY. By H. F. TOZER, M.A.
  GEOGRAPHY. By Sir G. GROVE, D.C.L. Maps.
  ROMAN ANTIQUITIES. By Prof. WILKINS, Litt.D. Illustrated.

  ANALYSIS OF ENGLISH HISTORY. By Prof. T. F. TOUT, M.A.
  INDIAN HISTORY: ASIATIC AND EUROPEAN. By J. TALBOYS WHEELER.
HOLE.—A GENEALOGICAL STEMMA OF THE KINGS OF ENGLAND AND
  FRANCE. By Rev. C. HOLE. On Sheet. 1s.
JENNINGS.—CHRONOLOGICAL TABLES. A synchronistic arrangement of
  the events of Ancient History (with an Index). By Rev. ARTHUR C.
  JENNINGS. 8vo. 5s.
LABBERTON.—NEW HISTORICAL ATLAS AND GENERAL HISTORY. By
  R. H. LABBERTON. 4to. New Ed., revised and enlarged. 15s.
LETHBRIDGE.—A SHORT MANUAL OF THE HISTORY OF INDIA. With
  an Account of INDIA AS IT IS. The Soil, Climate, and Productions; the
  People, their Races, Religions, Public Works, and Industries; the Civil
  Services, and System of Administration. By Sir ROPER LETHBRIDGE, Fellow
  of the Calcutta University. With Maps. Cr. 8vo. 5s.

MAHAFFY.—GREEK LIFE AND THOUGHT FROM THE AGE OF ALEX-ANDER TO THE ROMAN CONQUEST. By Rev. J. P. MAHAFFY, D.D., Fellow of Trinity College, Dublin. Cr. 8vo. 12s. 6d.

THE GREEK WORLD UNDER ROMAN SWAY. From Plutarch to Polybius. By the same Author. Cr. 8vo. 10s. 6d.

MARRIOTT.—THE MAKERS OF MODERN ITALY: MAZZINI, CAVOUR, GARI-BALDI. Three Lectures. By J. A. R. MARRIOTT, M.A., Lecturer in Modern History and Political Economy, Oxford. Cr. 8vo. 1s. 6d.

MICHELET.—A SUMMARY OF MODERN HISTORY. By M. Michelet. Trans-lated by M. C. M. SIMPSON. Gl. 8vo. 4s. 6d.

NORGATE.—ENGLAND UNDER THE ANGEVIN KINGS. By KATE NORGATE. With Maps and Plans. 2 vols. 8vo. 32s.

OTTÉ.—SCANDINAVIAN HISTORY. By E. C. OTTÉ. With Maps. Gl. 8vo. 6s.

SEELEY.—Works by J. R. SEELEY, M.A., Regius Professor of Modern History in the University of Cambridge.

THE EXPANSION OF ENGLAND. Crown 8vo. 4s. 6d.

OUR COLONIAL EXPANSION. Extracts from the above. Cr. 8vo. Sewed. 1s.

*TAIT.—ANALYSIS OF ENGLISH HISTORY, based on Green's "Short History of the English People." By C. W. A. TAIT, M.A., Assistant Master at Clifton. Revised and Enlarged Ed. Cr. 8vo. 4s. 6d.

WHEELER.—Works by J. TALBOYS WHEELER.

*A PRIMER OF INDIAN HISTORY. Asiatic and European. 18mo. 1s.

*COLLEGE HISTORY OF INDIA, ASIATIC AND EUROPEAN. With Maps. Cr. 8vo. 3s.; sewed, 2s. 6d.

A SHORT HISTORY OF INDIA AND OF THE FRONTIER STATES OF AFGHANISTAN, NEPAUL, AND BURMA. With Maps. Cr. 8vo. 12s.

YONGE.—Works by CHARLOTTE M. YONGE.

CAMEOS FROM ENGLISH HISTORY. Ex. fcap. 8vo. 5s. each. (1) FROM ROLLO TO EDWARD II. (2) THE WARS IN FRANCE. (3) THE WARS OF THE ROSES. (4) REFORMATION TIMES. (5) ENG-LAND AND SPAIN. (6) FORTY YEARS OF STUART RULE (1603-1643). (7) REBELLION AND RESTORATION (1642-1678).

EUROPEAN HISTORY. Narrated in a Series of Historical Selections from the Best Authorities. Edited and arranged by E. M. SEWELL and C. M. YONGE. Cr. 8vo. First Series, 1003-1154. 6s. Second Series, 1088-1228. 6s.

THE VICTORIAN HALF CENTURY—A JUBILEE BOOK. With a New Portrait of the Queen. Cr. 8vo. Paper covers, 1s. Cloth, 1s. 6d.

ART.

*ANDERSON.—LINEAR PERSPECTIVE AND MODEL DRAWING. A School and Art Class Manual, with Questions and Exercises for Examination, and Examples of Examination Papers. By LAURENCE ANDERSON. Illustrated. 8vo. 2s.

COLLIER.—A PRIMER OF ART. By the Hon. JOHN COLLIER. Illustrated. 18mo. 1s.

COOK.—THE NATIONAL GALLERY, A POPULAR HANDBOOK TO. By EDWARD T. COOK, with a preface by JOHN RUSKIN, LL.D., and Selections from his Writings. 3d Ed. Cr. 8vo. Half-morocco, 14s.

. Also an Edition on large paper, limited to 250 copies. 2 vols. 8vo.

DELAMOTTE.—A BEGINNER'S DRAWING BOOK. By P. H. DELAMOTTE, F.S.A. Progressively arranged. New Ed., improved. Cr. 8vo. 3s. 6d.

ELLIS.—SKETCHING FROM NATURE. A Handbook for Students and Amateurs. By TRISTRAM J. ELLIS. Illustrated by H. STACY MARKS, R.A., and the Author. New Ed., revised and enlarged. Cr. 8vo. 3s. 6d.

GROVE.—A DICTIONARY OF MUSIC AND MUSICIANS. A.D. 1450-1889. Edited by Sir GEORGE GROVE, D.C.L. In four vols. 8vo. Price 21s. each. Also in Parts.

Parts I.-XIV., Parts XIX.-XXII., 3s. 6d. each. Parts XV., XVI., 7s. Parts XVII., XVIII., 7s. Parts XXIII.-XXV. (Appendix), 9s.

A COMPLETE INDEX TO THE ABOVE. By Mrs. E. WODEHOUSE. 8vo. 7s. 6d.

HUNT.—TALKS ABOUT ART. By WILLIAM HUNT. With a Letter from Sir J. E. MILLAIS, Bart., R.A. Cr. 8vo. 3s. 6d.

MELDOLA.—THE CHEMISTRY OF PHOTOGRAPHY. By RAPHAEL MELDOLA, F.R.S., Professor of Chemistry in the Technical College, Finsbury. Cr. 8vo. 6s.

TAYLOR.—A PRIMER OF PIANOFORTE-PLAYING. By FRANKLIN TAYLOR. Edited by Sir GEORGE GROVE. 18mo. 1s.

TAYLOR.—A SYSTEM OF SIGHT-SINGING FROM THE ESTABLISHED MUSICAL NOTATION; based on the Principle of Tonic Relation, and Illustrated by Extracts from the Works of the Great Masters. By SEDLEY TAYLOR. 8vo. 5s. net.

TYRWHITT.—OUR SKETCHING CLUB. Letters and Studies on Landscape Art. By Rev. R. ST. JOHN TYRWHITT. With an authorised Reproduction of the Lessons and Woodcuts in Prof. Ruskin's "Elements of Drawing." 5th Ed. Cr. 8vo. 7s. 6d.

DIVINITY.

ABBOTT.—BIBLE LESSONS. By Rev. EDWIN A. ABBOTT, D.D. Cr. 8vo. 4s. 6d.

ABBOTT—RUSHBROOKE.—THE COMMON TRADITION OF THE SYNOPTIC GOSPELS, in the Text of the Revised Version. By Rev. EDWIN A. ABBOTT, D.D., and W. G. RUSHBROOKE, M.L. Cr. 8vo. 3s. 6d.

ARNOLD.—Works by MATTHEW ARNOLD.

A BIBLE-READING FOR SCHOOLS,—THE GREAT PROPHECY OF ISRAEL'S RESTORATION (Isaiah, Chapters xl.-lxvi.) Arranged and Edited for Young Learners. 18mo. 1s.

ISAIAH XL.-LXVI. With the Shorter Prophecies allied to it. Arranged and Edited, with Notes. Cr. 8vo. 5s.

ISAIAH OF JERUSALEM, IN THE AUTHORISED ENGLISH VERSION. With Introduction, Corrections and Notes. Cr. 8vo. 4s. 6d.

BENHAM.—A COMPANION TO THE LECTIONARY. Being a Commentary on the Proper Lessons for Sundays and Holy Days. By Rev. W. BENHAM, B.D. Cr. 8vo. 4s. 6d.

CASSEL.—MANUAL OF JEWISH HISTORY AND LITERATURE; preceded by a BRIEF SUMMARY OF BIBLE HISTORY. By Dr. D. CASSEL. Translated by Mrs. H. LUCAS. Fcap. 8vo. 2s. 6d.

CHURCH.—STORIES FROM THE BIBLE. By Rev. A. J. CHURCH, M.A. Illustrated. Cr. 8vo. 5s.

*CROSS.—BIBLE READINGS SELECTED FROM THE PENTATEUCH AND THE BOOK OF JOSHUA. By Rev. JOHN A. CROSS. 2d Ed., enlarged, with Notes. Gl. 8vo. 2s. 6d.

DRUMMOND.—INTRODUCTION TO THE STUDY OF THEOLOGY. By JAMES DRUMMOND, LL.D., Professor of Theology in Manchester New College, London. Cr. 8vo. 5s.

FARRAR.—Works by the Venerable Archdeacon F. W. FARRAR, D.D., F.R.S., Archdeacon and Canon of Westminster.

THE HISTORY OF INTERPRETATION. Bampton Lectures, 1885. 8vo. 16s.

THE MESSAGES OF THE BOOKS. Being Discourses and Notes on the Books of the New Testament. 8vo. 14s.

*GASKOIN.—THE CHILDREN'S TREASURY OF BIBLE STORIES. By Mrs. HERMAN GASKOIN. Edited with Preface by Rev. G. F. MACLEAR, D.D. 18mo. 1s. each. Part I.—OLD TESTAMENT HISTORY. Part II.—NEW TESTAMENT. Part III.—THE APOSTLES: ST. JAMES THE GREAT, ST. PAUL, AND ST. JOHN THE DIVINE.

GOLDEN TREASURY PSALTER.—Students' Edition. Being an Edition of "The Psalms chronologically arranged, by Four Friends," with briefer Notes. 18mo 3s. 6d.

GREEK TESTAMENT.—Edited, with Introduction and Appendices, by Bishop WESTCOTT and Dr. F. J. A. HORT. Two Vols. Cr. 8vo. 10s. 6d. each. Vol. I. The Text. Vol. II. Introduction and Appendix.

SCHOOL EDITION OF TEXT. 12mo. Cloth, 4s. 6d. ; Roan, red edges, 5s. 6d. 18mo. Morocco, gilt edges, 6s. 6d.

*GREEK TESTAMENT, SCHOOL READINGS IN THE. Being the outline of the life of our Lord, as given by St. Mark, with additions from the Text of the other Evangelists. Arranged and Edited, with Notes and Vocabulary, by Rev. A. CALVERT, M.A. Fcap. 8vo. 2s. 6d.

*THE GOSPEL ACCORDING TO ST. MATTHEW. Being the Greek Text as revised by Bishop WESTCOTT and Dr. HORT. With Introduction and Notes by Rev. A. SLOMAN, M.A., Headmaster of Birkenhead School. Fcap. 8vo. 2s. 6d.

THE GOSPEL ACCORDING TO ST. MARK. Being the Greek Text as revised by Bishop WESTCOTT and Dr. HORT. With Introduction and Notes by Rev. J. O. F. MURRAY, M.A., Lecturer at Emmanuel College, Cambridge. Fcap. 8vo.
[In preparation.

*THE GOSPEL ACCORDING TO ST. LUKE. Being the Greek Text as revised by Bishop WESTCOTT and Dr. HORT. With Introduction and Notes by Rev. JOHN BOND, M.A. Fcap. 8vo. 2s. 6d.

*THE ACTS OF THE APOSTLES. Being the Greek Text as revised by Bishop WESTCOTT and Dr. HORT. With Explanatory Notes by T. E. PAGE, M.A., Assistant Master at the Charterhouse. Fcap. 8vo. 3s. 6d.

GWATKIN.—CHURCH HISTORY TO THE BEGINNING OF THE MIDDLE AGES. By H. M. GWATKIN, M.A. 8vo. [In preparation.

HARDWICK.—Works by Archdeacon HARDWICK.

A HISTORY OF THE CHRISTIAN CHURCH. Middle Age. From Gregory the Great to the Excommunication of Luther. Edited by W. STUBBS, D.D., Bishop of Oxford. With 4 Maps. Cr. 8vo. 10s. 6d.

A HISTORY OF THE CHRISTIAN CHURCH DURING THE REFORMA-TION. 9th Ed. Edited by Bishop STUBBS. Cr. 8vo. 10s. 6d.

HOOLE.—THE CLASSICAL ELEMENT IN THE NEW TESTAMENT. Considered as a proof of its Genuineness, with an Appendix on the Oldest Authorities used in the Formation of the Canon. By CHARLES H. HOOLE, M.A., Student of Christ Church, Oxford. 8vo. 10s. 6d.

JENNINGS and LOWE.—THE PSALMS, WITH INTRODUCTIONS AND CRITICAL NOTES. By A. C. JENNINGS, M.A.; assisted in parts by W. H. LOWE, M.A. In 2 vols. 2d Ed., revised. Cr. 8vo. 10s. 6d. each.

KIRKPATRICK.—THE MINOR PROPHETS. Warburtonian Lectures. By Rev. Prof. KIRKPATRICK. [In preparation.

THE DIVINE LIBRARY OF THE OLD TESTAMENT. By the same. [In prep.

KUENEN.—PENTATEUCH AND BOOK OF JOSHUA: An Historico-Critical Inquiry into the Origin and Composition of the Hexateuch. By A. KUENEN. Translated by P. H. WICKSTEED, M.A. 8vo. 14s.

LIGHTFOOT.—Works by the Right Rev. J. B. LIGHTFOOT, D.D., late Bishop of Durham.

ST. PAUL'S EPISTLE TO THE GALATIANS. A Revised Text, with Introduc-tion, Notes, and Dissertations. 10th Ed., revised. 8vo. 12s.

ST. PAUL'S EPISTLE TO THE PHILIPPIANS. A Revised Text, with Intro-duction, Notes, and Dissertations. 9th Ed., revised. 8vo. 12s.

ST. PAUL'S EPISTLES TO THE COLOSSIANS AND TO PHILEMON. A Revised Text, with Introductions, Notes, and Dissertations. 8th Ed., revised. 8vo. 12s.

THE APOSTOLIC FATHERS. Part I. ST. CLEMENT OF ROME. A Revised Text, with Introductions, Notes, Dissertations, and Translations. 2 vols. 8vo. 32s.

THE APOSTOLIC FATHERS. Part II. ST. IGNATIUS—ST. POLYCARP. Revised Texts, with Introductions, Notes, Dissertations, and Translations. 2d Ed. 3 vols. 8vo. 48s.

THE APOSTOLIC FATHERS. Abridged Edition. With short Introductions, Greek Text, and English Translation. 8vo. 16s.

ESSAYS ON THE WORK ENTITLED "SUPERNATURAL RELIGION." (Reprinted from the *Contemporary Review*.) 8vo. 10s. 6d.

MACLEAR.—Works by the Rev. G. F. MACLEAR, D.D., Warden of St. Augustine's College, Canterbury.

ELEMENTARY THEOLOGICAL CLASS-BOOKS.

*A SHILLING BOOK OF OLD TESTAMENT HISTORY. With Map. 18mo.

*A SHILLING BOOK OF NEW TESTAMENT HISTORY. With Map. 18mo. These works have been carefully abridged from the Author's large manuals.

*A CLASS-BOOK OF OLD TESTAMENT HISTORY. Maps. 18mo. 4s. 6d.

*A CLASS-BOOK OF NEW TESTAMENT HISTORY, including the Connection of the Old and New Testaments. With maps. 18mo. 5s. 6d.

AN INTRODUCTION TO THE THIRTY-NINE ARTICLES.   [In the Press.

*AN INTRODUCTION TO THE CREEDS. 18mo. 2s. 6d.

*A CLASS-BOOK OF THE CATECHISM OF THE CHURCH OF ENGLAND. 18mo. 1s. 6d.

*A FIRST CLASS-BOOK OF THE CATECHISM OF THE CHURCH OF ENGLAND. With Scripture Proofs. 18mo. 6d.

*A MANUAL OF INSTRUCTION FOR CONFIRMATION AND FIRST COMMUNION. WITH PRAYERS AND DEVOTIONS. 32mo. 2s.

MAURICE.—THE LORD'S PRAYER, THE CREED, AND THE COMMANDMENTS. To which is added the Order of the Scriptures. By Rev. F. D. MAURICE, M.A. 18mo. 1s.

THE PENTATEUCH AND BOOK OF JOSHUA: An Historico-Critical Inquiry into the Origin and Composition of the Hexateuch. By A. KUENEN, Professor of Theology at Leiden. Translated by P. H. WICKSTEED, M.A. 8vo. 14s.

PROCTER.—A HISTORY OF THE BOOK OF COMMON PRAYER, with a Rationale of its Offices. By Rev. F. PROCTER. 18th Ed. Cr. 8vo. 10s. 6d.

PROCTER and MACLEAR.—AN ELEMENTARY INTRODUCTION TO THE BOOK OF COMMON PRAYER. Rearranged and supplemented by an Explanation of the Morning and Evening Prayer and the Litany. By Rev. F. PROCTER and Rev. Dr. MACLEAR. New Edition, containing the Communion Service and the Confirmation and Baptismal Offices. 18mo. 2s. 6d.

THE PSALMS, CHRONOLOGICALLY ARRANGED. By Four Friends. New Ed. Cr. 8vo. 5s. net.

THE PSALMS, WITH INTRODUCTIONS AND CRITICAL NOTES. By A. C. JENNINGS, M.A., Jesus College, Cambridge; assisted in parts by W. H. LOWE, M.A., Hebrew Lecturer at Christ's College, Cambridge. In 2 vols. 2d Ed., revised. Cr. 8vo. 10s. 6d. each.

RYLE.—AN INTRODUCTION TO THE CANON OF THE OLD TESTAMENT. By Rev. H. E. RYLE, M.A., Hulsean Professor of Divinity in the University of Cambridge. Cr. 8vo.   [In preparation.

SIMPSON.—AN EPITOME OF THE HISTORY OF THE CHRISTIAN CHURCH DURING THE FIRST THREE CENTURIES, AND OF THE REFORMATION IN ENGLAND. By Rev. WILLIAM SIMPSON, M.A. 7th Ed. Fcap. 8vo. 3s. 6d.

ST. JAMES' EPISTLE.—The Greek Text, with Introduction and Notes. By Rev. JOSEPH MAYOR, M.A., Professor of Moral Philosophy in King's College, London. 8vo.   [In the Press.

ST. JOHN'S EPISTLES.—The Greek Text, with Notes and Essays. By Right Rev. B. F. WESTCOTT, D.D., Bishop of Durham. 2d Ed., revised. 8vo. 12s. 6d.

ST. PAUL'S EPISTLES.—THE EPISTLE TO THE ROMANS. Edited by the Very Rev. C. J. VAUGHAN, D.D., Dean of Llandaff. 5th Ed. Cr. 8vo. 7s. 6d.

THE TWO EPISTLES TO THE CORINTHIANS, A COMMENTARY ON. By the late Rev. W. KAY, D.D., Rector of Great Leghs, Essex. 8vo. 9s.

THE EPISTLE TO THE GALATIANS. Edited by the Right Rev. J. B. LIGHTFOOT, D.D. 10th Ed. 8vo. 12s.

THE EPISTLE TO THE PHILIPPIANS. By the Same Editor. 9th Ed. 8vo. 12s.

THE EPISTLE TO THE PHILIPPIANS, with Translation, Paraphrase, and Notes for English Readers. By the Very Rev. C. J. VAUGHAN, D.D. Cr. 8vo. 5s.

THE EPISTLE TO THE COLOSSIANS AND TO PHILEMON. By the Right Rev. J. B. LIGHTFOOT, D.D. 8th Ed. 8vo. 12s.

THE EPISTLES TO THE EPHESIANS, THE COLOSSIANS, AND PHILE-
MON; with Introductions and Notes, and an Essay on the Traces of Foreign
Elements in the Theology of these Epistles. By Rev. J. LLEWELYN DAVIES,
M.A. 8vo. 7s. 6d.

THE EPISTLE TO THE THESSALONIANS, COMMENTARY ON THE GREEK
TEXT. By JOHN EADIE, D.D. Edited by Rev. W. YOUNG, M.A., with Preface
by Prof. CAIRNS. 8vo. 12s.

THE EPISTLE TO THE HEBREWS.— In Greek and English. With Critical and
Explanatory Notes. Edited by Rev. F. RENDALL, M.A. Cr. 8vo. 6s.

THE ENGLISH TEXT, WITH COMMENTARY. By the same Editor. Cr.
8vo. 7s. 6d.

THE GREEK TEXT. With Notes by C. J. VAUGHAN, D.D., Dean of Llandaff.
Cr. 8vo. 7s. 6d.

THE GREEK TEXT. With Notes and Essays by the Right Rev. Bishop
WESTCOTT, D.D. 8vo. 14s.

VAUGHAN.—THE CHURCH OF THE FIRST DAYS. Comprising the Church
of Jerusalem, the Church of the Gentiles, the Church of the World. By C. J.
VAUGHAN, D.D., Dean of Llandaff. New Ed. Cr. 8vo. 10s. 6d.

WESTCOTT.—Works by the Right Rev. BROOKE FOSS WESTCOTT, D.D., Bishop of
Durham.

A GENERAL SURVEY OF THE HISTORY OF THE CANON OF THE NEW
TESTAMENT DURING THE FIRST FOUR CENTURIES. 6th Ed. With
Preface on "Supernatural Religion." Cr. 8vo. 10s. 6d.

INTRODUCTION TO THE STUDY OF THE FOUR GOSPELS. 7th Ed.
Cr. 8vo. 10s. 6d.

THE BIBLE IN THE CHURCH. A Popular Account of the Collection and
Reception of the Holy Scriptures in the Christian Churches. 18mo. 4s. 6d.

THE EPISTLES OF ST. JOHN. The Greek Text, with Notes and Essays.
2d Ed., revised. 8vo. 12s. 6d.

THE EPISTLE TO THE HEBREWS. The Greek Text, with Notes and Essays.
8vo. 14s.

SOME THOUGHTS FROM THE ORDINAL. Cr. 8vo. 1s. 6d.

WESTCOTT and HORT.—THE NEW TESTAMENT IN THE ORIGINAL
GREEK. The Text, revised by the Right Rev. Bishop WESTCOTT and Dr.
F. J. A. HORT. 2 vols. Cr. 8vo. 10s. 6d. each. Vol. I. Text. Vol. II.
Introduction and Appendix.

SCHOOL EDITION OF TEXT. 12mo. 4s. 6d.; Roan, red edges, 5s. 6d. Fcap.
8vo. Morocco, gilt edges, 6s. 6d.

WRIGHT.—THE COMPOSITION OF THE FOUR GOSPELS. A Critical En-
quiry. By Rev. ARTHUR WRIGHT, M.A., Fellow and Tutor of Queen's College,
Cambridge. Cr. 8vo. 5s.

WRIGHT.—THE BIBLE WORD-BOOK: A Glossary of Archaic Words and
Phrases in the Authorised Version of the Bible and the Book of Common
Prayer. By W. ALDIS WRIGHT, M.A., Vice-Master of Trinity College, Cam-
bridge. 2d Ed., revised and enlarged. Cr. 8vo. 7s. 6d.

'YONGE.—SCRIPTURE READINGS FOR SCHOOLS AND FAMILIES. By
CHARLOTTE M. YONGE. In Five Vols. Ex. fcap. 8vo. 1s. 6d. each. With
Comments. 3s. 6d. each.

FIRST SERIES.—GENESIS TO DEUTERONOMY. SECOND SERIES.—FROM JOSHUA TO
SOLOMON. THIRD SERIES.—THE KINGS AND THE PROPHETS. FOURTH SERIES.
—THE GOSPEL TIMES. FIFTH SERIES.—APOSTOLIC TIMES.

ZECHARIAH—THE HEBREW STUDENT'S COMMENTARY ON ZECHARIAH,
HEBREW AND LXX. With Excursus on Syllable-dividing, Metheg, Initial
Dagesh, and Siman Rapheh. By W. H. LOWE, M.A., Hebrew Lecturer at
Christ's College, Cambridge. 8vo. 10s. 6d.

The English Illustrated Magazine.

Each Volume Complete in Itself.

Volume for 1884.

Containing 792 pages, with 428 Illustrations. Price 7s. 6d.

The Volume contains the following Complete Stories and Serials :—

The Armourer's 'Prentices. By C. M. YONGE. An Unsentimental Journey through Cornwall. By Mrs. CRAIK. Julia. By WALTER BESANT. How I became a War Correspondent. By ARCHIBALD FORBES. The Story of a Courtship. By STANLEY J. WEYMAN, etc.

Volume for 1885.

Containing 840 pages, with nearly 500 Illustrations. Price 8s.

The Volume contains the following Complete Stories and Serials :—

A Family Affair. By HUGH CONWAY. Girl at the Gate. By WILKIE COLLINS. The Path of Duty. By HENRY JAMES. Schwartz. By D. CHRISTIE MURRAY. A Ship of '49. By BRET HARTE. That Terrible Man. By W. E. NORRIS. Interviewed by an Emperor. By ARCHIBALD FORBES. In the Lion's Den. By the Author of "John Herring," etc.

Volume for 1886.

Containing 832 pages, with nearly 500 Illustrations. Price 8s.

The Volume contains the following Complete Stories and Serials :—

Kiss and be Friends. By the Author of "John Halifax, Gentleman." Aunt Rachel. By D. CHRISTIE MURRAY. A Garden of Memories. By MARGARET VELEY. My Friend Jim. By W. E. NORRIS. Harry's Inheritance. By GRANT ALLEN. Captain Lackland. By CLEMENTINA BLACK. Witnessed by Two. By Mrs. MOLESWORTH. The Poetry did It. By WILKIE COLLINS. Dr. Barrere. By Mrs. OLIPHANT. Mere Suzanne. By KATHARINE S. MACQUOID. Days with Sir Roger de Coverley, with pictures by HUGH THOMSON, etc.

Volume for 1887.

Containing 832 pages, with nearly 500 Illustrations. Price 8s.

The Volume contains the following Complete Stories and Serials :—

Marzio's Crucifix. By F. MARION CRAWFORD. A Secret Inheritance. By B. L. FARJEON. Jacquetta. By the Author of "John Herring." Gerald. By STANLEY J. WEYMAN. An Unknown Country. By the Author of "John Halifax, Gentleman." With Illustrations by F. NOEL PATON. A Siege Baby. By J. S. WINTER. Miss Falkland. By CLEMENTINA BLACK, etc.

Volume for 1888.

Containing 832 pages, with nearly 500 Illustrations.   Price 8s.

Among the chief Contents of the Volume are the following Complete Stories and Serials :—

Coaching Days and Coaching Ways. By W. O. TRISTRAM. With Illustrations by H. RAILTON and HUGH THOMSON. **The Story of Jael.** By the Author of "Mehalah." **Lil : a Liverpool Child.** By AGNES C. MAITLAND. **The Patagonia.** By HENRY JAMES. **Family Portraits.** By S. J. WEYMAN. **The Mediation of Ralph Hardelot.** By Prof. W. MINTO. **That Girl in Black.** By Mrs. MOLESWORTH. **Glimpses of Old English Homes.** By ELIZABETH BALCH. **Pagodas, Aurioles, and Umbrellas.** By C. F. GORDON CUMMING. **The Magic Fan.** By JOHN STRANGE WINTER.

Volume for 1889.

Containing 900 pages, with nearly 500 Illustrations.   Price 8s.

Among the chief Contents of the Volume are the following Complete Stories and Serials :—

Sant' Ilario. By F. MARION CRAWFORD. **The House of the Wolf.** By STANLEY J. WEYMAN. **Glimpses of Old English Homes.** By ELIZABETH BALCH. **One Night—The Better Man.** By ARTHUR PATERSON. **How the "Crayture" got on the Strength.** And other Sketches. By ARCHIBALD FORBES. **La Belle Americaine.** By W. E. NORRIS. **Success.** By KATHARINE S. MACQUOID. **Jenny Harlowe.** By W. CLARK RUSSELL.

Volume for 1890.

Containing 900 pages, with nearly 550 Illustrations.   Price 8s.

Among the chief Contents of the Volume are the following Complete Stories and Serials :—

The Ring of Amasis. By the EARL OF LYTTON. **The Glittering Plain : or, the Land of Living Men.** By WILLIAM MORRIS. **The Old Brown Mare.** By W. E. NORRIS. **My Journey to Texas.** By ARTHUR PATERSON. **A Glimpse of Highclere Castle—A Glimpse of Osterley Park.** By ELIZABETH BALCH. **For the Cause.** By STANLEY J. WEYMAN. **Morised.** By the MARCHIONESS OF CARMARTHEN. **Overland from India.** By Sir DONALD MACKENZIE WALLACE, K.C.I.E. **The Doll's House and After.** By WALTER BESANT. **La Mulette, Anno 1814.** By W. CLARK RUSSELL.

Volume for 1891.

Containing 900 pages, and about 500 Illustrations.   Price 8s.

Among the chief Contents of the Volume are the following Complete Stories and Serials :—

The Witch of Prague. By F. MARION CRAWFORD. **The Wisdom Tooth.** By D. CHRISTIE MURRAY and HENRY HERMAN. **Wooden Tony.** By Mrs. W. K. CLIFFORD. **Two Jealousies.** By ALAN ADAIR. **Gentleman Jim.** By MARY GAUNT. **Harrow School. Winchester College. Fawsley Park. Ham House. Westminster Abbey. Norwich. The New Trade-Union Movement. Russo-Jewish Immigrant. Queen's Private Garden at Osborne.**

MACMILLAN AND CO., LONDON.

vi.50.9.91.

www.ingramcontent.com/pod-product-compliance
Lightning Source LLC
Chambersburg PA
CBHW021940220326
41599CB00011BA/925